On Biomineralization

On Biomineralization

HEINZ A. LOWENSTAM
California Institute of Technology

STEPHEN WEINER
Weizmann Institute of Science

New York Oxford
OXFORD UNIVERSITY PRESS
1989

Oxford University Press

Oxford New York Toronto
Delhi Bombay Calcutta Madras Karachi
Petaling Jaya Singapore Hong Kong Tokyo
Nairobi Dar es Salaam Cape Town
Melbourne Auckland

and associated companies in
Berlin Ibadan

Published by Oxford University Press, Inc.,
200 Madison Avenue, New York, New York 10016
Oxford is a registered trademark of Oxford University Press

Library of Congress Cataloging-in-Publication Data

Lowenstam, Heinz A. (Heinz Adolf), 1912– On biomineralization.
Bibliography: p. Includes index.
1. Biomineralization. I. Weiner, Stephen. II. Title
QH512.L68 1989 574.19′29 88-15119
ISBN 0-19-504977-2 (U.S.)

9 8 7 6 5 4 3 2 1

Printed in the United States of America

Preface

Biomineralization is a field that has its origins in the last century and over the years thousands of scientific papers have been written on the subject. However, it is a field that is still very much in the process of gathering basic information, and synthesis of the information on hand is difficult, to say the least. In spite of this we feel that a book that brings together information from a wide variety of sources, compares and contrasts mineralization processes from bacteria to man, and makes an attempt to pinpoint some of the underlying processes can contribute to the field. This book is the product of our efforts in this regard.

Although we each have different scientific backgrounds and research interests, in many respects we view the field of biomineralization from similar vantage points. We have made strenuous efforts to integrate our different approaches and in this way have broadened the book's perspectives to encompass a wide range of topics. All chapters are, therefore, the products of our joint efforts. We emphasize that this is not a book for geologists or for biologists, but for all those involved in one way or another with biomineralization and to avoid a misunderstanding in this regard, we wish to note that even the order that we list our names on the title page is alphabetical!

We would like to acknowledge first and foremost the help and support of our respective families in this endeavor. In particular H.A.L. acknowledges his mother Frieda Lowenstam, his grandfather Emil Bindseil, and his uncle Carl Bindseil for encouraging and helping him enter the field of natural sciences. S.W. acknowledges his father Motty Weiner and wife Nomi Weiner, both of whom have always unquestionably and enthusiastically supported his efforts to pursue a scientific career.

We are particularly grateful to Janice Lester for her enormous help in producing this manuscript. We also thank friends and colleagues who have contributed ideas, comments, and criticisms. These include in particular L. Addadi, S. Bengtson, J. L. Kirschvink, and W. Traub. We also thank B. R. Constantz, R. N. Ginsburg, P. B. Kaufman, J. J. Lee, M. D. Ross, R. Trench, and J. Wattendorf for providing data and helpful suggestions.

We acknowledge the financial help of the Weizmann Institute of Science, where this book was written.

Pasadena, Calif. H.A.L.
Rehovot, Israel S.W.
May 1988

Contents

CHAPTER 1 INTRODUCTION, 3

CHAPTER 2 MINERALS AND MACROMOLECULES, 7

The Minerals, 7

Impact of Biomineralization on the Biosphere, 18

The Macromolecules, 20

CHAPTER 3 BIOMINERALIZATION PROCESSES, 25

Controlled and Uncontrolled Biomineralization Processes, 26

Biologically Induced Mineralization, 26

Biologically Controlled Mineralization, 27

Space Delineation, 28
The Preformed Organic Matrix Framework, 29
Setting up the Saturated Solution, 30
Control over Nucleation, 32
Control over Crystal Growth, 38
Cessation of Crystal Growth, 39

The Real World, 41

CHAPTER 4 PROTOCTISTA, 50

Diatoms (Bacillariophyta), 54

Ultrastructure of Valve Formation, 56
Valve Formation, 56
Uptake, Transport, and Deposition of Silicon, 58

Foraminiferida, 60

Agglutinating Foraminifera, 63
Miliolids, 63
Rotaline Foraminifera, 65

Haptophyta (Coccolithophoridae), 67

Intracellular Coccolith Formation, 69
Extracellular Holococcolith Formation, 72
Non-Coccolith-Associated Mineralization, 72
Silicification of Cysts, 73

CHAPTER 5 CNIDARIA, 74

Spicules, 77

Spicule Aggregates, 79

Fused Spicular Aggregates, 79

Massive Skeletons: The Scleractinian Corals, 81

Larval Scleractinian Skeleton, 82
Adult Scleractinian Skeleton, 82
Processes of Scleractinian Coral Mineralization, 83

CHAPTER 6 MOLLUSCA, 88

Aplacophora, 89

Monoplacophora, 89

Scaphopoda, 94

Polyplacophora: Tooth Formation, 94

Cephalopoda, Bivalvia, and Gastropoda: Shell Formation, 99

The Mantle, 99
The Periostracum, 101
The Shell, 103
The Zone between the Mantle and the Shell, 109
Shell Dissolution and Remodeling, 109

CHAPTER 7 ARTHROPODA, 111

Arthropod Cuticle, 115

The Mineralized Crustacean Cuticle, 117

Moulting and Mineralization in the Crustacea, 120

CHAPTER 8 ECHINODERMATA, 123

Spicule Formation in Sea Urchin Larvae, 127

Mineralization in Adult Sea Urchins, 130

The Nature of the Mineral Phase, 132

CHAPTER 9 CHORDATA, 135

Ascidiacea, 140

Craniata (Vertebrates), 144

Bone, 144

Molecular Organization of Bone, 149
The Mineral, 149
The Organic Matrix, 152
Collagen–Crystal Relations, 155
Stages of Bone Mineralization, 162

Cartilage, 167

Cartilage in the Unmineralized Form, 168
Mineralized Cartilage, 169

Enamel and Enameloid, 175

Enameloid, 180

Enamel, 182

The Crystals, 183
The Organic Maxtrix, 184
Maturation, 185

A Perspective, 187

CHAPTER 10 SOME NONSKELETAL FUNCTIONS IN BIOMINERALIZATION, 189

Gravity Perception, 190

Functions of Biologically Formed Magnetite Crystals, 196

Ferritin: An Iron Storage Macromolecule, 202

Biological Control over Ice Formation, 204

Induction of Ice Crystals by Certain Plant Bacteria, 204
Inhibition of Ice Crystal Formation by Glycoproteins from Polar Fish Blood, 205

CHAPTER 11 ENVIRONMENTAL INFLUENCES ON BIOMINERALIZATION, 207

Increase in the Amount of Biogenic Mineral Formed in Marine Warm Waters as Compared to Cold Waters, 208

Different Minerals Formed in Response to Environmental Changes, 210

Environmental Influences on Trace Element and Oxygen Isotopic Composition, 217

Trace Element Contents, 218
The Environment and Stable Oxygen Isotopes, 221

Environmental Influence on Skeletal Growth, 223

CHAPTER 12 EVOLUTION OF BIOMINERALIZATION, 227

The Early Evolution of Biomineralization, 228

Biologically Induced Mineralization in the Early Precambrian, 229
Biologically Controlled Mineralization in the Precambrian, 229
The Advent of Composite Skeletal Formation, 232

Evolution of Carbonate Biomineralization, 232

The Deposition of Aragonite or Calcite, 235
The Increase of Biogenic Carbonate Formation during the Phanerozoic, 238

Evolution of Phosphate Mineralization, 240

Evolution of Silicification, 244

The Precambrian–Cambrian Boundary Zone: The Evolution of Composite Mineralized Skeletons, 247

REFERENCES, 252

INDEX, 309

On Biomineralization

1

Introduction

Biomineralization refers to the processes by which organisms form minerals. It is, therefore, by definition a true multidisciplinary field that spans both the inorganic and the organic world. Although the vast majority of organisms do not form mineralized deposits, the phenomenon is still extremely widespread. All five kingdoms contain members that mineralize and these are distributed among no less that 55 phyla. These organisms are capable of forming some 60 different minerals and it is patently clear that the true diversity of the field is still far from having been ascertained. Some biogenic minerals are formed on such a huge scale in the biosphere that they have a major impact on ocean chemistry and are important components of marine sediments and ultimately of many sedimentary rocks as well. One of the functions of biogenic minerals is to provide mechanical strength to skeletal hard parts and teeth. The resultant materials often have remarkable mechanical properties and are of interest in their own right. When organisms evolved the ability to form mineralized hard parts, they provided themselves with a major adaptational advantage and their durable skeletons constituted the basis for a more complete record of life on earth in the form of their fossilized skeletal remains. The vertebrate skeleton, in particular, fulfills a variety of functions and this brings with it a multitude of health-related problems that plague our own species, such as dental caries, bone fractures, mineral loss from bone, and kidney stones. Biomineralization, therefore, is an unusual field in that it lies at the center of many other disciplines.

Figure 1.1 is an imaginary wheel showing on its rim some of the disciplines that overlap the field of biomineralization. There are many scientists in each of these disciplines who have more than a passing interest in biomineralization and one objective of this book is to provide them with easier access to the field. The center of the wheel contains a partial list of some of the fields within biology and chemistry that have a major contribution to make toward a more complete understanding of the processes involved in biomineralization. Within each of these fields

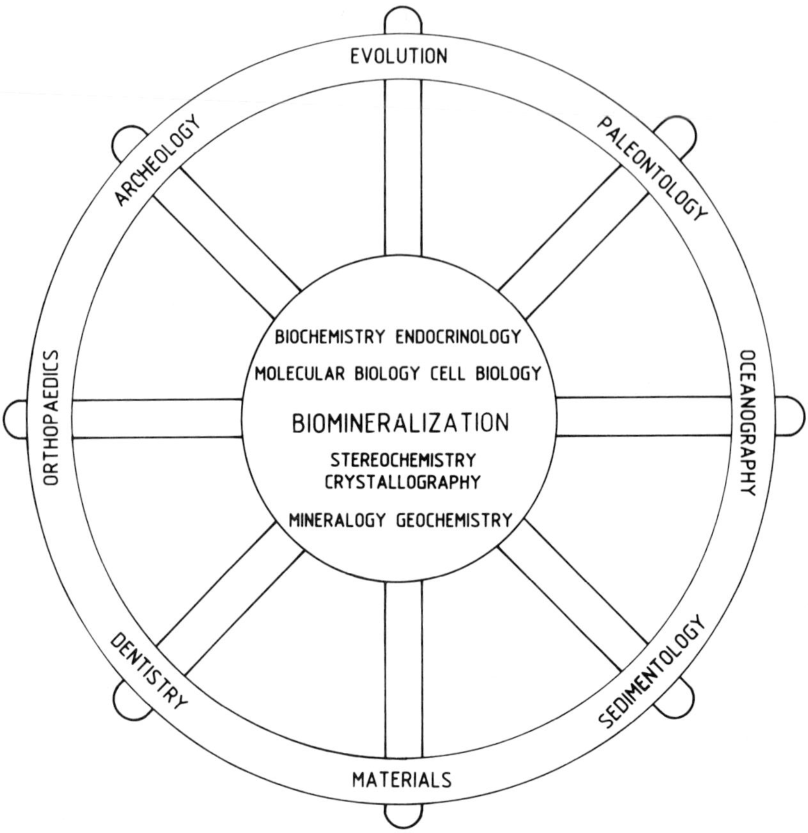

Figure 1.1. An imaginary wheel showing on its rim some of the scientific disciplines that overlap the field of biomineralization and at its center the disciplines upon which an understanding of biomineralization is based.

are investigators who focus much or all of their efforts on biomineralization. They too have very diverse backgrounds, and use different methodologies and even different vocabulary to describe their observations, with the result that effective communication within the field is a serious obstacle to progress. Hopefully this book will help in that regard as well. Finally a situation has arisen within the field in which even investigators with similar backgrounds and interests, but working on different mineralization processes, simply do not communicate. By bringing together under one cover a sampling of the major mineralization processes, we sincerely hope that not only will communication improve, but investigators in this field will again begin to more fully exploit the very powerful tool of comparative biology. In this regard it is of interest to put the current spectrum of activities in the biomineralization field into an historic perspective.

The field has its roots in the second half of the last century and the early part of this century when many organisms were being discovered and systematically described. The mineralized hard parts, of course, are conspicuous and scientists such as Haeckel, Gegenbauer, Grobben, Hatschek, Huxley, Lankester, Lamarck,

Bronn, and Bütschli, with their great attention to detail and accuracy, provided the field of biomineralization with a very solid foundation. The major tool available to these early investigators was the light microscope, which they used primarily for studying histological sections. Although somewhat crude by todays standards, the results achieved were sometimes remarkable. Mineral identification, for example, was a major undertaking in itself and yet by careful and often ingenious means, minerals were not only identified, but identified correctly, something that even today is not always the case. A good example is the identification by Bethe in 1895 of fluorine in mysid statoliths. He added acid to a known quantity of statoliths and then estimated the amount of fluorine gas evolved by the extent to which a glass slide was etched. He calibrated the assay by repeating it with known amounts of inorganic fluorite.

This period culminated in the 1920s with the publication of three major works in the field. W. J. Schmidt's (1924) volume, "Die Bausteine des Tierkörpers in polarisiertem Lichte," represents to this day an invaluable collection of data that encompasses much of what is known in the field. F. W. Clarke and W. C. Wheeler (1922) published the first comprehensive and accurate listing of the elemental compositions of many biogenic minerals. O. B. Bøggild's (1930) monumental work focused primarily on mollusk shell ultrastructure and was a milestone in integrating observations on the mineralized products of living organisms with their fossil ancestors.

The modern era of biomineralization research has its beginnings in the 1930s with the introduction of such powerful tools as X-ray diffraction. Improved optical microscopes and histological techniques provided better access to the cells and the tissue as a whole. The comparative approach still prevailed, although the trend was to focus more and more on vertebrates with their potential for providing solutions to the pressing problems of mineral-related diseases. A major change occurred in the early 1960s when, as far as we can ascertain, many investigators came to realize that the mineralization processes in various invertebrates and the so-called more primitive vertebrates are not relatively simple when compared to mammals, just different. This conclusion led them to focus their efforts on a few complex problems, and, for obvious reasons, they chose the medically more relevant ones. The result was a significant narrowing of the scope of activities in this field. This coincided with the development of many new techniques in biochemistry along with an exciting new molecular understanding of the basics of biology. The field of biomineralization also benefitted tremendously, albeit a decade or so later, when the important macromolecules associated with the minerals began to be investigated in earnest. In addition, the discovery of an array of hormones that are involved in regulating mineralization stimulated a second exciting area of research. Today the field still seems to be trying to solve many of these problems and perhaps at last the all-important cells responsible for the whole process are beginning to receive well-deserved attention. In the near future, we think investigators will by necessity be much more concerned with integrating the system into a whole, and we hope that better utilization will be made once again of the tremendous resource that nature has provided in the form of diverse mineralization processes.

The scope of the book is limited primarily by our own inabilities to comprehend this broad field. We, therefore, arbitrarily focus on the endproducts of mineralization: the minerals, the macromolecules, and how they are organized. We are concerned less with events at the cellular level and even less with the factors that regulate the cells' activities and the ways in which the ions that end up in the minerals are supplied. This is by no means a judgment of importance. We are fully aware that all these elements are essential for the system to work and all must be studied for a more complete understanding of the phenomenon. We have also arbitrarily decided not to discuss pathologic or regenerative forms of mineralization.

The book is organized at a number of levels with different reader's requirements in mind. At the most detailed level we have tried to be encyclopedic in terms of documenting what is known about the deposition of biogenic minerals among all organisms. We have listed all the cases of biogenically formed minerals known to us together with their literature citations. We have restricted this list, for the most part, to minerals whose actual identities have been ascertained and have excluded reports such as "calcareous substances" or even those that include elemental analyses of the mineral without its identity being determined. This is the first time such a listing with literature citations has been made. We emphasize that it is far from complete, in part because of our limitations, but even more, because so little is still known about the true extent of biomineralization products in the biosphere. Table 2.1 lists the known minerals and their distribution among the phyla. The key to Table 2.1 lists the literature citations or refers the reader to other tables in the book that do.

Chapters 2 and 3 provide the reader with an overview of the minerals and the macromolecules known to be involved in biomineralization and with the types of processes that are responsible for their formation. These chapters also contain numerous references to more detailed discussions in other chapters and can be used as a basis for more discriminating reading. Chapters 4 through 9 are written for the reader who is familiar with mineralization processes in some taxa, but would like to learn more about others. They describe mineralization processes within a kingdom (the Protoctista, Chapter 4) or within individual phyla (the Cnidaria, Chapter 5; Mollusca, Chapter 6, Arthropoda, Chapter 7; Echinodermata, Chapter 8; and Chordata, Chapter 9). The phyla we have chosen to discuss contain many members that form mineralized deposits; in addition, their processes of formation have been studied in some detail. It is, therefore, no coincidence that Chapter 9 on the Chordata, which includes, of course, the vertebrates, is the longest chapter in the book! Chapters 10, 11, and 12 discuss three different topics that we are personally interested in. We realize that many other topics, possibly more "important," could have been included.

We have also had to make some arbitrary decisions about terminology and taxonomy. We always refer to the process by which organisms form minerals as "biomineralization" and not "calcification," because a quick perusal of Table 2.1 will show that less than half the minerals are calcium minerals. The term *mineral* itself includes both crystalline and amorphous forms and we have taken the liberty of also adding to this the so-called organic minerals formed by organisms. The common names often used to describe the minerals and their chemical formula are listed in Table 2.2. We follow the terminology of Ferraiola (1982). Taxonomy is a more controversial issue. We have chosen to follow the outline of Margulis and Schwartz (1988), which is both convenient and updated. Within some phyla we have opted to use the classification of Barnes (1980).

2

Minerals and Macromolecules

Biomineralization is a diverse, widespread, and common phenomenon. This statement is based on our current knowledge of the known diversity of biogenic mineral types, the taxonomic affinities of the organisms that form these minerals, and their abundance in the biosphere. In this chapter, we present an updated compilation of biogenic mineral types and the organisms that form them. We also briefly discuss aspects of their impact on the environment. In addition, we list the basic types of macromolecules that are often, but by no means always, associated with biogenic minerals. Information of this type is invaluable for gaining an overall perspective of the subject, for beginning to identify any common trends and strategies, and eventually for determining whether or not different organisms use similar underlying principles for forming their minerals. It is also one of the only means available for roughly assessing what proportion of mineralizing organisms has been discovered to date and what proportion still remains to be discovered.

2.1 The Minerals

Table 2.1 lists the known biogenic mineral types and the taxonomic affinities of the organisms that form them at the phylum level. This compilation differs from earlier published versions in that the extensive key to the table allows the reader to identify the literature sources upon which the data are based. Table 2.2 lists the common names and chemical formulas of known biogenic minerals.

Table 2.1 lists almost 60 different biogenic minerals! In 1963 only 10 different mineral types had been identified (Lowenstam 1963); this increased to 19 mineral types by 1974 (Lowenstam 1974), 30 by 1981 (Lowenstam 1981), and 39 by 1983 (Lowenstam and Weiner 1983). Thus, there is no indication that the rate at which new minerals are being discovered is slowing down, and that we are even close to discovering the true diversity of biogenic mineral types, let alone the identities of

Table 2.1a. Distribution of Biogenic Minerals in the Monera and Protoctista (see key for identification of numbers and letters).

		KINGDOM	MONERA												PROTOCTISTA																
		PHYLUM NUMBER	1	2	3	4	5	6	7	8	9	10	11	12	13	14	15	16	17	18	19	20	21	22	23	24	25	26	27	28	29
	MINERAL	MAJOR																													
CARBONATES	Calcite	Ca	a											a	a	a	a	a	a	a	a	a	a								
	Aragonite	Ca	b	a										b				a			a	a	a	a							
	Vaterite	Ca																													
	Monohydrocalcite	Ca	c											c																	
	Protodolomite	Ca, Mg																													
	Amorph hydrous carbonate	Ca	b																												
	Hydrocerussite	Pb																													
PHOSPHATES	Hydroxyapatite indet.[2]	Ca					a						a																		
	Octacalcium phosphate (OCP)	Ca																													
	Francolite	Ca												d																	
	Dahllite	Ca			a											a															
	$Ca_3Mg_3(PO_4)_4$	Ca, Mg																													
	Whitlockite	Ca, Mg				a																									
	Struvite	Mg		b	b	a	b																								
	Brushite	Ca																													
	Amorph pyrophosphate	Ca																													
	Amorph calcium phosphate (ACP) indet.	Ca															a														
	ACP (dahllite precursor)[3]	Ca																													
	ACP (brushite precursor)[3]	Ca																													
	ACP (whitlockite precursor)[3]	Ca																													
	ACP (francolite precursor)[3]	Ca																													
	Amorph Mg Ca phosphate	Mg, Ca																													
	Amorph hydrous Fe^{3+} phosphate	Fe, Ca		c[4]			c[4]	a[4]	a[4]																						
	$K\ Na_3\ (Fe_{15}Mg_25)\text{-}(PO_4)_3(OH_3)^3$	Fe										a																			
	Vivianite $Fe_3^{2+}(PO_4)_2 \cdot 8H_2O$	Fe										b																			
HALIDES	Fluorite	Ca																													
	Amorph fluorite	Ca																													
	Hieratite	K																													
SULFATES	Gypsum	Ca																							a						
	Celestite	Sr																								a					
	Barite	Ba														a	a						a		a						
	Jarosite	K, Fe									a																				
SILICA	Opal	Si															a	a	a	a	a					a	a	a	a	a	

IRON OXIDES	Magnetite	Fe					d																a								a
	Goethite	Fe																													
	Lepidocrocite	Fe																													
	Ferrihydrite	Fe					e																								
	Amorph iron oxide	Fe																													
	Amorph "ilmenite"	Fe, Ti												e																	
MANGANESE OXIDES	Todorokite	Mn												e																	
	Birnessite	Mn								a																					
SULFIDES	Pyrite	Fe										c																			
	Hydrotroilite	Fe										d																			
	Sphalerite	Zn										c																			
	Wurtzite	Zn										e																			
	Galena	Pb										f																			
	Greigite	Fe										c																			
	Mackinawite	Fe										c																			
METALS	Sulfur	–																													
CITRATE	Earlandite	Ca																													
OXALATES	Whewellite	Ca																					a								
	Weddelite	Ca																													
	Glushinskite	Mg																													
	$Mn_2C_2O_4 \cdot 2H_2O$	Mn																													
	$Cu\ C_2O_4 \cdot nH_2O$	Cu																													
	Ca oxalate indet.	Ca																					a		a						
OTHER ORGANIC CRYSTALS	Sodium urate	Na																													
	Uric acid	–																													
	Paraffin hydrocarbon	–																													
	Wax (long chain)	–																													
	Ca tartrate	Ca																													
	Ca malate	Ca																													

1 Taxonomic assignments are arbitrarily based on Margulis and Schwartz (1988)
2 Hydroxyapatite is often loosely used for apatite minerals that also contain carbonate and/or fluorine. We do not imply that the designations shown signify that these organisms form hydroxyapatite and not one of the other forms.
3 The term "precursor" refers to an amorphous phase which upon heating to 500°C converts to the designated crystalline phase.
4 Found in bacterial ferritin

Table 2.1b. Distribution of Biogenic Minerals in the Fungi, Plantae, and Animalia (see key for identification of numbers and letters).

		KINGDOM	FUNGI					PLANTAE									ANIMALIA													
		PHYLUM NUMBER	30	31	32	33	34	35	36	37	38	39	40	41	42	43	44	45	46	47	48	49	50	51	52	53	54	55	56	57
	MINERAL	MAJOR																												
CARBONATES	Calcite	Ca				a		a		a				a			a	a	a		a	a	a	a	a	a		a	a	
	Aragonite	Ca												b			b	a			b		b	a	a	b			a	
	Vaterite	Ca												c										a	a				a	
	Monohydrocalcite	Ca																						a					a	
	Protodolomite	Ca, Mg																										a		
	Amorph hydrous carbonate	Ca												d					b					a	a				a	
	Hydrocerussite	Pb				b																								
PHOSPHATES	Hydroxyapatite indet.[2]	Ca																					c						a	
	Octacalcium phosphate (OCP)	Ca																											a	
	Francolite	Ca																				b		a					a	
	Dahllite	Ca			a																			a					a	
	$Ca_3Mg_3(PO_4)_4$	Ca, Mg																					d							
	Whitlockite	Ca, Mg																												
	Struvite	Mg																											a	
	Brushite	Ca																						a						
	Amorph pyrophosphate	Ca																						a						
	Amorph calcium phosphate (ACP) indet.	Ca																		a				a	a				a	
	ACP (dahllite precursor)[3]	Ca																			c									
	ACP (brushite precursor)[3]	Ca																												
	ACP (whitlockite precursor)[3]	Ca																	c											
	ACP (francolite precursor)[3]	Ca																												
	Amorph Mg Ca phosphate	Mg, Ca																a												
	Amorph hydrous Fe^{3+} phosphate	Fe, Ca																					e	a				a		
	K $Na_3(Fe_{15}Mg_25)\cdot(PO_4)_3(OH_3)^3$	Fe																												
	Vivianite $Fe_3^{2+}(PO_4)_2\cdot 8H_2O$	Fe																												
HALIDES	Fluorite	Ca																						a	a					
	Amorph fluorite	Ca																						a					a	
	Hieratite	K															c													
SULFATES	Gypsum	Ca																a												
	Celestite	Sr																												
	Barite	Ba																						a						
	Jarosite	K, Fe																												
SILICA	Opal	Si						b	a	b	a			e		a	d						f	a	a			a		

IRON OXIDES	Magnetite	Fe																						a					a	
	Goethite	Fe				c																		a					a	
	Lepidocrocite	Fe															e							a						
	Ferrihydrite	Fe				d	a							f										a					a	
	Amorph iron oxide	Fe																					g							
	Amorph "ilmenite"	Fe, Ti																												
MANGANESE OXIDES	Todorokite	Mn																												
	Birnessite	Mn																												
SULFIDES	Pyrite	Fe																												
	Hydrotroilite	Fe																												
	Sphalerite	Zn																												
	Wurtzite	Zn																												
	Galena	Pb																												
	Greigite	Fe																												
	Mackinawite	Fe																												
METALS	Sulfur	–																									a			
CITRATE	Earlandite	Ca												g										a						
OXALATES	Whewellite	Ca		a		e								h											a				a	
	Weddelite	Ca	a	b		f	b							i										a	a			a	a	
	Glushinskite	Mg				g																								
	$Mn_2C_2O_4 \cdot 2H_2O$	Mn				h																								
	$Cu\ C_2O_4 \cdot nH_2O$	Cu				i																								
	Ca oxalate indet.	Ca						c		c	b	a	a		a										a					
OTHER ORGANIC CRYSTALS	Sodium urate	Na																											a	
	Uric acid	–																							a					
	Paraffin hydrocarbon	–																							a					
	Wax (long chain)	–																							a					
	Ca tartrate	Ca												j																
	Ca malate	Ca												j																

1 Taxonomic assignments are arbitrarily based on Margulis and Schwartz (1988)
2 Hydroxyapatite is often loosely used for apatite minerals that also contain carbonate and/or fluorine. We do not imply that the designations shown signify that these organisms form hydroxyapatite and not one of the other forms.
3 The term "precursor" refers to an amorphous phase which upon heating to 500°C converts to the designated crystalline phase.
4 Found in bacterial ferritin

Key to Table 2.1

Phylum number	Kingdom	Phylum	*a*	*b*	*c*	*d*	*e*	*f*	*g*	*h*	*i*	*j*
1	Monera	Cyanobacteria	Schönleber (1936) Friedmann (1979)	Golubic and Campbell (1980)	Krumbein (1975, 1979)							
2		Pseudomonads	Greenfield (1963)	Rivadeneyra et al. (1983)	Harrison et al. (1987)							
3		Actinobacteria	Ennever and Takazoe (1973)	Rivadeneyra et al. (1983)								
4		Fermenting bacteria	Roth and Calmes (1981) Lowenstam (unpublished)									
5		Omnibacteria	Boyan et al. (1984)	Rivadeneyra et al. (1983)	Bauminger et al. (1980)	Blakemore (1975)	Frankel et al. (1985)					
6		N_2-fixing aerobic bacteria	Stiefel and Watt (1979)									
7		Aphragmabacteria	Bauminger et al. (1980)									
8		Aeroendospore	Northfield et al. (unpublished)									
9		Chemoautotrophic bacteria	Lazaroff et al. (1982)									
10		Thiopneutes	Hallberg and Wadston (1980)	Ivarson and Hallberg (1976)	Hallberg (1972)	Hallberg (1965)	Leleu et al. (1975)	Leleu and Goni (1974)				
11		Micrococci	Boyan et al. (1984)									
12		Undetermined	Morita (1980)	Krumbein (1974)	Krumbein (1979)	O'Brien et al. (1981)	Lowenstam (unpublished)					

13	Protoctista	Myxomycota	Table 4.1							
14		Ciliophora	Table 4.1							
15		Rhizopoda	Table 4.1							
16		Foraminifera	Table 4.1							
17		Dinoflagellata	Table 4.1							
18		Zoomastigina	Table 4.1							
19		Haptophyta	Table 4.1							
20		Rhodophyta	Table 4.1							
21		Chlorophyta	Table 4.1							
22		Phaeophyta	Table 4.1							
23		Gamophyta	Table 4.1							
24		Actinopoda	Table 4.1							
25		Bacillariophyta	Table 4.1							
26		Xanthophyta	Table 4.1							
27		Pyrrhophyta	Table 4.1							
28		Chyrsophyta	Table 4.1							
29		Euglenophyta	Table 4.1							
30	Fungi	Ascomycota	Horner et al. (1983)							
31		Basidiomycota	Graustein et al. (1977)	Graustein et al. (1977)	Horner et al. (1983)					
32		Deuteromycota	Ennever and Summers (1975)							
33		Mycophycophyta	Arnott and Pautard (1970) Pobeguin (1954)	Jones et al. (1982)	Ascaso et al. (1976) Galvan et al. (1981)	Jackson and Keller (1970) Jones et al. (1980)	Erdman et al. (1977) Jones and Wilson (1986)	Wadsten and Moberg (1985)	Wilson et al. (1980)	Wilson and Jones (1984)	Purvis (1984)
34		Zygomycota	Peat and Banbury (1968) David and Easterbroek (1971)	Jones et al. (1976) Urbanus et al. (1978)						

Key to Table 2.1 (*Continued*)

Phylum number	Kingdom	Phylum	*a*	*b*	*c*	*d*	*e*	*f*	*g*	*h*	*i*	*j*
35	Plantae	Bryophyta	Pobeguin (1954)	In Voronkov et al. (1975)	Arnott (1973)							
36		Sphenophyta	Witty and Knox (1964) Kaufman et al. (1971)									
37		Filicinophyta	Pobeguin (1954)	Riquier (1960)	Arnott (1973)							
38		Coniferophyta	Brydon et al. (1963)	Wattendorf and Meier (1970)								
39		Gnetophyta	Scurfield et al. (1973)									
40		Ginkgophyta	Franceschi and Horner (1980)									
41		Angiospermaphyta	Pobeguin (1954)	Swineford and Franks (1959) Smith et al. (1971)	Arnott (1980)	Pobeguin (1954)	Frey-Wyssling (1930) Kaufman et al. (1981) Sangster and Parry (1981)	Hyde et al. (1963) Roberts and Humpherson (1967)	Arnott (1973)	Arnott et al. (1965)	Franseschi and Horner (1980)	In Arnott and Pautard (1970)
42		Cycadophyta	In Napp-Zinn (1966)									
43		Lycopodophyta	In Voronkov et al. (1975)									

44	Animalia	Porifera	Haeckel (1872)	Hickson (1911) Hartman and Goreau (1970)	Gregson et al. (1979)	Thoulet (1884)	Towe and Rützler (1968)		
45		Cnidaria	Table 5.1						
46		Platyhelminthes	von Brand et al. (1965)	von Brand et al. (1965)	Nieland and von Brand et al. (1965)				
47		Nemertina	Stricker and Weiner (1985)						
48		Ectoprocta	Kelly (1901)	Lowenstam (1954a)	Hunt (1972)	Neff (1971)			
49		Brachiopoda	Sorby (1879)	McConnell (1963) Lowenstam (1972b)					
50		Annelida	Lowenstam (1954)	Lowenstam (1954)	Neff (1971)	Neff (1971)	Lowenstam (1972b)	Lowenstam and Rossman (1975)	Lowenstam (1972b)
51		Mollusca	Table 6.1						
52		Arthropoda	Table 7.1						
53		Sipuncula	Lowenstam (unpublished) Rice (1969)						
54		Pogonophora	Jones (1981)						
55		Echinodermata	Table 8.1						
56		Chordata	Table 9.1						

Table 2.2 The Common Names of Biologically Formed Minerals and Their Chemical Formulas[a]

Name	Chemical formula
Calcite	$CaCO_3$
Aragonite	$CaCO_3$
Vaterite	$CaCO_3$
Monohydrocalcite	$CaCO_3 \cdot H_2O$
Protodolomite	$Ca\ Mg(CO_3)_2$
Hydrocerussite	$Pb_3(CO_3)_2(OH)_2$
Hydroxylapatite	$Ca_5(PO_4)_3(OH)$
Octacalcium phosphate	$Ca_8H_2(PO_4)_6 \cdot SH_2O$
Fluorapatite (francolite)	$Ca_5(PO_4)_3F$
Carbonate-hydroxylapatite (dahllite)	$Ca_5(PO_4,CO_3)_3(OH)$
Whitlockite	$Ca_{18}H_2(Mg,Fe)_2^{2+}(PO_4)_{14}$
Struvite	$Mg(NH_4)(PO_4) \cdot 6H_20$
Brushite	$Ca(HPO_4) \cdot 2H_2O$
Vivianite	$Fe_3^{2+}(PO_4)_2 \cdot 8H_2O$
Fluorite	CaF_2
Hieratite	K_2SiF_6
Gypsum	$CaSO_4 \cdot 2H_2O$
Celestite	$SrSO_4$
Barite	$BaSO_4$
Jarosite	$KFe_3^{3+}(SO_4)_2(OH)_6$
Opal	$SiO_2 \cdot nH_2O$
Magnetite	$Fe^{2+}Fe_2^{3+}O_4$
Goethite	α-FeO(OH)
Lepidocrocite	γ-FeO(OH)
Ferrihydrite	$5Fe_2O_3 \cdot 9H_2O$
Todorokite	$(Mn^{2+}Ca\ Mg)Mn_3^{4+}O_7 \cdot H_2O$
Birnessite	$Na_4\ Mn_{14}O_{27} \cdot 9H_2O$
Pyrite	FeS_2
Hydrotroilite	$FeS \cdot nH_2O$
Sphalerite	ZnS
Wurtzite	ZnS
Galena	PbS
Greigite	$Fe^{2+}Fe_2^{3+}S_4$
Mackinawite	$(Fe, Ni)_9\ S_8$
Earlandite	$Ca_3(C_6H_5O_2)_2 \cdot 4H_2O$
Whewellite	$CaC_2O_4 \cdot H_2O$
Weddelite	$CaC_2O_4 \cdot (2 + X)H_2O\ (X < 0.5)$
Glushinskite	$MgC_2O_4 \cdot 4H_2O$

[a]The names and compositions given for the most part follow Ferraiolo (1982).

all the organisms that form them. In fact over the last few decades large numbers of organisms have been discovered that were previously not known to form minerals. We can only conclude that as our knowledge of the true extent of biomineralization in the biosphere is still far from complete, we can expect many surprises!

Part of the reason for this resides in the methodologies used for identifying biogenic mineral types. X-ray diffraction has been available since the early 1930s and has been extensively used since the late 1940s. Fairly large (milligram) samples are needed in powder form and the method is applicable only to crystalline minerals in which the crystals are fairly large (hundreds of Ångstroms). Electron diffraction requires only very small amounts of crystalline minerals. However, as most biologists routinely embed their specimens for transmission electron microscopy and use expensive knives for cutting thin sections, they usually take the precaution of first removing any mineral. Even if the mineral is left, the thin sections are difficult to cut and the mineral is often lost. Electron diffraction, however, can be conveniently used to identify the mineralogy of individual crystals isolated from the tissues provided they are not too thick. Energy dispersive X-ray spectrometer (EDS) attachments to scanning and transmission electron microscopes are convenient for elemental analyses but are not capable of identifying the mineral. The method is widely used, but, unfortunately in many cases, no attempt at identifying the mineral is made. The result is that the literature of the last decade or so is replete with not very helpful reports of elemental analyses without any mineralogical identification; this impedes progress in the field. We strongly recommend that EDS be used in conjunction with other mineralogical identification techniques. Infrared spectroscopy, in our view, is a most valuable technique for studying biogenic minerals. With the advent of Fourier transform infrared spectrometers, microgram samples can be conveniently analyzed and the infrared spectrum provides information on the mineralogical identity of both crystalline and amorphous phases, even if they end up in the same sample.

A study of the known biogenic minerals (Table 2.1) shows that approximately 80% are crystalline and 20% are amorphous. Wherever possible we arbitrarily designate a mineral as amorphous only when it shows no discernible X-ray diffraction pattern and its infrared spectrum shows broad absorption peaks. In practice this probably means that the amorphous phase is not ordered over distances greater than 10 or 15 Å [see the study by Mann et al. (1983b) on amorphous biogenic silica]. We suspect that many amorphous minerals remain undetected even in tissues that have been examined to date because only diffraction techniques have been used to identify their mineralogy. Calcium minerals constitute about 50% of all known biogenic minerals. The calcium ion certainly has a "privileged" status in biomineralization and in biology in general. The widespread use of calcium minerals is probably a direct result of the fact that organisms very early in the evolution of life developed the means to manipulate this ion. The widely used term *calcification* reflects the fact that calcium minerals are predominant and abundant. However, they still constitute only half the known minerals and, for this reason, we studiously avoid using this term. We also note that some 60% of the known biogenic mineral types contain hydroxyl groups and/or bound water molecules. This, in part, is a consequence of many organisms having specifically evolved the ability to form relatively soluble solid phases, so that these can be redissolved with minimal energy expense.

Phosphates constitute about 25% of the known biogenic mineral types. The apatitic crystal lattice for one reason or another is particularly amenable to the incorporation of various additional ions, such as fluoride, carbonate, and magnesium, over and above calcium and phosphate (Neuman and Neuman 1953). Mineralogical identification of apatites is, therefore, complicated. Furthermore, apatitic crystals tend to be small and even though they are crystalline, X-ray diffraction patterns are generally not very informative. Many investigators describe the apatitic mineral phase as "hydroxyapatite" in spite of the fact that they have not unequivocally determined that carbonate, for example, is absent. In Table 2.1 we list these so-called "hydroxyapatite" minerals, but concur with McConnell (1969) in suspecting that most of them are really dahllite (carbonate apatite) or francolite (carbonate fluorapatite), a fact that can easily be detected by infrared spectroscopy.

We note that more and more biogenic "organic crystals" are being identified. Table 2.1 lists no less than 12 different ones under the headings citrates, oxalates, and other organic crystals. They appear to fulfill functions analogous to minerals and, in our opinion, are an integral part of the field of biomineralization. We deftly ignore the contradiction between the terms *mineral* and *organic compounds* and suspect that in the future the study of the formation of organic crystals will become a prominent part of this field.

Organisms belonging to 55 phyla are now known to form minerals. The distribution of biogenic minerals between the 5 kingdoms shows that 37 are formed by animals, 10 by protoctists, 24 by monerans, 11 by vascular plants, and 10 by fungi. Carbonate minerals are fomed by species in 26 phyla, phosphates by species in 23 phyla, silica by species in 21 phyla, and iron oxides by species in 17 phyla. Thus, in terms of taxonomic distribution, carbonates are still the most widely utilized bioinorganic constituents. This updated compilation shows, for the first time, that phosphate, silica, and iron oxides are also extremely widely distributed. In terms of quantities formed, carbonates followed by opal are undoubtedly the most abundant. However, biogenic phosphates and iron oxides, particularly magnetite, are also formed biologically in huge amounts in the biosphere.

The list of minerals formed by organisms contains very few minerals that are not found somewhere on earth, having been formed by inorganic processes. Many of the biogenic minerals are, however, formed by organisms in environments in which their inorganic counterparts would not normally precipitate. Furthermore, the biogenic minerals are often very different in morphology from the inorganic ones and certainly the way they are ordered in biological systems (texture or ultrastructure) generally bears no resemblance to the way they form in the inorganic world.

One important conclusion to draw from Table 2.1 is that the large number of already identified biogenic mineral types precludes the possibility of invoking unique mechanisms of formation for each case (Wilbur 1984). It forces investigators in this field to search for basic underlying processes used by organisms even in different phyla. This, to us, is an almost self-evident conclusion, but has not obviously influenced the manner in which research is conducted in this field. Most investigators work on only one mineralized tissue. Topics covered at scientific conferences are generally restricted to the phosphatic mineralized tissues of vertebrates, the predominantly carbonatic invertebrate skeletons, or biological silicification. In this book we try to approach the subject as one entity with an emphasis on searching for basic common principles.

2.1.1 Impact of Biomineralization on the Biosphere

Biomineralization occurs in the oceans on such a huge scale that it influences many aspects of the chemistry of seawater and has a major affect on the nature of the sediments forming at the ocean bottom. Less appreciated is the influence that minerals formed by organisms can have on freshwater chemistry. For example, a freshwater protozoan belonging to the genus *Loxodes* forms intracellular spherical granules of barium sulfate. These organisms are very abundant in some lakes and may affect the whole biogeochemical cycle of barium in these freshwaters (Finlay et al. 1983). Silicification by diatoms is well known to directly affect the silica cycle in lakes (reviewed by Reynolds 1986). On land, bacteria and fungi are active in weathering processes of rocks and in soil formation (Ehrlich 1981). In some cases, such as lichens (Jones and Wilson 1986), biogenic minerals are formed, which are in essence just biologically mediated end-products of the weathering process. Certain plants form relatively large amounts of opal (Voronkov et al. 1975), which in local areas can accumulate in quite substantial amounts in the soils. Processes on land that were once thought to be purely inorganic, such as the formation of desert varnish, are in fact mediated by organisms that participate in the formation of iron and manganese minerals (Dorn and Oberlander 1982). The impact of biomineralization on the continental environment is poorly understood. In contrast, much more is known about the affect of biomineralization on the oceans. The subject has been reviewed by Lowenstam (1974) and Whitfield and Watson (1983).

Vast amounts of biogenic calcium carbonates are formed primarily in the surface waters of the open oceans. The organisms mainly responsible for this are the calcitic coccolithophoridae, the calcitic foraminifera, and the aragonitic pteropoda, which are nektonic gastropods. In spite of the fact that very large amounts of carbonate are formed biologically, the ocean surface waters are still supersaturated with respect to both calcite and aragonite. The major effect that the formation of these skeletons has on seawater chemistry is that as they sink through the water column to the ocean floor, they redistribute not only the calcium and bicarbonate ions, but also various associated trace elements, such as strontium and barium. Thus, the whole geochemical carbonate cycle is intimately dependent upon the rates of skeletal formation and the rates at which they dissolve (Whitfield and Watson 1983). It has been observed empirically that the calcitic coccoliths dissolve more slowly than the calcitic foraminiferal tests, even though the surface areas of the former are much larger than those of the latter (Honjo and Erez 1978). The reason for this is not known, but may well have to do with the skeletal ultrastructure, or even possibly occluded macromolecules inside the crystals themselves (see Chapters 3 and 8). It should also be noted that many of these biogenic carbonate skeletons accumulate on the ocean bottom and are eventually incorporated into the sediments and ultimately into the sedimentary rocks. These in turn make up a very large proportion of the continents themselves!

Biogenic carbonates are formed in huge quantities by reef-building communities widely distributed throughout the world's tropical oceans. Unlike the skeletons formed by the open ocean planktonic organisms previously discussed, which for the most part dissolve and are recycled back into the system, the biogenic reef carbonates are semipermanent features of the earth's sedimentary record. They thus represent an enormous "sink" of calcium carbonate extracted from seawater. Smith (1978) has estimated that some 50% of all the calcium carbonate entering the oceans mainly through rivers is taken up by reef-building communities.

The second major effect biomineralization processes have on ocean chemistry is on the silica cycle. The oceans are undersaturated with respect to opaline silica because organisms such as diatoms, radiolaria, silicoflagellates, and siliceous sponges form opaline skeletons. In fact, there is good reason to believe that before these organisms evolved, the oceans were supersaturated with respect to opaline silica, and as each of the major silica-forming taxonomic groups evolved, the degree of saturation decreased (Chapter 12). Most of the biogenic opal dissolves either in the water column or on the ocean bottom. Major accumulations of siliceous skeletons on the ocean bottom occur primarily under areas in which the surface waters have very high productivity (Broecker 1974). Antarctica is a good example. The transport of opaline skeletons from the surface waters, where they are primarily formed by diatoms and to a lesser extent by radiolaria, to the deep oceans also results in a redistribution of trace elements such as Ge and Zn, and possibly of Se, Cr, Be, Cu, and Ni (Whitfield and Watson 1983).

There are other, less dramatic ways in which biomineralization affects the marine environment. Some organisms form minerals that would not otherwise normally form in the oceans. This includes not only the carbonate minerals discussed above, but also, for example, magnetite formed by magnetotactic bacteria and various invertebrates, as well as fluorite formed by mysid shrimps and gastropods (reviewed by Lowenstam 1981). The planktonic Acantharia form skeletons of celestite (strontium sulfate) that would never otherwise form in the oceans. However, upon death, they immediately dissolve, and, as a result, have very little long-lasting effect on the ocean chemistry of strontium and certainly not of sulfate (Whitfield and Watson 1983).

Thus, biomineralization products have a major impact on the chemistry of seawater, on the sediments that form on the ocean floor, and through them on the sedimentary geological record itself. Much less is known about the impact of biomineralization on freshwater and continental environments.

2.2 The Macromolecules

Macromolecules are probably not associated with all biologically formed minerals. In general mineralization processes that are controlled in some way by the organism tend to have associated macromolecules (see Chapter 3). There is no doubt that these macromolecules fulfill important functions, both during the formation of the tissue (Chapter 3) and in their contribution to the biomechanical properties of the mature product—a topic not dealt with in this book, but described in the book by Wainwright et al. (1976). Almost all currently available information on the biochemistry of these macromolecules is restricted to carbonate- and phosphate-bearing mineralized tissues. Furthermore within this group, relatively few tissues have been studied in any detail [bone, dentin, and enamel in the chordata (Chapter 9), echinoderm skeletons (Chapter 8), and certain types of mollusk shells (Chapter 6)]. These mineralized tissues contain tens of different associated macromolecules, but a more or less complete inventory of the quantitatively abundant constituents is available only for bone (Tracy et al. 1987). In this section we briefly summarize the information available on the types of macromolecules found in mineralized tissues from different phyla. (See also review by Weiner et al. 1983a and Weiner 1984.)

A study of the macromolecules of a mineralized tissue almost always begins by removing the mineral phase. The usual procedure for carbonates and phosphates is to use the calcium chelating agent ethylenediaminetetracetic acid (EDTA) to dissolve the mineral at or close to neutral pH. The EDTA is removed with great difficulty we caution (Worms and Weiner 1986), and the presence of macromolecules both in solution and as an insoluble phase is checked. In *almost* all cases macromolecules are found in solution and, on further examination, many of them are found to have one particular characteristic: they are highly acidic. The proteins are rich in aspartic and/or aspartic and glutamic acids, and, in addition, some may have phosphorylated serine and threonine residues. They often have covalently bound polysaccharides that are also acidic, being rich in carboxylate groups and sometimes sulfate as well. For convenience we refer to all these macromolecules as *acidic macromolecules.* After demineralization there is often an insoluble fraction as well. Biochemical characterization shows that these macromolecules extracted from different tissues have little in common with each other, except that they are usually cross-linked and relatively hydrophobic when compared to the acidic constituents. As they are almost always quite distinct from the acidic constituents, we refer to them as *framework* macromolecules. This alludes to the fact that in many tissues they are the quantitatively abundant constituents that make up the mold or framework in which mineral forms (Chapter 3).

The distinction between these two major classes of macromolecules is empirical in that it depends upon their solubility first in EDTA and then when the EDTA is removed, in water. Unfortunately not all biogenic minerals dissolve in EDTA. This includes the quantitatively very important opaline silica, which dissolves only in hydrofluoric acid. Other biogenic minerals, such as the oxides, also have to be dissolved in acid. Thus, even extracting these macrocmolecules in their intact state is a problem, let alone comparing them with the material obtained by EDTA dissolution. This is one reason almost nothing is known about the macromolecules in noncarbonatic and nonphosphatic mineralized hard parts. It should be noted, however, that exposure to acidic conditions does not necessarily degrade macromolecules. Butler et al. (1981), for example, used 0.1 *M* formic acid for preparing dentin acidic glycoproteins. Furthermore King (1975) has shown that cold hydrofluoric acid, necessary for dissolving opaline skeletons, does not cause detectable hydrolysis of the protein constituents of radiolarian skeletons or the extracted proteins from calcitic foraminifera, which were used as a control. These demineralization procedures should be used, albeit with caution, as we really need biochemical information on the macromolecules associated with nonphosphatic and noncarbonatic skeletons.

Table 2.3 lists the macromolecules in terms of "framework" and "acidic" constituents for most of the mineralized tissues for which data are currently available. All the tissues contain acidic glycoproteins and/or proteoglycans. One possible exception in which they may be absent is the calcareous alga, *Halimeda* (Wilbur et al. 1969; Nakahara and Bevelander 1978; Borowitzka 1982), which forms its aragonitic crystals under poorly controlled conditions (Weiner 1986; Chapter 3). All the remaining tissues listed in Table 2.3 are formed under relatively well-controlled

Table 2.3 Some of the Major Classes of Macromolecules Identified to Date in Various Skeletal Hard Parts[a]

Phylum	Skeletal hard part	Major "framework" macromolecules	Major acidic macromolecules
Chordata			
Craniata	Bone and dentin (dahllite)	Collagen, type I	Glycoproteins, proteoglycans, Gla-containing proteins
	Tooth enameloid (francolite, dahllite)	Collagen	Glycoproteins
	Tooth enamel (dahllite)	Amelogenin?	Glycoproteins
	Cartilage (dahllite)	Absent?	Glycoproteins, proteoglycans
Echinodermata			
Echinoidea	Skeletal test, spines, spicules (calcite)	Absent?	Glycoproteins, proteoglycans
Mollusca	Shell (calcite, aragonite)	β-Chitin–silk–fibroinlike protein complex, absent in some	Glycoproteins, proteoglycans?
Arthropoda	Cuticle (calcite)	α-Chitin–protein complex	Glycoproteins?
	Granules (ACP)	Absent?	Glycoproteins
Cnidaria	Skeleton, spicules (aragonite, calcite)	Absent?; protein or chitin complexes in some	Glycoproteins
Annelida			
Serpulida	Tubes (calcite, aragonite)	α-Chitin–protein complex (Bubel et al. 1983)	Glycoproteins (Mitterer 1971)
Brachiopoda			
Inarticulata	Shell (francolite)	Chitin–protein complex (Kelley et al. 1953)	Proteins (Tuross and Fisher 1988)
Articulata	Shell (calcite)	Glycine-rich protein (Jope 1967)	No data
Ectoprocta	Cuticle (calcite)	Chitin-protein complex (Banta 1968; Hunt 1972)	Present, but poorly defined (Banta 1968)
Porifera	Spicule (opal)	Filamentous protein (Shore 1972)	No data
Radiolaria	Skeleton (opal)	No data	Glyco(?)proteins (King 1974, 1975)
Foraminifera	Shell (calcite)	No data	Glycoproteins, proteoglycans? (Weiner and Erez 1984)
Coccolithophoridae	Coccoliths (calcite)	Absent?	Acidic polysaccharides, proteins? (de Jong et al. 1976)

[a]References are given only for phyla that are not reviewed in more detail in other chapters.

conditions and acidic macromolecules are present. This observation strongly supports the notion that these acidic macromolecules fulfill important functions in biomineralization. The fact that they are so highly charged in itself implies that they are actively involved in regulating mineral formation. Furthermore, in the few cases in which their locations in the tissue are known (Chapter 3), they are closely associated with the mineral. We do not know enough about these macromolecules or about the evolution of biomineralization (Chapter 12) to determine whether

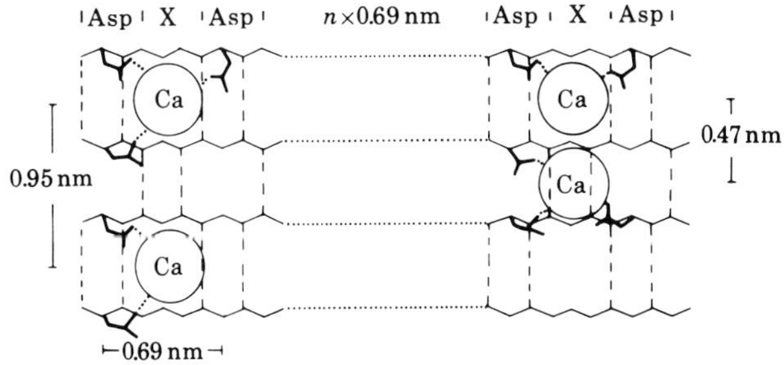

Figure 2.1. Schematic illustration of the possible modes of calcium binding to the surface of a β-sheet containing aspartic acid residues. The calcium ions are liganded to two or three aspartic acid residues. Reproduced from Weiner and Traub (1984) by courtesy of the Royal Society, London.

these molecules have their origin in a common ancestor and still share various structural properties, or were selected from different sources to fulfill these mineralizing functions.

Very little is known about the secondary conformations of the acidic macromolecules. We do note, however, that acidic glycoproteins rich in aspartic and/or glutamic acids from mollusks (Worms and Weiner 1986), echinoderms (Berman et al. 1988), tooth enamel (Little 1959; Jodaikin et al. 1986), and bone (Renugopalakrishnan et al. 1986b) all partially adopt the β-sheet conformation in vitro in the presence of calcium. Furthermore, in vitro experiments show that acidic glycoproteins can interact specifically with certain crystal faces and not others, provided they are in the β-sheet conformation (Addadi and Weiner 1985). The β-sheet is thus expected to be an important factor in determining the manner in which these proteins function. The β-sheet conformation results in the carboxylate groups of the aspartic and glutamic acid residues being perpendicular to the plane of the sheet and, depending upon the amino acid sequence, spaced at some specific distances from each other in a two-dimensional plane (Fig. 2.1). The distance between adjacent polypeptide chains is fixed at about 4.7 Å by hydrogen bonding. At least two carboxylate groups are needed to bind a calcium ion. In Figure 2.1 we schematically illustrate how this *may* occur.

The distribution of major framework macromolecules between phyla (Table 2.3) presents an entirely different picture as compared to the acidic macromolecules. The framework macromolecular types vary considerably from tissue to tissue and a number of mineralized hard parts from various phyla do not *appear* to have any framework macromolecules at all. In Table 2.3, a question mark is placed after each notation of "absent," as it is not clear whether framework-type macromolecules are really absent or present in only small amounts, as they might fulfill an essential mineralizing function. One such role could be as a substrate for the acidic macromolecules. In mollusk shells, they do perform this function (Chapter 6) and in vitro experiments (Greenfield et al. 1984; Addadi and Weiner 1985) demonstrate the importance of such substrates for the functioning of the acidic macromolecules.

The diversity of framework macromolecular types and the fact that in some tissues they are present in very small amounts or are absent suggest that their major function is not in the mineralization per se of the tissue, but in contributing to the mechanical properties of the formed product. This is best illustrated by the two mollusks, *Tridacna* (Bivalvia) and *Strombus* (Gastropoda). They both form very large robust shells and even though they are taxonomically quite different, they both have no detectable amounts of insoluble framework-type macromolecules (Weiner et al. 1983a). In contrast, all "normal"-sized mollusk shells we have examined do contain considerable quantities of framework constituents. This suggests that these large mollusks achieve their optimal mechanical properties by shear mass, rather than by constructing a material with advantageous properties. The shell of the pearly *Nautilus* presents the opposite situation; it is thin and is subjected to very large hydrostatic pressures as the animal descends into the ocean depths with gas inside its shell chambers. It is interesting to note, therefore, that *Nautilus* shell contains more framework constituents than any other known marine mollusk (Hare and Abelson 1965).

In this chapter we show that the minerals and macromolecules known to be utilized by organisms to form mineralized hard parts are almost overwhelmingly diverse. Furthermore, it seems that we are not even close to finding out the real extent of their diversity. The challenge, therefore, is to try to delve below the surface and understand biomineralization at a level that is common and relevant to many different mineralization processes. This is the major theme of the next chapter.

3

Biomineralization Processes

The large number of different minerals formed by organisms from almost 50 different phyla described in Chapter 2 should in itself discourage anyone from searching for *the* mechanism of biomineralization. On the other hand, the survey of macromolecules used by many organisms to control mineralization (Chapter 2), even though limited primarily to carbonate- and phosphate-bearing mineralized hard parts, shows that similar and rather unusual acidic glycoproteins and proteoglycans are widely utilized in biomineralization. This raises the possibility that many organisms may have adopted common approaches or strategies for regulating mineral formation. We do not know whether this arose as a result of divergence from a common ancestor or is a product of convergent evolution in which many different phyla independently began utilizing similar macromolecules for controlling mineralization (see Chapter 12). Either way we view the diversity in biomineralization as the product of a very broad and almost continuous spectrum of processes that organisms use to control mineralization. This ranges from no apparent control at one end to, it seems, control over every detail at the other. However, this is achieved by a fairly limited number of different basic processes used in various combinations and ways to produce a unique final product. This last statement is, we readily admit at this point in time, more an act of faith than an established fact. In this chapter we will try to identify and/or speculate about some of these basic processes.

We will draw upon material from many different sources, and, in particular, we will refer whenever possible to the more detailed descriptions of mineralization processes given in the chapters that follow. As a consequence, this chapter may also be used by the reader as a guide toward more discriminating reading on selected topics in the remainder of the book.

3.1 Controlled and Uncontrolled Biomineralization Processes

The spectrum of biomineralization processes can in principle be easily divided into cases in which control is exercised in some way over mineralization and those in which it is not. In practice the differentiation is not that simple as all organisms do exercise some control at one level or another, even if it simply involves, for example, removing from the cell some undesirable metabolic end-product or ion that combines with another ion in the external medium and precipitates. To differentiate between processes that are not specifically designed for mineralization, but do in fact result in mineral being formed, and processes in which a specific "machinery" is set up for the purpose of mineralization, Lowenstam (1981) coined the phrase "biologically induced mineralization" for the former situation. The latter is more aptly referred to as "biologically controlled mineralization," a term proposed by Mann (1983), rather than "organic-matrix-mediated mineralization" (Lowenstam 1981), which is restrictive in that it refers more specifically to mineralization within a preformed framework.

3.2 Biologically Induced Mineralization

Mineral formation in many of the aqueous environments in which organisms live is not inherently difficult to achieve. Even a relatively minor perturbation, such as the introduction of biologically produced metabolic end-products, the release of particular cations by the cell, or even the construction of a charged surface such as a cell wall, will under certain circumstances induce minerals to precipitate. When this occurs, we refer to it as "biologically induced mineralization." It is often, but not always obvious that the organism derives any benefit from the process; in fact, not surprisingly, many forms of pathologic mineralization occur under similar circumstances.

There are a number of different criteria that are useful for recognizing biologically induced as opposed to biologically controlled mineralization. Mineralization occurs in the open environment and not in a space particularly delineated for this purpose. No specialized cellular or macromolecular "machinery" is set up to induce mineralization. The minerals themselves, if they are crystalline, have habits that are the same or very similar to those formed by their inorganic counterparts; they have a wide range of sizes and are usually random aggregates. One particularly characteristic feature of biologically induced mineralization is that the type of mineral formed is a function just as much of the environmental conditions in which the organism lives as it is of the biological processes involved in its formation. The same organism in different environments can form different minerals. We caution that these criteria do not define a particular process, but are useful as guidelines for recognizing a broad and heterogeneous spectrum of biologically induced biomineralization processes.

Biologically induced mineralization appears to be the predominant process among the monerans and fungi and it occurs fairly frequently in the protoctists (Chapter 4). It is by no means uncommon among animals. Some examples of biologically induced mineralization in animals are the calcite crystals formed in the axial skeleton of the pennatulid sea fan (Ledger and Franc 1978) (Chapter 5) and the weddelite and brushite crystals that form between the layers of biologically controlled deposits of calcite in the mandible or beak of *Nautilus* (Lowenstam et al. 1984).

The sulfate-reducing bacteria release hydrogen sulfide into the environment where it is free to react with any metal ion that happens to be present in high enough concentration in solution. *Desulfovibrio* in the phylum Thiopneutes has been shown in laboratory culture to form no less than seven different sulfide minerals (see Table 2.1). Similar processes are thought to occur in nature as well. Some of the variability simply reflects the environmental pH of the aqueous medium. This in part controls the formation and stabilization of greigite (Fe_3S_4) and mackinawite [$(Fe, Ni)_9 S_8$]. The mineral can form at some distance from the bacterium itself depending upon the concentrations of the ions and molecules involved and their rates of diffusion. The cell wall is not necessarily involved in mineralization.

Crustose lichens can chemically decompose the rocky substrates on which they grow (reviewed by Jones and Wilson 1986). The fungal partner of the lichen ("mycobiont") is principally responsible for the decomposition process that is accomplished by acid dissolution. The acids so far known to be used are a group of polyphenolic compounds and oxalic acid. It is not surprising that depending on the mineralogy of the substrate, lichens can induce the formation of several different minerals (see Table 2.1) including a variety of oxalates. For example, the species *Pertusaria corallina* forms weddellite, or an unnamed manganese oxalate, but when it encrusts iron-rich volcanic rocks, it forms ferrihydrite (Jones and Wilson 1986).

The calcareous algae are a diverse group of mineralizers (reviewed by Borowitzka 1982). They utilize a fairly broad spectrum of processes that appear to span the (ill-defined) boundary between biologically induced and biologically controlled mineralization. The so-called brown algae, such as *Padina,* form their aragonite crystals in the open environment without any apparent control over the process. In the red algae *Galaxaura* and *Liagora* and particularly the green algae *Halimeda,* the crystals form mainly between cells in a semiconfined space that is, however, still partially open to the environmental seawater (Hillis 1959). Their crystals form in close association with the cell wall. In *Chara,* the crystals form in regions of the cell wall that correspond to areas that are more alkaline (Spear et al. 1969). In the green algae, such as *Penicillus, Rhipocephalus,* and *Acetabularia,* crystal growth occurs in an extracellular wall layer or "sheath" (Böhm et al. 1978) and the crystals are slightly out of equilibrium with respect to trace element and stable isotopic composition (Lowenstam and Epstein 1957; Kinsman 1969), indicating that processes responsible for their formation are more appropriately classified as "biologically controlled," albeit in a very simple manner.

3.3 Biologically Controlled Mineralization

Much of the research in biomineralization is in one sense ultimately aimed at understanding the control processes that organisms have evolved, and that differentiate mineralization in biology from mineralization in the inorganic world.

Although we are only just beginning to identify some of these processes, it is already apparent that many organisms can combine different control processes and end up with a unique final product. It is necessary to identify and understand the various processes, but not to lose sight of the fact that they act in concert.

3.3.1 Space Delineation

One of the most distinctive features of biologically controlled mineralization is that the site at which mineral forms is first sealed off from the environment by a barrier through which ions cannot freely diffuse. Space delineation (Wilbur 1984; Simkiss 1986) is a fundamental part of the "machinery" that cells use to control mineralization. Suprisingly, organisms, as far as we know, do not use a wide variety of methods for delineating space.

Lipid bilayers, either in the cell wall membrane or as part of vesicles located within the cell or outside the cell, are the most common means of space delineation. The other less widely used material for space delineation is composed of water-insoluble macromolecules (proteins and/or polysaccharides) that polymerize to form impervious sheets. The best characterized of these is periostracin (Waite et al. 1979), the protein that forms the main constituent of the outer organic membrane of mollusk shells (described in some detail in Chapter 6). Some organisms use both membranes and polymerized macromolecules for space delineation. Several of the more common combinations are schematically illustrated in Figure 3.1 and the legend refers to some examples in nature in which they occur.

A highly specialized form of space delineation occurs in ferritin. Ferritin is composed of a multisubunit protein shell, so-called apoferritin. The iron in some ferritins is deposited as paracrystalline ferrihydrite, whereas in others (e.g., certain bacteria) it is an iron phosphate mineral. Organisms ranging from bacteria to mammals use ferritin for temporarily storing iron (see Chapter 10). The use of a protein shell to create the space for mineral deposition appears to be unique.

Space delineation as a means of exerting control over mineralization probably arises from the necessity to carefully control the composition of the mother liquor from which the mineral forms. In some organisms the barriers used for space delineation also participate in mineral formation itself. For example, some of the lipids in the vesicle membranes are acidic and may also contain membrane proteins that participate directly in controlling aspects of mineralization (Boskey 1986) (Chapter 9). It has been proposed that the mollusk periostracum and the arthropod cuticle both facilitate mineral formation as a consequence of their demonstrated semiconducting properties (Digby 1965, 1968). Note that substrates often play a very important part in controlling mineralization (see following), but these are usually the acidic glycoproteins and proteoglycans that constitute part of the matrix framework rather than the "membrane" used for space delineation.

For the purposes of describing and understanding the controls exerted over mineralization once the space is delineated, it is helpful to divide the entire process into stages. The stages, which will be discussed separately, are: subdividing the delineated space, setting up the saturated solution from which the mineral will precipitate, the nucleation events, modulation of the mineral as it is forming, and cessation of mineral formation.

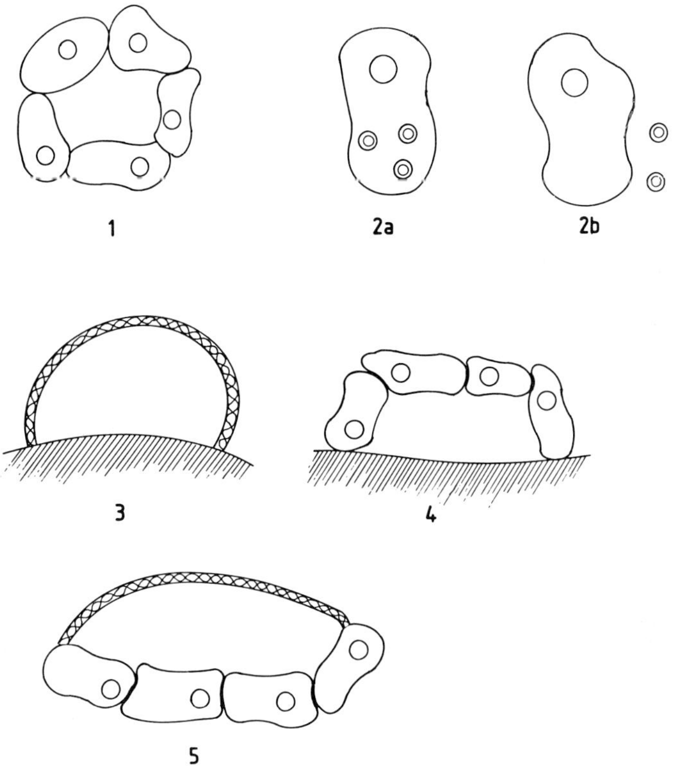

Figure 3.1. Schematic illustrations of the common modes of space delineation in biomineralization. (1) Space formed by cells with tight junctions. (2) Vesicles bounded by lipid bilayers located (a) intracellularly and (b) extracellularly. (3) Space formed by a "cyst" of polymerized macromolecules on a substrate such as an already formed part of the skeleton. (4) Space formed by cells and a substrate. (5) Space formed by cells and polymerized macromolecules. An example of (1) is the syncytium of echinodermata (Chapter 8), of (2a) intracellular calcium storage granules (see Brown 1982), of (2b) the cartilage matrix vesicles (Chapter 9), of (3) the rotaline foraminifera (Chapter 4), and of (4) and (5) the inner and outer shell layers of mollusks respectively (Chapter 6).

3.3.2 The Preformed Organic Matrix Framework

In many mineralization processes, particularly those occurring in most skeletal hard parts, the space set up by cells and/or polymerized macromolecules is further subdivided prior to mineralization by an assemblage of so-called organic-matrix macromolecules. The term *organic matrix* was coined by Le Gros Clark (1945) for bone. The cells involved in space delineation synthesize an array of macromolecules, which are then secreted into the extracellular space, where they self-assemble into a three-dimensional framework. These phenomena have been particularly well studied in the forming dentin of continuously growing rat incisors (Weinstock and Leblond 1973; see Chapter 9) in which the framework proteins (collagen type I) are first secreted and then, only some 24 hours after they have self-assembled, are some of the acidic proteins secreted by a different pathway directly into the site of mineralization (Maier et al. 1983). In dentin the preformed organic matrix framework is formed just ahead of the mineralization front. In the teeth of chitons (Chapter 6) and fish (Chapter 9), the whole tooth structure is preformed prior to any mineral being introduced. The tooth is surrounded by epithelial cells that set up the matrix framework and supply the ions for mineralization. We can only surmise that the organic matrix structures are permeable so that the ions can diffuse through them to the nucleation sites.

We emphasize that not all spaces delineated for controlled mineralization are subdivided by matrix macromolecules. This is often the case in vesicle mineralization, including possibly the syncytium in which echinoderms deposit their calcitic skeletons (Chapter 8). On the other hand, there are also a few known vesicle-bound mineralization processes in which the space is indeed subdivided by matrix macromolecules (see section on Octacorallia in Chapter 5).

The term *organic matrix* is used in a wide variety of contexts. We prefer to refer to an organic matrix as a continuous sheetlike structure that subdivides the mineralization space. The fact that organic matrixlike macromolecules are extracted out of a mineral phase does not necessarily imply that they exist as continuous "membranes," as they could also be dispersed monomers or discontinuous aggregates. The latter situation has been proposed for echinoderm skeletons by Berman et al. (1988) (Chapter 8).

3.3.3 Setting Up the Saturated Solution

It is obvious that a precondition for mineralization is that the solution from which the mineral is to precipitate must be saturated and all the thermodynamic and kinetic considerations that apply to mineral formation from saturated solutions in a beaker also apply to the spaces delineated by organisms for mineralization. Nancollas (1977) and Mann (1986) provide information about the chemical control of biological mineralization processes.

The formation of the mother liquor in biologically controlled mineralization is dominated by the cell or cells responsible for orchestrating the entire mineralization process. The cells are either part of the boundaries of the space set up for this purpose or enclose or secrete the vesicles in which mineralization is to occur. They are usually thoroughly equipped with the means to actively pump ions of choice into the mineralizing compartment or, when appropriate, allow passive diffusion of specific ions. Ion-specific pumps and channels are part of the machinery of all cells and they are no doubt fully utilized in controlled biomineralization to determine the order in which the ions are introduced into the mineralization compartment, the ionic composition, and the degree of saturation to be achieved.

Potentially one of the most powerful tools cells have for controlling mineralization is the ability to introduce ions into the solution in a controlled sequence. This phenomenon has been well documented in the case of endochondral cartilage formation in which calcium first enters the extracellular matrix followed by phosphate after a significant lapse in time (Shapiro and Boyde 1984; see Chapter 9). Another strategy is to keep the major cations and anions separate until mineralization is about to be initiated; this phenomenon is known to occur in certain benthonic foraminifera (ter Kuile and Erez 1988). Although documentation of this phenomenon is still sparse, we suspect that it is widespread. This is based on the rather circumspect reasoning that all acidic matrix glycoproteins examined to date, essential apparently for regulating mineralization, need to first associate with calcium ions to fold into their ordered conformations (Worms and Weiner 1986; Stetler-Stevenson and Veis 1987). This, in turn, is almost certainly a prerequisite for matrix self-assembly. Thus, a compartment, which is subdivided by a matrix framework, is already partially loaded up with calcium ions at an early stage. The subsequent introduction of the anion will trigger mineral nucleation. We wonder whether this is not one reason for the widespread intimate association of anion manipulating enzymes such as alkaline phosphatase and carbonic anhydrase with mineralization processes.

The composition of the ions that actually enter into the mineralization compartment is determined by the manner in which they are extracted from the environment, transported through the tissues, and introduced into the mineralization site. No attempt is made to review this very important aspect of biomineralization here. The fact is that many organisms are well equipped to determine which ions will enter into the mineralization compartment. There are many examples of biogenic minerals that are formed by organisms that live in environments in which that same mineral would never form inorganically. One of the best known is the protoctists, Acantharia (Chapter 4). These single-celled marine animals form hard parts composed of celestite or strontium sulfate. (Table 2.1) This mineral is highly undersaturated in the oceans (Odum 1951) and, upon death, the shells rapidly dissolve. Skeleton-containing animals that live in freshwater or on land have to expend a large amount of energy sequestering ions for setting up saturated solutions for mineralization. It is, therefore, quite surprising that organisms that live in water that is supersaturated with respect to calcite and aragonite, namely the shallow water marine environment, generally do not simply fill up their mineralization compartments with "filtered" seawater. In Chapter 11 we describe cases that are often thought to be the exceptions to this rule, namely those organisms whose mineralization products are directly influenced by the environment in which they live. Upon closer examination, however, we note that most of the observed cases are also genetically controlled. It seems that the vast majority of biogenic minerals formed under controlled conditions precipitate from solutions whose compositions are very closely monitored by the responsible cells. The large body of information showing that the trace element and stable isotopic compositions of many mineralized hard parts are not in equilibrium with the environmental water attests to this (Chapter 11).

The composition of the mother liquor can also be influenced in a rather dramatic manner by the subtle effects of so-called *additives* or molecules present in small amounts in the solution that interact with the incipient solid phase. A good example of such an effect in biomineralization is a case in which amorphous calcium phosphate (ACP) is stabilized and prevented from spontaneously converting to the crystalline phase. The ACP granules present in the mitochondria of hepatopancreas cells of the blue crab *Callinectes sapidus* contain significant amounts of adenosine triphosphate (ATP) and Mg (Becker et al. 1974). In vitro experiments demonstrate that both delay the conversion of ACP to hydroxyapatite, but together

ATP and Mg act synergistically and reduce the conversion rate by about an order of magnitude (Posner et al. 1978). The mechanism proposed for this effect is that during the precipitation process the additives poison the growth of embryonic hydroxyapatite nuclei and these redissolve preventing hydroxyapatite crystal growth and in effect stabilize the amorphous phase.

In echinoderms, the skeleton is composed of high Mg calcite. If the surfaces of the mineral phase are cleaned and washed and the mineral phase is then dissolved, an assemblage of acidic glycoproteins is released (Weiner 1985; Benson et al. 1986). In vitro experiments in which calcite crystals are grown from a saturated solution that contains low concentrations (micrograms per milliliter) of some of these acidic macromolecules (the "additives") result in calcite crystals forming with some of the proteins inside the crystal (Berman et al. 1988). The presence of the protein radically changes the mechanical properties of the crystal and it behaves in fracture somewhat like the sea urchin skeleton. The strategy that the sea urchins have employed, apparently, is to dope their saturated solutions of mother liquor with low concentrations of these proteins, so as to affect the nature of the mineral phase formed. This example also serves to illustrate the fact that in cases in which these interactions occur, the mineral phase should contain a record of these types of interactions and, thus, much more could be learned about such processes from careful analysis of the trace constituents, both organic and inorganic, present in the mineral phase.

We note that the ability of additives to influence crystal nucleation and growth in synthetic solutions is a well-known phenomenon and indeed many industrial crystallization processes utilize additives. A detailed understanding of the basis of these effects on organic crystals grown in the presence of small amounts of additives in solution has been achieved by Lahav, Leiserowitz, Addadi, and their colleagues, and is reviewed by Addadi et al. (1985). Highly specific stereochemical interactions occur between the additives and the surfaces of the growing crystals that depend to a large extent on the molecular structures of the crystal surfaces and the additives, as well as on the solvent itself. Subtle changes, such as the differing position of a hydrogen bond in chiral molecules, for example, result in large differences in the morphologies of the crystals formed (Addadi et al. 1982). The additives influence crystal morphology if their structures are such that they interact preferentially with certain crystal faces and not others. The interaction during crystal formation retards growth on these faces causing them to grow more slowly and, as a result, the affected faces have relatively large surface areas. Alternatively, the additive can interact with the incipient crystal nuclei and, again depending upon the stereospecificity between surfaces and molecules involved, can inhibit the formation of certain nuclei and allow others to form. We suspect that these types of effects are widely utilized by organisms for controlling biological mineralization, even though, to date, not much documentation is available.

3.3.4 Control over Nucleation

There are good indications from a wide variety of mineralization processes that many organisms do exploit the nucleation event to exert control over mineral formation. We still understand very little about the ways in which this occurs, but comparisons of mineralization processes from different phyla show that the extent of control can vary enormously, from no control to highly controlled (Weiner 1986). Although we are concerned specifically with the subject of "control over nucleation," we note that it is not uncommon for organisms to exert no apparent control over the nucleation event, but to still control other aspects of mineralization. The formation of some amorphous mineral phases may belong in this category.

Control over nucleation can be indirect in the sense that molecules in solution may specifically inhibit the formation of nascent nuclei of one mineral phase and, in so doing, allow another to form. As noted, adenosine triphosphate may cause amorphous calcium phosphate to form by inhibiting the nucleation of the crystalline phase (Posner et al. 1978).

Direct control over nucleation in biology generally occurs on specially created solid surfaces. Nucleation off solid surfaces rather than particles in solution in general is most effective, in part because the movements of the ions that have to come together to form a stable cluster ("critical nucleus") are restricted when bound to solid surfaces. This was demonstrated empirically for calcite crystal nucleation in vitro in the presence of acidic glycoproteins extracted from mollusk shells. The glycoproteins in solution are potent inhibitors of calcite nucleation, whereas adsorbed on a solid substrate they become more efficient nucleators than their counterparts in solution are inhibitors (Addadi and Weiner 1985).

To really understand controlled biological nucleation, it is necessary to have a good understanding of the molecular structure of the site itself and the organization of the first layers of ions that form over the site. Unfortunately techniques for fully resolving this problem currently do not exist, except possibly for the scanning tunneling electron microscope. In the interim, we can learn something about nucleation, albeit indirectly, from the formed crystals themselves, the molecular organization of the substrate, and particularly from the spatial relations between them. The types of substrates most likely to be involved in nucleation are the lipid bilayer membranes of vesicles and the acidic glycoproteins and proteoglycans located on the surfaces of organic matrix frameworks.

Mineralization within lipid bilayer-bound vesicles is extremely common. It does not, however, necessarily follow that the vesicle walls are always actively involved in nucleation. Observations of the stages of mineral formation in intracellular vesicles from different animals show that many of them form by accretion of mineral around a node at the center of the vesicle [reviewed in Simkiss (1976) and Brown (1982)], implying that the vesicle walls are not the sites of nucleation. In other cases, vesicles actually contain organic macromolecules, which provide the substrates for nucleation [e.g., the base plate in the coccolith vesicles of *Emiliania huxleyi* (Westbroek et al. 1984) (Chapter 4)] and here, too, the vesicle membranes are not involved in nucleation. To date, the best documented cases in which vesicle membranes do appear to be directly involved in nucleation are the extracellular "matrix vesicles" in which hydroxyapatite forms. These have been well studied in cartilage and other vertebrate mineralized tissues (Chapter 9). One other example in which hydroxyapatite crystals appear to be actively nucleated on a membrane occurs in a bacterium, *Bacterionema matruchotii,* extracted from the oral cavity (Vogel and Smith 1976).

The most convincing evidence that these membranes are directly involved in nucleation is that they are relatively rich in a multimolecular complex composed of a proteolipid and phospholipids, which is able to strongly bind calcium ions (Boyan and Boskey 1984). These complexes in vitro efficiently induce hydroxyapatite formation (Boskey and Posner 1977). It has also been shown that if these complexes are implanted back into an animal, they induce hydroxyapatite formation in a dose-dependent manner (Raggio et al. 1983; Raggio 1986). One complicating fact, however, is that these very same membrane complexes are actively involved in regulating calcium transport through the membrane (Yaari et al. 1984). Although it has been proposed that the complex fulfills a dual function, more evidence, particularly studies of the crystals themselves that are formed in these vesicles, is needed to really prove that they indeed nucleate on the membrane itself. Do the crystals, for example, have unique habits that arise because they nucleate on a membrane?

Membrane-induced nucleation is particularly amenable to study in vitro, because the lipid bilayers can easily be reconstituted. Most of these studies to date have focused on mineral formation inside vesicles and have provided some very interesting observations and insights (Eanes 1986; Mann et al. 1986a). It is still very difficult to really demonstrate that nucleation actually occurs on the membrane itself, let alone how it occurs. This, however, has been conclusively demonstrated using Langmuir monolayers. Small changes in the chirality of the monolayer structure result in crystals of the amino acid glycine being nucleated off two different faces (Landau et al. 1985). Similar results were obtained for sodium chloride crystals grown under monolayers (Landau et al. 1986). This type of experimental design can be used to elucidate basic mechanisms of crystal nucleation on membranes including those more relevant to biomineralization. These experiments do show that membranes can function as highly specific nucleating substrates.

Specific macromolecular substrates are, it seems, the most common means organisms use to nucleate minerals under controlled conditions. The actual macromolecules involved have not been identified, let alone characterized, and we can only surmise that they are a subset of the class of acidic glycoproteins found in many mineralized tissues (Chapter 2) and are intimately associated with the mineral phase. In the absence of detailed information on the identity of nucleating macromolecules, their biochemical properties, and molecular structures, it is not surprising that very little is understood about this type of controlled biological nucleation in vivo. Almost all the currently available information is indirect; most of it is derived from the mineral phase itself, some from the spatial relations between the mineral phase and the major macromolecular constituents of the substrate, and only a few scraps of information are available about the nucleation site itself. Information on the in vitro nucleation capability of acidic glycoproteins extracted from mineralized tissues has also provided some fairly detailed insights into what may be happening in vivo.

The ultrastructure of a mineralized tissue can in itself provide some clues about nucleation. It is often self-evident from the tissue fabric whether the nucleation site induces the formation of a single crystal or multiple crystals. Single crystal nucleation events occur in bone (Chapter 9), mollusk nacre, and prismatic layers (Chapter 6) and in the formation of an entire larval spicule of the sea urchin (Chapter 8). Spherulitic structures found in scleractinian corals (Chapter 5), egg shells (Simkiss 1968; Silyn-Roberts and Sharp 1986) and certain mollusks (Carter and Clark 1985) are the products of many crystals being nucleated at one site. The sizes of the spherulitic clusters or the single crystals in the mature tissue are often directly determined by the distances separating nucleation sites. This is particularly well illustrated in the fabric of avian egg shells.

Information about the relative orientations of the macromolecules in neighboring nucleation sites can be inferred from the orientations of the crystallographic axes of the minerals that form at these sites. In gastropod nacre the aragonitic crystals are all randomly oriented about their *c* axes (Wise 1970). The *a* and *b* axes of the crystals in the nacre of the bivalve *Mytilus californianus* and the cephalopod *Nautilus repertus* show a preferred orientation, whereas those in the bivalves *Neotrigonia margaritacea* and *Pinctada margaratifera* are perfectly aligned over distances as large as a few millimeters (Weiner et al. 1983b). These differences reflect the relative orientations of neighboring nucleation sites, which in turn must presumably represent the products of the activities of individual mantle epithelial cells. In *N. margaritacea* and *P. margaratifera* thousands of cells construct nucleation sites that are all in perfect alignment.

The type of mineral nucleated can conceivably reflect the molecular structure of the nucleation site, although without knowing the actual structure of the site, this is almost impossible to prove. It has been proposed that the structure of the nucleation site is responsible for organisms forming aragonite and not calcite, and vice versa (e.g., Hare 1963). Examination of the lattice structure of these two polymorphs of $CaCO_3$ reveals that they are very similar in the *ab* plane, off which they nucleate. This would imply that the nucleation sites have specific and highly complex structures and may even be three-dimensional. However, other explanations involving variations in the composition of the mother liquor (Lahann 1978) also do not appear to be able to account for this obviously genetically controlled phenomenon. The "calcite–aragonite problem" remains unresolved and, in our minds, represents one of the most challenging problems in biomineralization. The solution is probably not super sophisticated or trivial, just elegant and simple!

More relevant, but difficult to obtain information on nucleation can be derived from the spatial relations at the molecular level between the crystal and the macromolecular substrate on which it forms. A well-defined spatial relation would strongly suggest that nucleation occurs by epitaxy. Epitaxy is defined as the "oriented overgrowth of one crystalline phase on the face of another" and occurs when the "atomic (lattice) arrangement in the substrate matches the lattice arrangement of the nucleating phase" (Posner and Betts 1981, p. 260). When nucleation is less controlled, the spatial relations between overgrowth phases and substrate will not be invariant, even though the crystals that nucleate off the substrate may show a well-defined orientation of one of their crystallographic axes relative to the substrate. An example would be the orientation of aragonitic *c* axes perpendicular to the nucleating surface, whereas the *a* and *b* crystallographic axes bear no specific relation to the substrate structure. Weiner (1986) referred to this as "stereochemically controlled nucleation." It is also quite possible that no spatial relation exists between mineral and substrate. Thus, identifying this property of a mineralized tissue may provide important information on nucleation. To date fairly detailed information on mineral–substrate spatial relations at the molecular level is available for only three mineralized tissues: mollusk nacre, vertebrate tooth enamel, and mineralized turkey tendon as an analog of bone. The known crystal–matrix relations of these three tissues are schematically illustrated in Figure 3.2.

In mollusk nacre the orientations of all three crystallographic axes of aragonite are known with respect to the chitin and silk-fibroinlike β-sheet proteins that constitute the matrix framework. The crystals nucleate on the matrix surface that is coated by acidic glycoproteins; their orientations are unfortunately not known and can be surmised only by extrapolation from the insoluble core. Nacre from all three of the major classes of mollusks show the same spatial relations: the *a* and *b* crystallographic axes of aragonite are aligned with the chitin fibers and protein polypeptide chains, respectively, and the *c* axis is perpendicular to the surface of nucleation (Weiner et al. 1983b; see Chapter 6). This well-defined spatial relation strongly suggests that nucleation from the matrix surface occurs by epitaxy. This cannot be proven until information is available on the molecular structure of the nucleation site to determine whether it in any way resembles the lattice structure of aragonite (Weiner and Traub 1984).

In vertebrate tooth enamel, the orientation of the acidic matrix proteins relative to the carbonate apatite crystals is known (Jodaikin et al. 1985; 1988). The acidic "enamelins" are largely in the β-sheet conformation and their polypeptide chains are predominantly oriented perpendicular to the *c* axes of the crystals. This arrangement raises the possibility that the enamelins constitute part or all of the tubule in which the crystals grow. Tubules are known to be present in forming enamel (Jessen 1968; Chapter 9). It is not clear whether the enamelins are involved in modulating crystal growth during elongation and/or fulfill a role in nucleation. In fact it is not known precisely where enamel crystals nucleate: close to the dentin layer or repeatedly as the crystal grows (Weiner 1986). Our current understanding of this observed spatial relation indicates only that enamelins must fulfill a very important function in crystal formation.

In bone the crystals are extremely small and the overall ultrastructure is complex. Thus, the task of determining their precise orientations with respect to the matrix in which they form is formidable. Mineralized tendons of large land birds are better organized and more easily studied, with the result that to date our most detailed information on collagen–apatite relations is from mineralized turkey tendon (Chapter 9). We do not know how good an analog it is of bone. We know the orientations of the crystals and therefore also the orientations of the mineral crystallographic axes with respect to the collagen fiber axis (Weiner and Traub 1986) and we have a fairly well-substantiated model of the sites that the crystals occupy inside the fibril (Hodge and Petruska 1963; Miller, 1984; Chapter 9). When the structure of the fibril is known in more detail, we will be in a position to compare the atomic arrangements of the crystal with those of the collagen that abut the crystal. An exciting propect indeed! It is not known precisely where the crystals nucleate, or whether this occurs directly on collagen or on some acidic noncollagenous macromolecule(s). The available spatial information on turkey tendon is in one respect analogous to that of enamel; it shows that the matrix framework is intimately involved in regulating crystal growth, but so far provides little definitive information on the mode of nucleation.

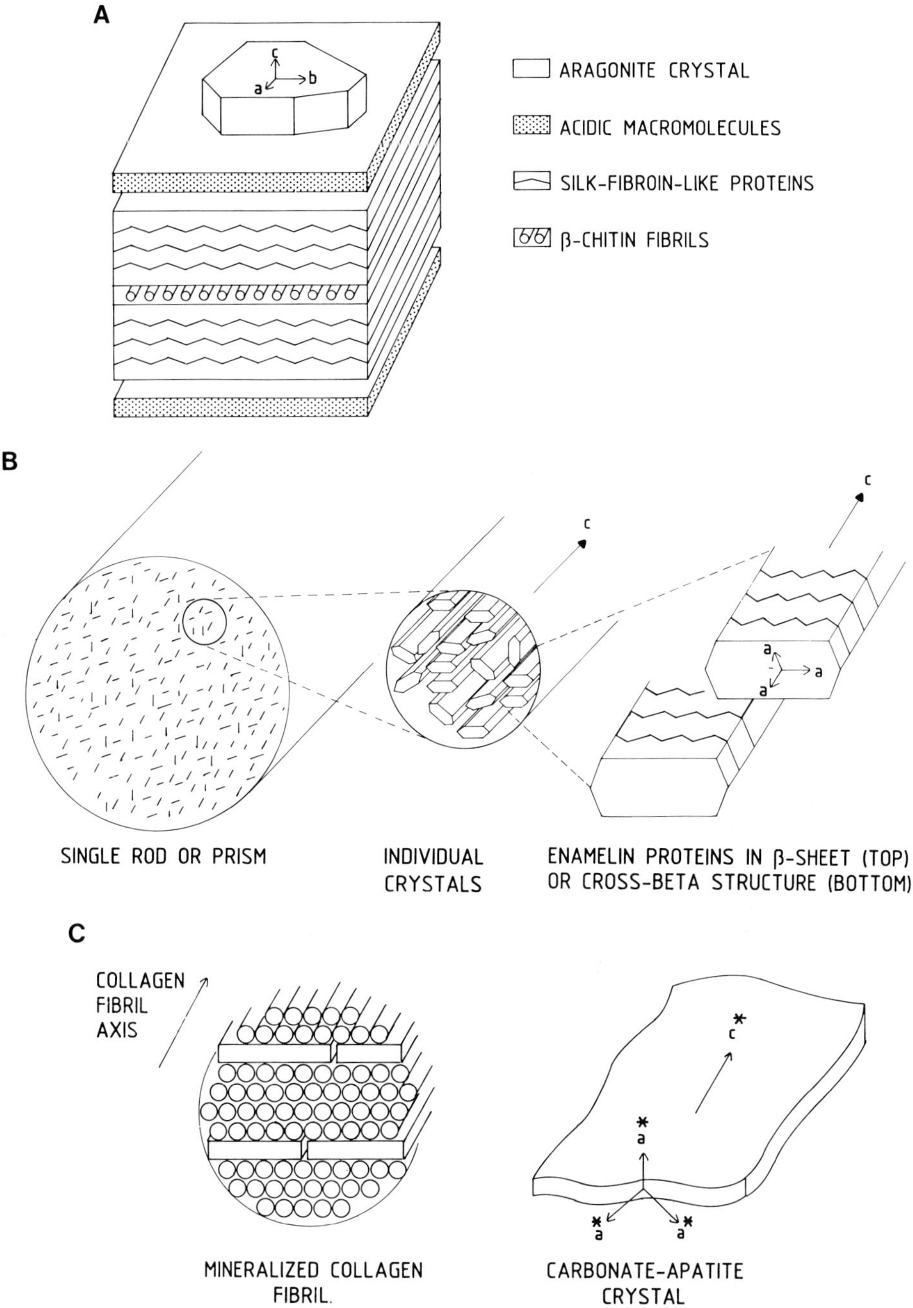

Figure 3.2. Schematic illustrations of the crystal–organic matrix relations in (a) mollusk shell nacreous layer, (b) mammalian tooth enamel, and (c) mineralized turkey tendon.

Hopefully we will soon begin to probe nucleation sites directly, both by isolating and characterizing molecules known to be involved in nucleation and by mapping the molecules in the site itself. To our knowledge the only direct information available on an actual nucleation site is from mollusk nacre. Wada (1980) has shown by electron probe that the site is rich in sulfur (proteoglycan?). In a landmark study, Crenshaw and Ristedt (1976) using histochemical techniques show specific calcium binding to the nucleation site of the nacre of *Nautilus pompilius,* as well as the presence of sulfated polysaccharides (Chapter 6).

It is clear from the above discussion that attempts to mimic in vivo nucleation in vitro can be very misleading. Numerous studies have been reported in which the effect of macromolecules from mineralized tissues on crystal nucleation and/or growth kinetics is examined. In most cases, these charged molecules inhibit crystal formation, as occurs with any nonspecific charged polymer (Birchall and Davey 1981). In a few cases (e.g., Nawrot et al. 1976) the macromolecules are reported to induce nucleation. It is not obvious that this behavior can be taken as proof of the molecule fulfilling a similar function in vivo.

Addadi et al. (1987) used a different in vitro approach to study nucleation by acidic matrix macromolecules. In a completely contrived situation that was not meant in any way to mimic in vivo nucleation, they showed that for oriented calcite crystal nucleation to occur, sulfonate ligands linked to a polystyrene substrate and carboxylate ligands of polyaspartate in the β-sheet conformation must cooperate to concentrate calcium over the ordered β-sheet nucleation site (schematically illustrated in Fig. 3.3). The latter is responsible for the nucleation event itself. They also showed that an apparently similar phenomenon occurs with actual mollusk shell glycoproteins in vitro, in which the polysaccharide moiety is heavily sulfated and the aspartic acid-rich proteins adopt the β-sheet conformation in the presence of calcium. Whether this actually occurs in vivo remains to be shown, although the widespread association of sulfate and carboxylate ligands in mineralized tissues is consistent with the hypothesis.

3.3.5 Control over Crystal Growth

Very little consideration has been given to the possibility that organisms can influence crystals as they are growing. The most direct evidence for this is the development of relatively unstable crystal faces during growth as a result of constituents in the solution specifically interacting with the face. Some stereochemical laws governing the process have been identified (Addadi et al. 1985) and it has been demonstrated that acidic glycoproteins from mineralized tissues are able to function in vitro in a similar manner (Addadi and Weiner 1985; Berman et al. 1988). A first hint that crystal growth modulation by acidic glycoproteins does occur in vivo comes from a study of the sea urchin *Paracentrotus lividus.* Acidic glycoproteins extracted from the test interact in vitro specifically with a set of crystal faces of calcite that are almost parallel to the *c* axis. The crystal overgrows the adsorbed proteins and it has been conclusively shown that the proteins are occluded inside the formed crystal. These calcite crystals fracture quite differently from normal calcite crystals, and, in fact, resemble the manner in which the sea urchin calcitic skeleton itself fractures. This suggests that this type of growth modification process does actually occur in vivo in echinoderm mineralization (Berman et al. 1988).

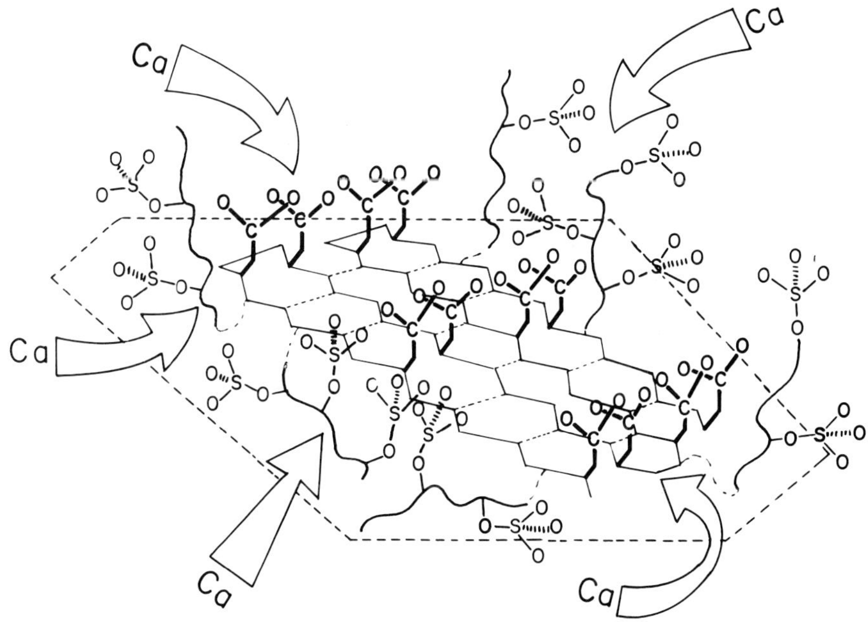

Figure 3.3. A schematic illustration of a possible nucleation site for mollusk shell comprising protein in the β-sheet conformation adsorbed onto a solid substrate and with covalently bound sulfated polysaccharides (following the study by Addadi et al. 1987). The arrows show the imagined flux of calcium ions toward the protein nucleation site. Courtesy of L. Addadi and S. Weiner.

The significance of crystal morphology in terms of reconstructing events that occurred during crystal growth has not been well appreciated. Constantz (1986) pointed out that the aragonitic crystals of different genera of scleractinian corals have characteristic morphologies. These could conceivably result from specific interactions between the acidic proteins associated with the coral skeleton (Chapter 5) and the growing crystals. Another example in which morphologies of growing crystals may be determined by interactions with acidic glycoproteins is the mollusk shell nacreous layer. Figure 3.4 shows that the forming nacreous aragonite crystals from two bivalve species have quite different morphologies. It will be interesting to discover whether these crystals also have occluded glycoproteins as was found in the sea urchin.

3.3.6 Cessation of Crystal Growth

In terms of the final mineralized product, control over the locations at which mineralization ceases often determines the overall shape of the material. If the mineral phase is amorphous, its final form is essentially the same as that of the space in which it precipitates. The incredibly intricate morphological forms of siliceous organisms (diatoms, radiolaria, silicoflagellates, and so on) are unparalleled in the field of biomineralization (Chapter 4). These structures form inside a so-called silica-depositing vesicle that is responsible for their final shape (see also chapters in

a

b

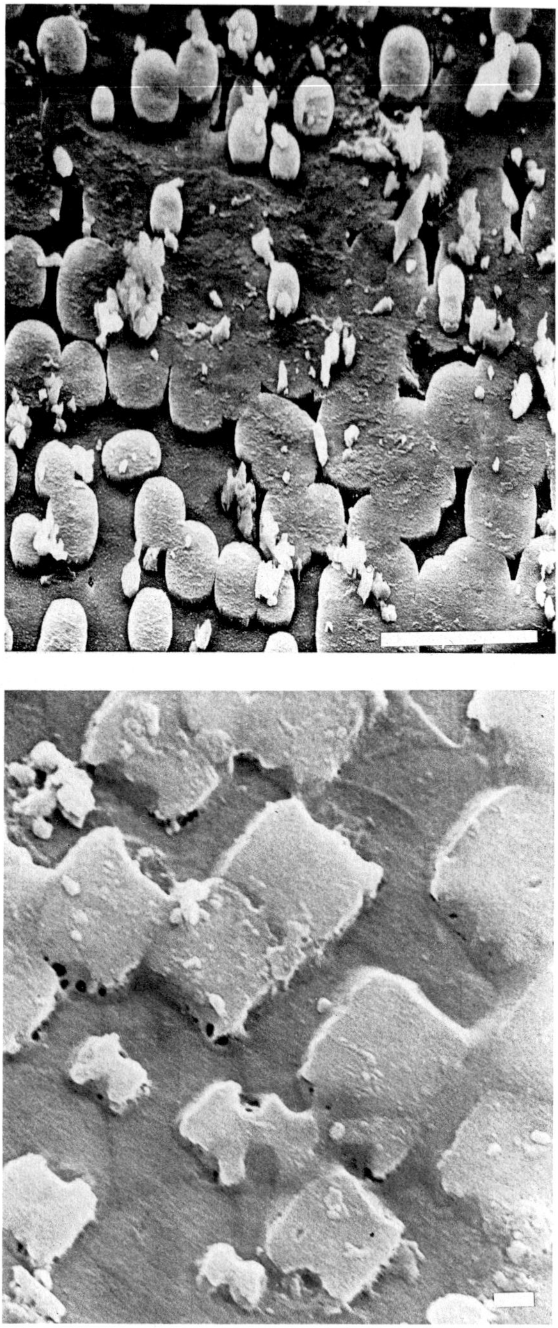

Figure 3.4. Scanning electron micrographs of the growing nacreous layer tablets of the bivalves (a) *Neotrigonia margaritacea* and (b) *Brachiodontus variabilis*. Scale bar in (a) 10 μm and in (b) 1.0 μm.

Simpson and Volcani 1981). Intuitively it is expected that the vesicle will be in its final shape even before mineralization starts. This is the case in the protistan flagellate, *Synura petersenii* (Leadbeater 1984), but in diatoms the vesicle apparently expands like a stocking as the mineral accumulates (Li and Volcani 1984). Unfortunately very little is known about the molecular structure of this vesicle. A similar process occurs in the coccolithophorida, *Emiliania huxleyi*, a marine protoctist that forms crystalline calcite structures (coccoliths) (Westbroek et al. 1984). Here too the intracellular vesicle expands with the mineral. Significantly the external morphology of coccoliths is also smooth and curved into intricate patterns (Chapter 4). A third example of this type of mineral cessation process occurs in the Echinodermata (Chapter 8). In this case it is also a crystalline mineral that forms in a vesicle and the mature product again takes on the shape of the vesicle including its smooth and curved surfaces.

Crystals that form in compartments delineated by matrix macromolecules also stop growing when they contact the preformed surface of the matrix. Our overall impression is that the morphologies of the mature crystals that form in this way are far more regular and generally do not have the curved surfaces of vesicle-bound mineralized deposits. The crystals of bone and enamel (Chapter 9) are good examples. In mollusk nacre (Chapter 6) the aragonite crystals are also thought to form in three-dimensional preformed compartments (Bevelander and Nakahara 1969a; Nakahara 1983). However, the fact that the lateral dimensions and hence shapes of these uniformly thick polygons vary significantly suggests an alternative possibility, namely that their lateral boundaries may simply be determined by the points at which neighboring crystals contact each other (see Weiner 1986). Growth cessation as a result of crystal–crystal contacts is quite common. In spherulitic structures, for example, the crystals clearly stop growing when they come into contact with crystals emanating from a neighboring nucleation site (Silyn-Roberts and Sharp 1986).

We know very little about the structures of lipid bilayers or macromolecular sheets that are responsible for crystal growth inhibition. One of the best studied examples is an elaborately structured polysaccharide present in the coccoliths of *Emiliania huxleyi* (Fichtinger-Schepman et al. 1981) and in the vesicle membrane (van Emburg et al. 1986), where it may be involved in crystal growth inhibition (Westbroek et al. 1984). We know nothing about the structures of the surface acidic matrix macromolecules that inhibit mineral growth. Weiner and Traub (1981) speculated that inhibition might occur as a result of lattice mismatches between the crystal and matrix surfaces. Another common cause of crystal growth cessation, although one that is difficult to prove, is the simple lack of supply of ions to the mineralization site. This is probably the main reason why crystals formed by biologically induced processes stop growing.

3.4 The Real World

So far in this chapter our attempts to understand biomineralization processes have been reductionist to the extreme. We have broken down one aspect of the process, albeit an important one, into semimanageable stages and described these stages by combining observations from many different sources. Contemplation of the in vivo world of a single biomineralization process and all the factors that operate together to form the final product presents an entirely different reality in which the unknown far outweighs the known! In this final section on biomineralization processes, we deliberately try to return the reader to the "real world" of biomineralization, which includes not only ions, macromolecules, and minerals, but the all-important cells that orchestrate the whole phenomenon and coordinate their activities with the rest of the organism. An enormous amount still remains to be learned. This is illustrated by presenting several unrelated but fundamental topics that are relatively poorly understood.

In the previous sections of this chapter we described biomineralization in terms of space delineation, formation of preformed organic matrices, mineral nucleation, modulation, and cessation of mineral formation. Within each category different strategies used for controlling the particular process were differentiated. In reality of course a given mineralization process involves only some of these phenomena. As an exercise and being fully aware of the poor quality of data available, we have made *educated guesses* at the various types of basic processes that may be involved in the formation of six different mineralized tissues, all of which involve crystalline, rather than amorphous mineral formation. This is presented in Table 3.1. The table also serves to illustrate a few important points. A tissue is the product of a combination of different processes at each stage of mineralization. Even though the final product is unique, the specific processes involved at each stage may be common to many different taxonomic groups. An important objective of research in biomineralization should be to identify and understand these underlying processes. Table 3.1 also shows that in a specific tissue there may be no required combinations of basic processes. For example, there is no indication that a crystal nucleated by epitaxy will grow under a certain set of controlled conditions and not under other conditions. Only one trend is apparent in Table 3.1; processes that involve little or no control over nucleation also involve little or no control over any other aspect of mineralization. This is exemplified in Table 3.1 by the bacterium *Desulfovibrio.* In actual fact it is this trend that caught Lowenstam's (1981) attention when he proposed distinguishing mineralization processes involving little or no control (so-called biologically induced mineralization) and all the rest that involve some degree of control.

We have limited the discussion in this chapter so far to mineralization processes in which the initial mineral formed is the same as the mineral in the mature end-product. This is by no means always the case. Table 3.2 lists eight examples known to date, in which the precursor mineral is different from the mature form. Bearing in mind the great difficulty in identifying transient precursor forms, the phenomenon may well be widespread. Table 3.2 shows that it is found in prokaryotes as well as eukaryotes and in biologically induced as well as biologically controlled mineralization processes.

In all known instances the precursor phase is thermodynamically less stable than the mature form. It can be amorphous, paracrystalline (ferrihydrite), or crystalline (vaterite). Thus, all eight cases conform to the Ostwald–Lussac Law of stages, according to which "the phase with the highest solubility forms preferentially during a sequential precipitation" (Nancollas 1982, p. 86). Nancollas (1982)

Table 3.1 Basic Processes that May Be Involved in the Formation of Six Different Mineralized Tissues[a]

			Nucleation			Growth modulation		Cessation of growth	
	Space delineation	Preformed organic matrix	Nonspecific	Stereochemical	Epitaxial	Absent	Interactive	Crystal–crystal contact	Preformed structures
Mollusk nacre	PM/cell membranes	X	—	—	X	—	P	X	X
Mammalian bone	Cell membranes	X	—	—	X	—	?	X	X
Mammalian enamel	Cell membranes	X	—	X	P	—	P	X	X
Echinoid tests	Lipid bilayer	?	—	P	X	—	X	—	X
Radial foraminifera	PM	?	—	X	P	X	—	X	—
Desulfovibrio bacteria	Absent	—	X	—	—	X	—	—	—

[a]X, Most probable process; P, an alternate possibility; PM, polymerized macromolecules.

Table 3.2 In Situ Maturation of Biologically Formed Minerals

Phylum	Lower taxa	Mineral		References
		Precursor	Mature	
Monera				
Cyanobacteria	*Rivularia* sp.	Amorphous $CaCO_3$ →	aragonite	Golubic and Campbell (1980); Lowenstam (1986)
	Aquaspirillum magnetotacticum	Amorphous hydrous ferric oxide →	magnetite	Mann (1985)
Thiopneutes	*Desulfovibrio desulfuricans*	Hydrotroilite → greigite →	pyrite + S[a]; (greigite →) pyrite + mackinawite[b]	Hallberg (1972)
Animalia				
Mollusca	All polyplacophora Chitonidae	Ferrihydrite →	magnetite	Towe and Lowenstam (1967)
		Amorphous calcium phosphate →	dahllite	Lowenstam and Weiner (1985)
	Viviparus viviparus	Vaterite →	aragonite	Kessel (1933); Levetzow (1932)
Chordata	*Helix* sp.	Vaterite →	aragonite	Stolkowski (1951)
	Mammalian dental enamel, dentin, and bone	Octacalcium phosphate →	dahllite	Nelson et al. (1986)

[a]At pH 6.
[b]At pH 8.

noted that the Ostwald–Lussac sequence of stages is also the sequence of decreasing hydration for calcium phosphates formed in vitro and in vivo. This is in fact true of most of the other cases listed in Table 3.2 except vaterite as a precursor form of aragonite. The mechanisms of transformation may be different in each case. Octacalcium phosphate converts to hydroxyapatite by a hydrolysis reaction (Tung and Brown 1983). In the chiton teeth amorphous calcium phosphate (ACP) is thought to convert to dahllite by a dissolution and reprecipitation process (Lowenstam and Weiner 1985; see Chapter 6 for more details). We know nothing about the manner in which vaterite converts to aragonite in vivo; it could be by dissolution and reprecipitation or by a solid-state transformation process (also known as a single crystal to single crystal transformation), even though pseudomorphs of vaterite have not been observed. In the chiton teeth, the timing of the transformation of ACP to dahllite is well controlled (Chapter 6), and the dahllite crystals that form secondarily have a well-defined preferred orientation, suggesting the involvement of matrix surfaces as well (Lowenstam and Weiner 1985).

We are fully aware that in this chapter and to a great extent throughout this book we describe biomineralization processes with almost complete disregard of the cells and their roles in mineralization. This is not, as we emphasized at the outset, because we minimize their importance. On the contrary all the processes described are the direct consequence of the cell's activities. They control every stage of mineralization, only part of which is by means of the macromolecules and ions they introduce into the mineralization site. They are usually responsible for the timing of the mineralization process and for determining the rates at which mineral will be deposited. They coordinate these rates with the rest of the organism's growth. They often have a direct input into the overall spatial organization of the tissue. In the chapters that follow ample evidence will be provided for the importance of the cells in controlling mineralization. Here we highlight this topic by drawing attention to a phenomenon that is not widely appreciated in biomineralization: the fact that in some organisms the mineralizing cells can form different mineralized products at different stages of development or ontogeny. The cells or cellular organelles responsible for the mineralogic changes may in some cases be the same, whereas in others the mineral is, as a rule, formed by different cells or organelles. Table 3.3 lists the incidences known to us in which this occurs. This phenomenon is also almost certainly far more widespread, but has not been well investigated to date. Furthermore, the manner in which it manifests itself varies considerably from case to case.

In the Actinopoda, Zoomastigina, and Mollusca, different minerals are formed at different developmental stages. In the case of the oyster *Crassostrea virginica*, for example, the larval "shell gland" cells form aragonite, whereas in the adult the mantle cells form calcite. In the holothurian species of the *Molpadiidae*, the juveniles initially form spicules of calcite (Fig. 3.5a) that is the characteristic mineral of all echinoderm hard parts. Beginning with the late juvenile stage the mesodermal spicules, except in the oral and caudal region, are resorbed, although occasionally remnants do remain (Fig. 3.5b). They are replaced by granules of amorphous hydrous ferric phosphate together with opal (Fig. 3.5c). These minerals are the only mineral precipitates that form in this body segment (Fig. 3.5d) thereafter and throughout the adult stage. It is also interesting to note that this represents the only example that we are aware of in which two different mineral phases coexist at the

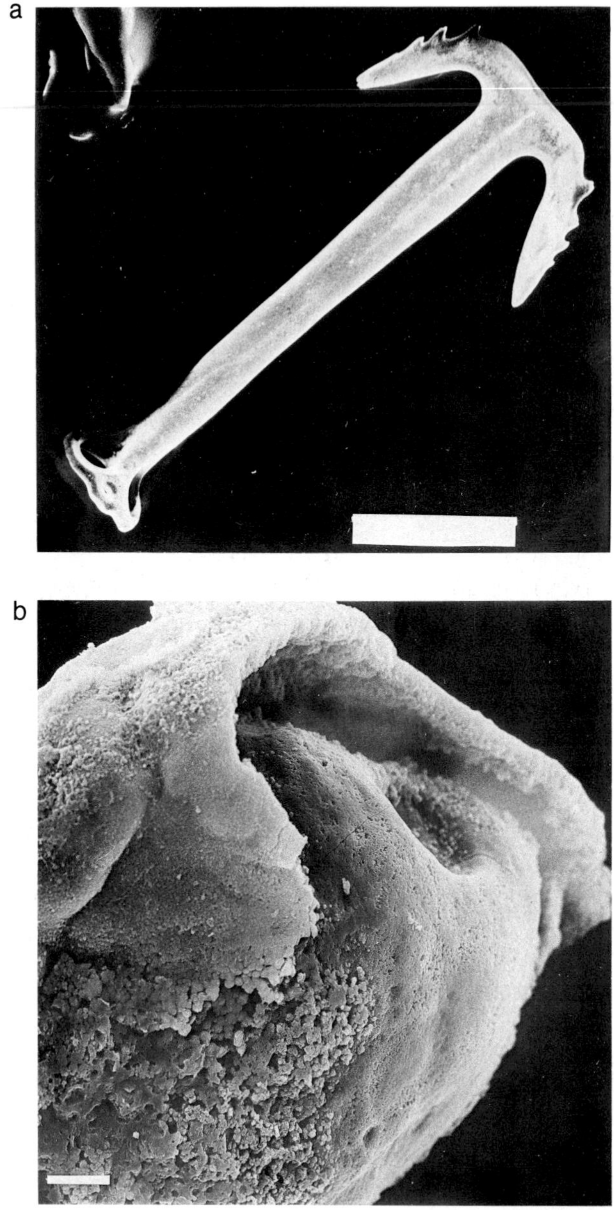

Figure 3.5. Scanning electron micrographs illustrating the stages of mineralization during development of the holothurian *Molpadia intermedia.* (a) An example of a calcitic spicule formed by a juvenile. Scale bar: 100 μm. (b) The spicule (center) is in the process of being resorbed and replaced by amorphous hydrous ferric phosphate (outer rim). Scale bar: 5 μm.

c

d

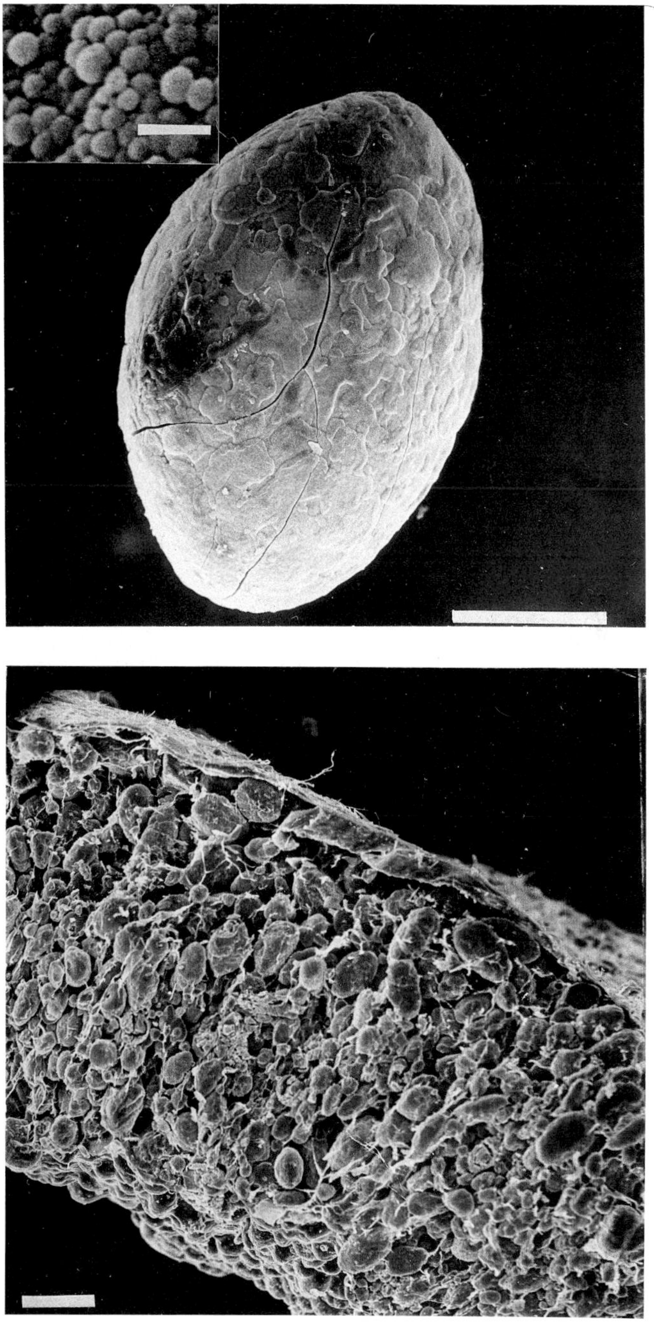

Figure 3.5. (*Continued*) (c) An amorphous hydrous ferric phosphate granule. Scale bar: 100 μm. Inset shows the spherical subunits that make up the granule. Scale bar: 0.45 μm. (d) The body segment containing granules. Scale bar: 650 μm.

Table 3.3 Ontogeny-Related Changes in Biomineralization Products

Phylum	Lower taxa	Ontogenic stage	Hard part	Mineralogy	References
Protoctista					
Actinopoda	Radiolaria	Isospore	Intracellular (crystal)	Celestite	Hollande and Martoja (1974)
		(Adult)	Skeleton	Opal (silica)	Anderson (1983)
Zoomastigina	*Pseudokephyrion*	Spore	Cyst	Opal	Tappan (1980)
		(Adult)	Lorica	Calcite	Tappan (1980)
Animalia					
Mollusca	*Crassostrea virginica*	Larva	Shell	Aragonite	Stenzel (1962, 1963, 1964)
		Adult	Shell[a]	Calcite	
Echinodermata	Molpadiidae	Juvenile	Mesodermal spicules	Calcite	Lowenstam and Rossman (1975)
		Adult	Mesodermal granules	Amorphous hydrous ferric phosphate and opal	Lowenstam and Rossman (1975)
Chordata	Elasmobranchii	Embryo	Otoconia	Amorphous calcium phosphate	Lowenstam and Fitch (1978); Lowenstam (1980)
	Prionace glauca	Juvenile	Otoconia	Amorphous calcium phosphate and aragonite	Lowenstam and Fitch (1978); Lowenstam (1980)
		Adult	Otoconia	Aragonite and trace of amorphous calcium phosphate	Lowenstam and Fitch (1978)
	Alopias volpinus	Embryo	Otoconia	Monohydrocalcite	Lowenstam and Fitch (unpublished)
		Juvenile	Otoconia	Monohydrocalcite and aragonite	Lowenstam and Fitch (unpublished)
		Adult	Otoconia	Aragonite	Lowenstam and Fitch (unpublished)

[a]Myostracum and ligament composed of aragonite.

exact same mineralization site. In the vestibulary apparatus of certain sharks, the otoconia of the embryos differ in mineralogy from those of the juveniles and adults. During subsequent development the otoconia formed by the embryos are not resorbed, but remain stable throughout life (Lowenstam 1980).

The manner in which the activities of the mineralizing cells are coordinated with the rest of the organism is an enormous subject in its own right. The importance of growth hormones, as well as hormones involved in regulating the uptake and transport of ions (for example, parathyroid hormone and vitamin D metabolites) are well appreciated and understood in vertebrates. It is not widely recognized that similar endocrinological controls exist in invertebrate animals (e.g., review by Joosse and Geraerts 1983) and this includes various vitamin D metabolites as well (Weiner et al. 1979; Hobbs et al. 1987). We make no attempt in this book to include this important aspect of the mineralization process.

It is well known that organisms may form different minerals at one deposition site. They are always, with the one known exception cited earlier, segregated into discrete microarchitectural units. Lowenstam and Weiner (1983) tabulated 22 known associations of different minerals at the *same* deposition site. Some organisms, however, form different minerals at *different* sites. As a final example of just how complicated these phenomena can be, we briefly describe the case of the pearly *Nautilus* (Lowenstam et al. 1984). This mollusk must qualify as one of the "superstars" of biomineralization. It forms mineralized hard parts at four different tissue sites: the shell, mandibles, vestibulary apparatus, and kidneys. It deposits no less than six different minerals: aragonite, calcite, brushite, amorphous calcium phosphate, weddelite, and another unidentified phosphatic mineral. The mandibles or jaws are in themselves a marvel of mineralization. They are composed predominantly of a structural organic complex of β-chitin and protein in the β-sheet conformation. The chitin fibrils are oriented perpendicular to the protein polypeptide chains. The mineral in contact with the chitin–protein complex is always aragonite, formed within a matrix under well-controlled conditions. Within the aragonite layers are isolated rosettes of brushite and it is suspected that they represent points of muscle attachment. The aragonite layers are in part overlain by thick segments of calcite. By fracturing the calcite, the growth surfaces are revealed and these often contain individual euhedral-shaped crystals of weddelite and brushite! These appear to have precipitated pseudoinorganically out of solution. Thus, within this one incredible animal, almost the whole spectrum of biomineralization processes is found, ranging from biologically induced to highly controlled organic-matrix mediated.

The real world of biomineralization covers a wide spectrum of biological disciplines that no single book can do justice to. In fact we hardly simplify matters by confining ourselves largely to the final acts of the mineralization process involving primarily the ions, macromolecules and minerals themselves. In the chapters that follow, we try to bring together information that is distributed throughout the literature and wherever possible we highlight some of the unifying concepts. It is our belief that as additional information becomes available, we will identify more of the underlying common principles and hopefully the field of biomineralization will gradually become simpler to understand. We are, however, still very much in the process of collecting the basic facts, without which the unifying principles are almost impossible to formulate.

4

Protoctista

This kingdom is defined by exclusion, in that its members are neither animals, plants, fungi, nor prokaryotes. They comprise eukaryotic microorganisms and their immediate descendants (Margulis and Schwartz 1988). Of the 27 phyla that make up this kingdom, no less than 17 contain members that form mineralized hard parts (Table 4.1). Although the vast majority of Protoctista are microorganisms, their smallness does not in any way imply an inability to control their biomineralization processes. On the contrary, many of the mineralizing Protoctista form very elaborate and complex structures. D'Arcy Thompson was one of many natural scientists who was both intrigued and fascinated by their skeletal morphologies. A perusal of his book *On Growth and Form* shows beautifully illustrated examples of protoctist skeletons and the text reveals a rare insight into some of the forces that govern their structure-forming processes. In the Radiolaria, for example, Thompson (1942) concludes that "the symmetry which the organism displays seems identical with that symmetry of forces which results from the play and interplay of surface-tensions in the whole system: this symmetry being displayed, in one class of cases, in a more or less spherical mass of froth, and in another class in a simpler aggregation of a few, otherwise isolated, vesicles" (p. 723). Although elegant and simple, physicochemical processes of interfacial chemistry are not sufficient to explain the complex, genetically controlled morphologies of many radiolarian species. Skeletal morphology is most likely the product of the delicate interplay between biologically controlled and physicochemically controlled processes (Anderson 1986). This is a recurring theme in biomineralization.

Not all the protoctists are expert mineralizers. In fact they exhibit the whole spectrum of mineralization processes, from uncontrolled to finely tuned. Within the foraminifera and testate amoeba, among the Rhizopoda, are examples in which this wide diversity is found even within an individual phylum. They both contain species that construct their tests entirely out of organic materials or organic materials reinforced with mineral grains scavenged from the environment. They also contain species in which the test is mineralized by the organism itself, and at least in the case of the foraminifera, this can occur both intracellularly and extracellularly (Lowenstam 1986). Thus, from the biomineralization point of view, the Protoctista display an almost endless variety of mineralization processes that offer many opportunities for studying basic underlying mechanisms.

Lowenstam (1986) made a comprehensive compilation of the known protoctist mineralization processes. Table 4.1 is an updated version of this compilation and the text below basically follows his discussion.

The most widely formed minerals among the protoctists are calcite and opal (silica). The calcitic tests of the Foraminifera and the Coccolithophoridae and the silica tests of the Bacillariophyta or diatoms are formed in such huge amounts in the world's oceans that they even affect many aspects of the chemistry of seawater. Furthermore the presence of abundant diatoms in freshwater bodies may influence their water chemistry as well (Chapter 12). Calcite is of course formed by many organisms other than protoctists (Table 2.1). Opal, on the other hand, although by no means exclusive to the protoctists, can be regarded as their "speciality." No less than 10 different protoctist phyla form siliceous mineralized hard parts, primarily, but not exclusively, for skeletal construction.

The protoctists, for reasons that remain obscure, form sulfate minerals more extensively than the other kingdoms (Table 2.1). The remarkable Acantharia build their skeletons out of strontium sulfate. Members of two different phyla use barium sulfate to form statoliths for gravity perception (Chapter 10). In fact they are the only known genera to use a noncalcium mineral for this purpose. Calcium sulfate (gypsum) crystals are formed by members of two different protoctist phyla, but their functions are not known. Given the fact that at least these protoctists do have a propensity to use sulfur, it is interesting that sulfur is used only in the oxidized form and not in the reduced form. The latter is the form commonly utilized by monerans (Table 2.1). This may hint at the possibility that these protoctists evolved in an aerobic atmosphere.

The majority of protoctists form only one mineral type. Lowenstam (1986), however, listed a number of cases in which two different minerals are formed by the same species (refer to Table 4.1). In one case both minerals are present at the same site or organelle. *Spirostomum* contains intracellular vacuoles with both dahllite and calcite (Pautard 1970). In *Spirogyra* and *Chara* the calcium oxalate and barium sulfate minerals are at two different anatomical sites. A third situation exists in which different minerals are formed during development. Species of *Pseudokephyrion* have a calcite-impregnated lorica, whereas the cysts are encased with silica (Tappan 1980). In the Radiolaria the isospores contain a vacuole that encloses a single celestite crystal, whereas the adults form a silica shell (Hollande and Martoja 1974; Anderson 1983). For additional examples and a more detailed discussion of mineralogic changes during ontogeny, see Chapter 3.

Given the facts that the kingdom of Protoctista is basically defined as a collection of phyla that are not obviously members of the other four kingdoms, we should not expect to find any mineralizing strategies or processes that are common to these organisms. It is, therefore, most surprising to discover that indeed six different protoctist phyla all have members that use a unique skeletal-forming process (Lowenstam 1986). Genera within the Haptophyta, Chrysophyta, Zoomastigina,

Table 4.1 The Diversity, Distribution, and Localization of Minerals Formed by the Protoctista[a]

Phylum[b]	Taxa reported	Mineralization site	Mineral	Reference
Myxomycota	*Didymium*	Peridium of spore coat	Calcite	Pobequin (1954)
Ciliophora	*Spirostomum*	Endodermal vacuoles	Calcite	Pautard (1970)
	Spirostomum	Endodermal vacuoles	Dahllite	Pautard (1970)
	Loxodes	Intracellular bodies	Barite	Hubert et al. (1975)
Rhizopoda	*Paraquadrula*	Exoskeleton	Calcite	Deflandre (1953)
	Cryptodifflugia	Exoskeleton	ACP	Ogden and Hedley (1980)
	Xenophyophora	Extracellular	Barite	Tendal (1972)
		Intracellular	Barite	Gooday and Nott (1982)
	Testacea (widespread)	Exoskeleton	Opal	Ogden and Hedley (1980)
Foraminifera	Almost all	Test or shell	Calcite	Blackmon and Todd (1959)
	Robertinacea	Test or shell	Aragonite	Blackmon and Todd (1959)
	Silicoloculina	Test or shell	Opal	Resig et al. (1980)
	Many arenaceous genera	Test or shell	Ferric oxides undefined	
Dinoflagellata	*Scrippsiella*	Resting cyst	Calcite	Wall et al. (1970)
	Actiniscaceae	Intracellular	Opal	Tappan (1980)
Zoomastigina	*Pseudokephyrion*	Lorica	Calcite	In Tappan (1980)
		Cyst	"Opal"	
	Loricate choanoflagellates (widespread)	Lorica	Opal	Leadbeater (1981)
Haptophyta	Coccolithophoridae	Coccoliths	Calcite	Isenberg et al (1963); Wilbur and Watabe (1963)
		Coccoliths	Aragonite	Manton and Oates (1980)
	Prymnesium	Cysts	Opal	Green et al. (1982)
Rhodophyta	Cryptonemiales (some)	Skeleton	Calcite	Borowitzka et al. (1974)
	Liagora, Galaxaura	Skeleton	Aragonite	Lowenstam (1955)
Chlorophyta	Most genera	Skeleton and gametangia	Calcite	In Tappan (1980)
	Chara	Statoliths	Barite	Schröter et al. (1975)
	Dasycladales (some)	Skeleton	Aragonite	Borowitzka et al. (1974)
	Penecillus, Udotea, Rhipocephalus,	Skeleton	Whewellite	Friedmann et al. (1972)

	Acetabularia, Bornetella,	Skeleton	Calcium oxalate	Arnott and Pautard (1970)
	Chlamydomonas	Intracellular	Magnetite	Lins de Barros et al. (1982)
	Chlamydomonas	Extracellular	Manganese oxides	Schultz-Baldes and Lewin (1975)
Phaeophyta	*Padina*	Surface deposit on thallus	Aragonite	Okazaki and Furuya (1977)
Gamophyta	*Oocardium*	Surface deposit	Calcium carbonate	Wallner (1933)
	Spirogyra	Intracellular statoliths	Calcium oxalate, barite	Arnott and Pautard (1970); Kreger and Boere (1969)
		Intracellular crystals	Aragonite	Mann et al. (1987)
	Closterium, Penium Pleurotaenium Telememorus	Intracellular	Gypsum	Fischer (1884)
Actinopoda	Acantharia	Skeleton	Celestite	Bütschli (1906); Odum (1951)
	Heliozoa	Exoskeleton	Opal	Bovee (1981)
	Radiolaria	Skeleton	Opal	Schröder (1901)
		Isospores	Celestite	Müller (1858); Hollande and Martoja (1974)
Bacillariophyta (diatoms)	Majority	Skeleton	Opal	Kützing in Ehrenberg (1834)
Xanthophyta	Few coccoid genera	Cysts	Opal	Round (1981)
?Eustigmatophyta	?Chlorobotrys	Cyst	Opal	Tappan (1980)
Chrysophyta	Synuracea,	Imbricated cell	Opal	McGrory and Leadbeater (1981), Bovee (1981)
	Aurosphaeraceae,	Covering scales	Opal	Tappan (1980)
	Silicoflagellates	Intracellular	Opal	Tappan (1980)
	Ebria, Hermesinum, Chrysococccus, Synura	Extracellular	Ferric oxides	Pringsheim (1946)
Euglenophyta	*Anisonema*	Intracellular crystal chains	Magnetite	Torres de Araujo et al. (1986)
	Siderophylic genera	Extracellular	Ferric oxides	Pringsheim (1946)

[a]Updated version of the table by Lowenstam (1986). Unlike the other tables of this kind, we do not have a "functions" column as, for the most part, the functions of protoctist mineralized hard parts are unknown.
[b]Phylum classification follows Margulis and Schwartz (1988).

Foraminifera, Actinopoda, Bacillariophyta, and Rhizopoda form mineralized skeletal elements intracellularly within specially designated vesicles. These structures are then extruded to the exterior of the cell where they are assembled to form the exoskeletons. The extrusion and assembly process is unique to these phyla and to the Protoctista. The reason this occurs or the evolutionary significance of this observation is not known. The processes are described in more detail below, particularly in the sections on diatom and coccolithophorid mineralization.

Although we deal in detail with mineralization in only three different phyla, we would like to draw the reader's attention to just a few of the many fascinating mineralization processes that occur among the protoctists. The lorica of the choanoflagellate *Stephanoeca diplocostata* is a basketlike structure constructed out of siliceous ribs or costae. The costae are formed intracellularly in vesicles and are then extruded out of the protoplast where they accumulate. They are assembled only when sufficient costae have accumulated to form one lorica. The assembly process involving many costae takes just a few minutes! (Leadbeater 1984; Leadbeater and Riding 1986).

Certainly one of the most exotic mineralized skeletons known is that of the Acantharia. Not only is it composed of strontium sulfate (Table 4.1), but the symmetry of the skeleton is quite extraordinary. The 20 spines are arranged in an extremely regular manner. The symmetry pattern was even formulated as a law by Müller (1858). Müller's law is discussed in some detail by Thompson (1942). The symmetry does not reflect the crystallography of celestite (strontium sulfate). In fact, it has been shown that each pair of opposite spicules constitutes a single crystal of celestite, and one of its crystallographic axes is parallel to the long axes of the spicules (Schreiber et al. 1959). It would be fascinating to find out the biological basis for the symmetry of these organisms.

A short chapter such as this cannot begin to do justice to the Protoctista in terms of describing their mineralization processes. Fortunately some excellent books are available on the subject and the reader is referred in particular to Simpson and Volcani (1981) on siliceous organisms, Anderson (1983) on Radiolaria, and Leadbeater and Riding (1986) for detailed descriptions of a wide variety of protoctist mineralization processes.

In the remaining sections of this chapter we describe, in some detail, the mineralization processes of the diatoms, foraminifera, and coccolithophoridae. Their quantitative abundance in the biosphere and the fact that their skeletal remains are often major constituents of marine sediments make them important in their own right. Furthermore, for the same reasons their mineralization processes have been relatively well studied as compared to other protoctists.

4.1 Diatoms (Bacillariophyta)

Diatoms are photosynthetic aquatic organisms living in hypersaline, marine, and freshwater environments. There are as many as 10,000 living diatoms and many more extinct species are known from the fossil record. They are very widely distributed and abundant in the photic zones of the world's oceans and in freshwater bodies. They are, therefore, an important group at the base of the aquatic food

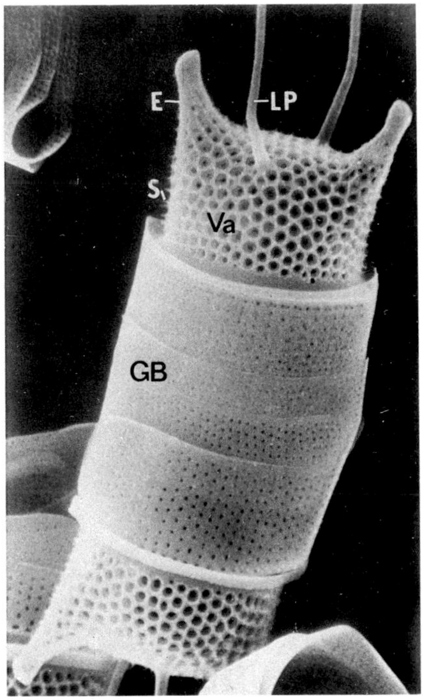

Figure 4.1. Scanning electron micrograph of the whole cell of the diatom *Odontella aurita* showing the valves (Va), and the girdle bands (GB). The maximum diameter of the girdle band is 28 μm. Reproduced from Li and Volcani (1985) by courtesy of Springer-Verlag New York, Inc.

chain. From our point of view their major interest lies in the fact that they form elaborate mineralized structures that enclose the entire organism. These structures, sometimes called frustules, comprise two interlocking valves as well as two or more ringlike girdles that surround the valves (Fig. 4.1). The mineral present in these structures is opal.

Opal is also often referred to as *silica.* It is an amorphous mineral that exists as a hydrated, covalent inorganic polymer with the general formula $[SiO_{2n/2}(OH)_{4-n})_m$, where n = 0–4 and m is a large number (Mann et al. 1983b). This formula indicates that it is not a stoichiometric mineral and varies both with respect to residual functional groups and degree of hydration. Transmission electron microscopy (TEM) examination shows that it is truly an amorphous material with no order being observed over distances greater than about 10 Å. ^{29}Si NMR studies confirm this observation and show that biogenic opal (the material obtained in these studies was actually from certain vascular plants) is very similar to inorganic silica gel (Mann et al. 1983b).

The terminology associated with silicification is confusing. Silicon (Si) is used to denote the element. Silicic acid $[Si(OH)_4]$, the nonionized form, predominates in aqueous solution at pH 8. Silicate ion (H_3SiO_4), the ionized form, comprises only about 2% of the dissolved fraction at pH 8.0. The mineral opal, as noted, is also sometimes referred to as "silica" (Simpson and Volcani 1981). For more information on silicon chemistry and biochemistry, see Iler (1979), Williams (1986), and Perry and Mann (1988).

Diatoms are the most intensely studied of all the siliceous organisms. Although their beautiful mineralized structures have been observed and described for more than a century, it was only in the 1960s that a series of pioneering studies triggered the start of the "modern" era of diatom silicification studies. Papers by Reimann (1964), Drum and Pankratz (1964), and Stoermer et al. (1965) described the sequence of events involved in the formation of new valves within a silica deposition vesicle. The membrane of this vesicle was termed the *silicalemma* by Reimann et al. (1966). Detailed biochemical studies of silicon uptake, transport, and its role in diatom metabolism were all made possible once Darley and Volcani (1969) devised the methods to grow diatoms in synchronous culture. Unfortunately to this date very little is known about the mineralization process itself. In this section, we briefly review the stages of valve formation and aspects of the uptake, transport, and deposition of silicon in diatom metabolism. We try to focus, wherever possible, on the many open questions relating to the mineralization process itself.

4.1.1 Ultrastructure of Valve Formation

Figure 4.2 is a schematic representation of the cell cycle with respect to the formation of the mineralized valves and girdle bands. It is interesting that one of the daughter cells is smaller than the parent, a fact first noted by Wallich in 1860. This does not occur in all diatoms, but when it does, restoration of maximum size at some stage is necessary. This occurs when the cells revert to a sexual mode of reproduction instead of the normal asexual mode (Drebes 1977).

4.1.2 Valve Formation

We will focus on valve formation, rather than girdle formation, and will basically follow the descriptions given in reviews by Li and Volcani (1984) and Crawford and Schmid (1986). The silica deposition vesicle starts to form at a discrete location and continues to grow by fusion with small vesicles (Dawson 1973). It is suspected that these vesicles also contain silicon in some precursor form and, upon fusion, the mineral enters the silica deposition vesicle (Schmid and Schulz 1979). The mineral forms long before the silica deposition vesicle adopts its final size and shape; in fact the silicalemma expands in conjunction with continued mineral accretion (Li and Volcani 1984). This does not occur in the silicoflaggelates, where the vesicle is preformed in its mature shape before mineralization is initiated (Leadbeater 1984).

The morphology of diatom valves and girdles is a marvel in itself and raises the question about the manner in which this intricate moulding is achieved. A series of elegant studies by Schmid (reviewed and illustrated schematically by Crawford and Schmid 1986) sheds some light on the subject by showing that the valve moulding machinery of *Coscinodiscus wailesii* is the product of highly complex and perfectly timed interactions of various cell constituents. These include the

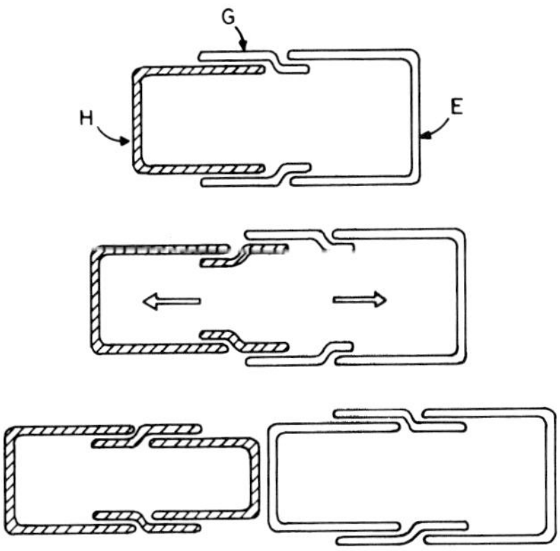

Figure 4.2. Schematic representation of a diatom cell cycle, showing the formation of the mineralized elements; E, epivalve; H, hypovalve; and G, girdle bands. (A) Nondividing cell. (B) Dividing cell showing the formation of a new set of girdle bands. (C) Completed cycle after the formation of two new valves. Note that the "shaded" daughter cell is smaller than the parent. Illustration is redrawn following Crawford (1981).

flattened saclike vesicles (cisternae) of the endoplasmic reticulum, certain large vesicles of unknown origin called the areolar vesicles, the external plasma membrane of the cell (plasmalemma), as well as various ill-defined fibrous elements. These vesicles and membranes arrange themselves into a specific pattern that in essence acts as the moulding of the about-to-be formed silica deposition vesicle. The latter initially adopts a tubular structure that is "squeezed" into the available space. It expands first in a two-dimensional plane and then in three dimensions in response to a precisely timed series of movements and changes in the shapes of the supporting vesicles. The forming silica deposition vesicle continuously fills up with mineral. Thus, views of the developing mineral phase of *C. wailesii* can graphically show the developmental stages of the silica deposition vesicle itself (Fig. 4.3). The sequence of events described by Schmid and colleagues illustrates not only the fine control exercised over every detail of the moulding process, but demonstrates that the highly curved and smooth morphology of the diatom valves is a direct result of the intimate involvement of the vesicular membranes in shaping hard part morphology. This type of morphology is characteristic not only of most siliceous protoctists, but of various other mineralized structures that form in vesicles. This even includes some composed of a crystalline mineral phase, such as the echinoderm skeleton (Chapter 8). The work of Schmid on *C. wailesii* provides, in our opinion, one of the most detailed insights into the manner in which these vesicles and membranes are directly responsible for determining morphology.

Morphogenetic studies of other diatoms (reviewed by Crawford and Schmid 1986) show that the detailed events vary from case to case, and also demonstrate the involvement of cytoskeletal elements, such as microtubules (Pickett-Heaps et al. 1979) and other organelles including the mitochondria.

As the silica deposition vesicle expands, mineral rapidly forms such that at any given time there is very little space between the forming surface of the mineral and

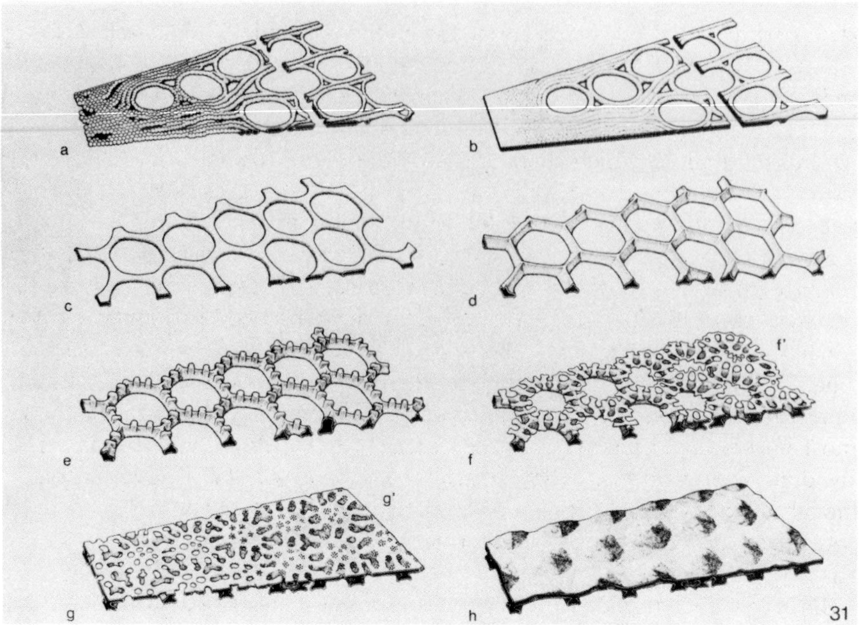

Figure 4.3. Schematic drawing of the valve development of *Coscinodiscus wailesii.* a–c shows the development of the base layer; d–e the development of the areolae walls; f–h the development of the outer layer. Reproduced from Schmid and Volcani (1983) by courtesy of the Phycological Society of America.

the silicalemma. The forming mineral surfaces have various morphologies—fibrous, spherical, and hexagonal columnar, depending upon the species of diatom examined (Li and Volcani 1984). These textures presumably reflect different modes of mineral accretion.

Valve formation, as well as girdle formation, occurs within vesicles inside the cell. The mature mineralized structures, however, end up outside the cell membrane. The manner in which this occurs in diatoms has not been unequivocally resolved. Four different models, critically reviewed by Crawford and Schmid (1986), have been proposed. They represent attempts to resolve an important problem, which is by no means unique to diatoms. Similar phenomena occur in a variety of protoctists, various octacorallia (Chapter 5), the echinodermata (Chapter 8), and the ascidians (Chapter 9). In none of these cases has this problem been fully resolved.

4.1.3 Uptake, Transport, and Deposition of Silicon

Diatoms are among the few mineralizing organisms that can be grown in synchronous culture. This was first achieved by Volcani and his colleagues and their studies of cultured diatoms have resulted in an impressive body of work on the uptake and transport of silicon. One of the highlights of these studies is the demonstration that silicon is an integral part of this organism's metabolism, necessary for activities of enzymes critical to DNA synthesis (Sullivan and Volcani 1973) as well as for the synthesis of various DNA-binding proteins (Oikita and Volcani 1980a) and acidic proteins (Oikita and Volcani 1980b). However, as Volcani (1981) himself points out, the actual manner in which the mineral phase, opal, is formed is still "largely *terra incognita* about which there are far more questions and speculations than answers" (p. 195). Sullivan (1986) has recently reviewed those aspects of diatom metabolism pertaining to silicification, which are briefly summarized here.

Silicon in the form of silicic acid [$Si(OH)_4$] is specifically bound by the diatoms. In vitro studies show that among the various membranes examined, the fractions enriched in plasmalemma as well as the one enriched in endoplasmic reticulum have the highest specific activities (Sullivan 1986). Silicic acid transport through the membrane is an energy-dependent process (Azam et al. 1974) requiring the presence of sodium in the uptake medium (Bhattacharyya and Volcani 1980). A most interesting observation is the presence of two silicate-specific ionophores in the diatom *Nitzschia alba* that are also dependent upon the presence of sodium. Although their biological function(s) has not yet been determined, they could be involved in active membrane transport of silicate (Bhattacharyya and Volcani 1983).

Silica is stored in one or more pools before it is deposited as a mineral in the silica deposition vesicle (Sullivan 1979). The concentration of silicon in these pools is extraordinarily high (around 500 mM Si per liter cell water) (Sullivan 1986), particularly in view of the fact (Sullivan 1986) that in vitro solutions with silicon concentrations greater than 3.5 mM Si per liter autopolymerize (Alexander et al. 1954). Sullivan (1986) thus hypothesizes that these silicon pools already contain oligomeric silica chains or macromolecular colloidal silica in addition to silicic acid. It is also conceivable that some of this silicon is bound to organic molecules. Thus, the polymerization process ultimately leading to opal formation may actually start outside the silica deposition vesicle. It also seems that these diatoms may well have evolved specific mechanisms to prevent autopolymerization at undesirable locations.

Unfortunately very little is known about the all-important silicalemma, the membrane of the silica deposition vesicle. It may have the capability of transporting silicon into the vesicle, and, in addition, may need to inhibit the continued formation of the mineral phase when it contacts its inner surface. Gaining insight into the actual mineralization process inside the silica deposition vesicle has proved to be very difficult. The mature cell wall of a diatom comprises the mineral phase, the silicalemma and an exterior organic casing. Ideally we would like to know if any macromolecules are intimately associated with the mineral phase inside the vesicle, as much as possible about the silicalemma itself, and about the structure of the casing. The latter may play some role in mineralization, but is more likely to be actively involved in protecting the mineralized fraction from dissolution, as evidenced by the fact that diatom frustules dissolve more rapidly from killed cells than from live cells (Lewin 1961).

No information is available on whether or not macromolecules are present within the mineral phase, even though conceptually at least it would seem possible to clean the valve surfaces with an oxidant, then dissolve the mineral phase and check to see whether macromolecules are present. It has not been possible to isolate the silicalemma itself from the casing and the only information available to date on its composition is inferred indirectly. Coombs and Volcani (1968) showed that the carbohydrate/protein ratio is sharply reversed during wall formation. The mature wall accretes primarily by the addition of carbohydrate, whereas about half of the macromolecular constitutents incorporated into the newly formed walls are proteins. They suggest that some of these proteins are part of the silicalemma, as at this stage it is a quantitatively major constituent of the fraction analyzed. These proteins are rich in glutamate, aspartate, serine, alanine, and glycine (Volcani 1981). (We note that this amino acid composition is reminiscent of most organic matrix proteins.) Similar observations suggest that certain carbohydrates are also part of the silicalemma. In contrast, analyses of the mature cell wall show that these proteins are rich in serine, threonine, and glycine and are depleted in acidic amino acids (Hecky et al. 1973). It is not clear what the functions of these hydroxyl-rich proteins really are, although their unusual amino acid composition does suggest some role in mineralization, possibly as a template for silica deposition, or as a means of protecting the mineralized hard parts from dissolution (Hecky et al. 1973).

4.1.4 Concluding Comment

Diatoms are just one of many organisms that form siliceous hard parts (Table 2.1). Even though they are the most thoroughly studied, we still have much to learn about frustule silicification. It is quite conceivable that other siliceous organisms will yield some of their secrets more easily than diatoms. Studies of silicification in vascular plants by Perry et al. (1984a,b) provide, for example, important information about the ultrastructural and molecular organization of the silicified deposits in relation to their environments of deposition. Clearly this process is also under strict biological control. It seems to us that the identification of other more convenient and amenable systems for studying silicification could help break the current impasse in this important field.

4.2 Foraminiferida

The planktonic foraminiferida, along with the coccolithophorida (next section) and the nektonic mollusks called pteropoda, account for most of the biogenic carbonate precipitated in the open oceans (Chapter 11). Foraminifera are widely distributed in the surface waters of all the world's oceans and accumulate in large quantities in the ocean bottom sediments. They are, therefore, a conspicuous component of many sedimentary rocks dating all the way back to the early Mesozoic. Because of their small size and widespread distribution, they have assumed great importance in biostratigraphy. They also have economic importance as they are used to determine the ages and environments of deposition of rocks extracted from drill holes. A large amount of taxonomic information is thus available on the foraminifera, most of which is contained in the Catalogue of Foraminifera (Ellis and Messina 1940). This now comprises more than 70 volumes (each about 6 cm thick!) and is continually updated. More than 34,000 species have been described, approximately

4,000 of which are living today. Foraminifera were first described by Strabo (63 B.C. to 20 A.D.) who observed specimens of the giant species, *Nummulites,* in the pyramids of Egypt (Boltovsky and Wright 1976). Most of the living species are not planktonic, but live on the ocean bottom in the benthonic habitat. Note that in spite of their tremendous diversity, the foraminifera, according to the more traditional classification systems, do not constitute a phylum. They are only an order in the class Granuloreticulosea, which is in the subphylum Sarcodina. This, in turn, belongs to the phylum Sarcomastigophora (Lee et al. 1985). Margulis and Schwartz (1988) on the other hand do regard them as constituting a phylum.

Fundamental taxonomic distinctions among foraminifera are based on the materials used for constructing their tests or shells (Sorby 1879; Wood 1949), a most convenient practice for those of us interested in biomineralization. The fact that this feature of foraminiferal biology provides a useful basis for taxonomic identification is as good an indication as any that shell formation is under strict genetic control and that the foraminifera use a variety of means for forming their shells. The Allogromiina have smooth, flexible, membranous tests composed of structural organic macromolecules and some agglutinated material picked up from the environment and "glued" onto the exterior test surface. Interestingly some species are very selective in their choice of materials. In one case only rather rare rutile mineral grains are chosen for this purpose (Dick 1928) and in another case only glauconite is used (Z. Reiss, personal communication). The manner in which they perform their mineralogical identifications remains a mystery. The Textulariina are similar to the Allogromiina except that their tests are more elaborately constructed and the adhering grains can be cemented together by a biologically precipitated mineralized cement (Bender and Hemleben 1988). The Fusulinina is an extinct group that constructed their tests out of so-called microgranular calcite in which the crystals can still be shown to have had unique morphologies (Green et al. 1980). Miliolina tests are made up of small elongated calcite crystals that for the most part are haphazardly arranged so as to give the shell a porcellaneous appearance in reflected light. Curiously, one miliolidlike foraminifera is known to form an opaline test (Resig et al. 1980). Finally, the Rotaliina have walls with a glassy appearance and contain many perforations. Most of the members of this important group form calcitic tests with their *c* axes oriented perpendicular to the inside surface of the wall. However, one taxonomically distinct group within the Rotaliina, the Robertinina, form aragonitic tests. A fascinating group within the Rotaliina, from the biomineralization point of view, is the Spirillinacea. They form calcitic tests in which in polarized light or by X-ray diffraction the whole test behaves as one single crystal (Towe et al. 1977; Bellemo 1978)! (Fig. 4.4). This situation is reminiscent of echinoderm skeletons (Chapter 8), but without more information on the manner in which the Spirillinacea crystal(s) form, the analogy could be only superficial.

Even this brief survey suffices to show that the foraminifera are an extremely diverse group of mineralizers. They can be kept alive in the laboratory (Adshead 1967) and can therefore be conveniently studied. Many foraminifera contain photosynthesizing symbionts (dinoflagellates, unicellular rhodophytes, chlorophytes, and a variety of pennate diatoms) that are an integral part of their metabolism. Sorting the relative contributions to the organisms' metabolism of the symbionts as opposed to the food intake [some foraminifera are carnivores and others are

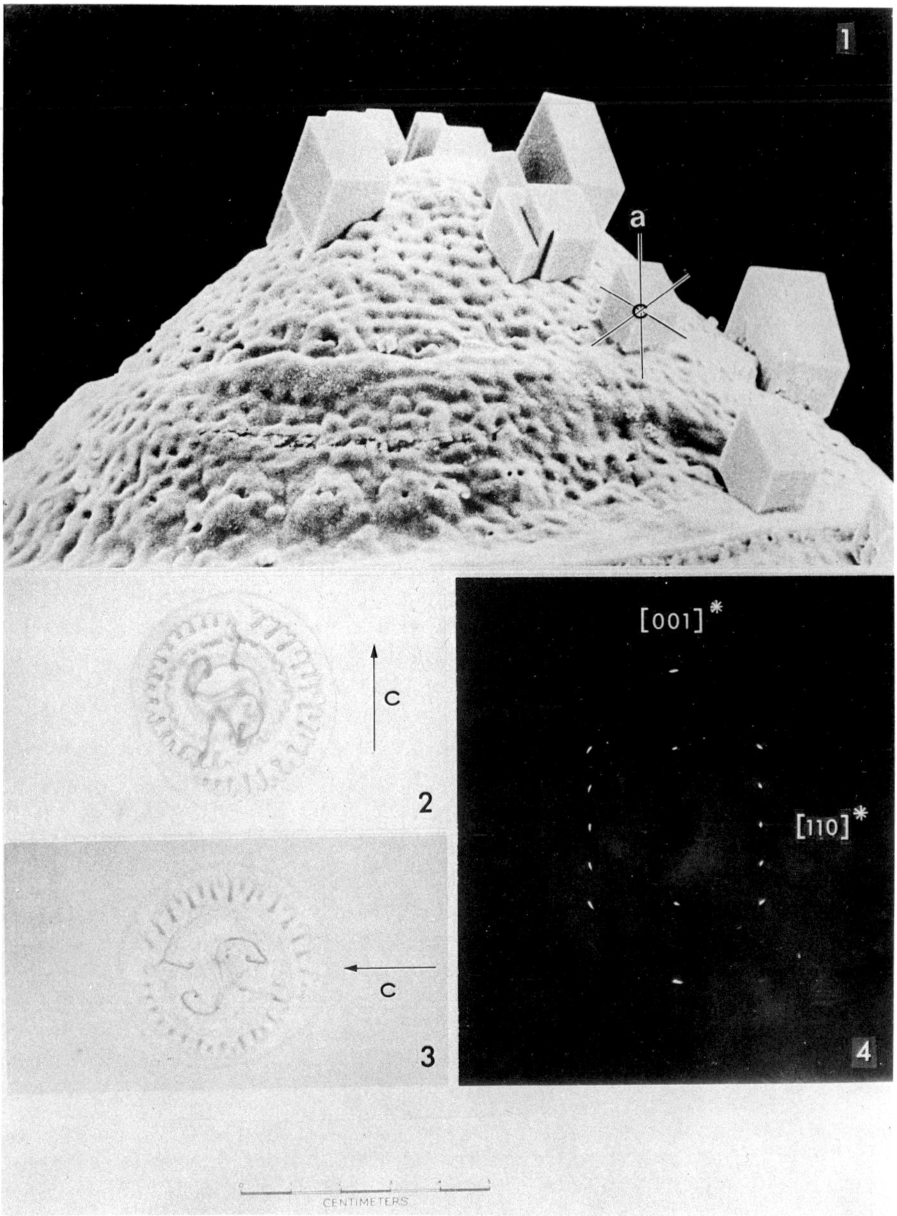

Figure 4.4. (1) Scanning electron micrograph of the test of *Patellina corrugata* secondarily overgrown with calcite crystals. (2, 3) An individual of *P. corrugata* viewed in polarized light with crossed-nicols and a gypsum plate. Both views are extinction positions at right angles showing maximum (2) and minimum relief (3). (4) Buerger precession X-ray photograph showing the single crystal pattern from *P. corrugata.* Reproduced from Towe et al. (1977) by courtesy of the Cushman Foundation.

herbivores (Spindler et al. 1984)], has given rise to some very interesting studies (Erez 1983; Spero and Parker 1985; ter Kuile and Erez 1987; 1988). The foraminifera, therefore, offer many exciting and challenging possibilities for understanding basic aspects of biomineralization. It is no surprise then that one of the most insightful papers written on biomineralization, in our opinion, is that by Towe and Cifelli (1967) on the foraminifera.

4.2.1 Agglutinating Foraminifera

Very little is known about the process of *agglutination*—the manner in which an organic or organic plus mineral cement is formed and how the particles are arranged. The very versatile pseudopodia are almost certainly involved in this process (Hottinger 1986). A pioneering study by Hedley (1963) identified the primary constituent of the cement as an acid mucopolysaccharide. A more detailed study of one particular species, *Allogromia laticollaris,* shows that the cement is mostly protein and that carbohydrates are a minor constituent (Schwab and Plapp 1983). We note that agglutination is by no means uncommon in the biomineralization world. One group of mollusks, the Xenophoridae, are superb "collectors" that arrange a motley array of objects on the exterior surface of their shells, often in neat and orderly ways (illustrated in Hyman 1967).

4.2.2 Miliolids

The porcellaneous tests of the miliolids have a most unusual ultrastructure (Fig. 4.5) in which elongated crystals of magnesium-calcite (Blackmon and Todd 1959) are somewhat haphazardly arranged. On the wall surfaces, however, they are aligned in the plane so as to form a smooth exterior. The mode of formation of these tests remained a complete mystery until Berthold (1976) showed in *Calcituba polymorpha* that bundles of crystals 1–2 μm long and 0.1 μm wide are formed intracellularly within vesicles. The bundles are composed of an oriented array of crystals. Each crystal is enveloped by organic material. The bundles are released to the cell exterior by exocytosis (Berthold 1976; Hemleben et al. 1986). Direct observation of the stages in chamber formation of the miliolid, *Spiroloculina hyalina,* shows that it begins with the foraminifer surrounding itself with a cyst of algal cells and detritus attached to a transparent organic membrane. The preformed crystals and matrix material begin to accrete on the surface of the underlying chamber. There is no evidence of a preformed framework or mould in which the crystals form. Pseudopodia are very active in the accretion and shaping of the newly formed wall (Angell 1980). Much still remains to be learned about this fascinating process. It would be particularly interesting to determine the exact morphology of the individual crystals. Are the surfaces rounded, implying perhaps that they grow into a preformed membrane-bound sheath, or are specific crystal faces expressed? Are there differences in crystal morphology between species? Are the envelopes surrounding the crystals composed of glycoproteins and are some of these perhaps within the crystals themselves? (See Chapter 8 on echinoderms.)

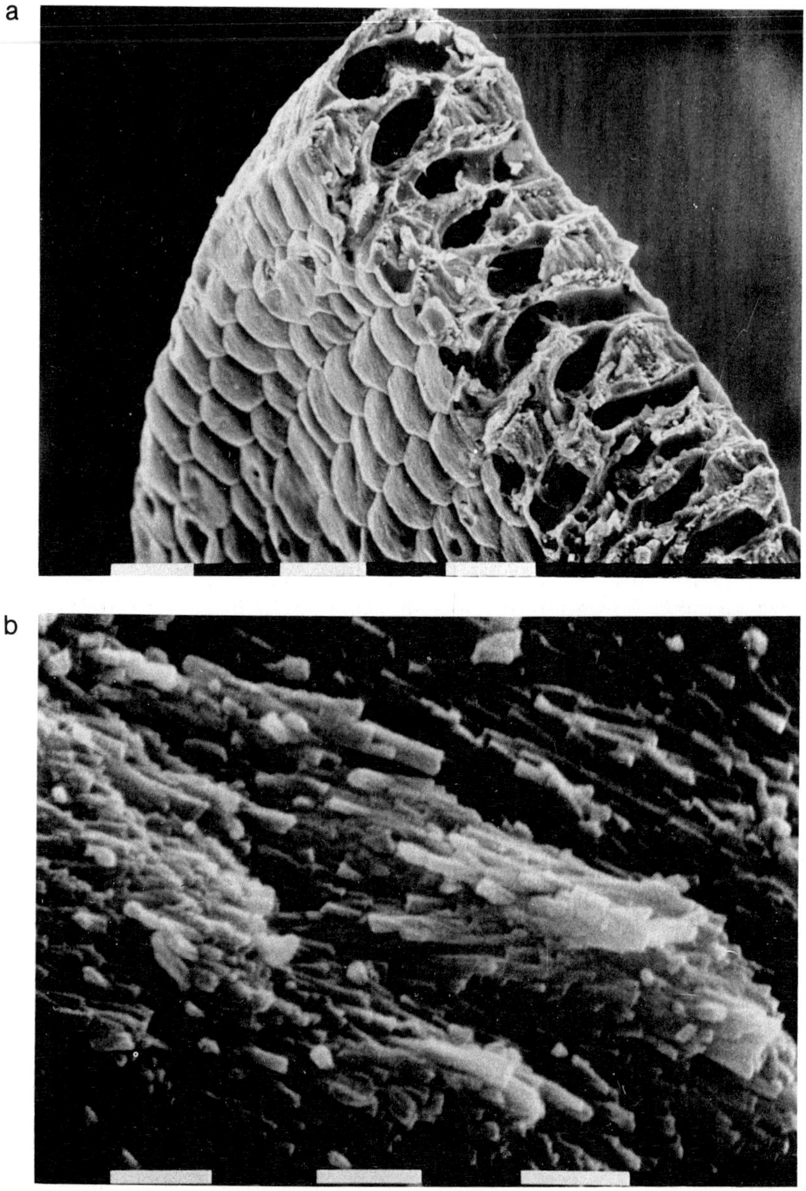

Figure 4.5. Scanning electron micrographs of (a) the broken test of the miliolid *Amphisorus hemprichii* (Elat). Scale bar: 0.1 mm. (b) Higher magnification of the test showing the individual calcite rod-shaped crystals. Scale bar: 1 μm.

4.2.3 Rotaline Foraminifera

By far the best studied foraminifera in terms of biomineralization are the Rotaliina. Their tests are calcitic, except for the aragonitic Robertinina. Different taxa have different magnesium contents (Blackmon and Todd 1959). A characteristic feature of many of their tests is that the crystallographic *c* axes are perpendicular to the test wall surface. This has given rise to the term "radial" to describe many of these wall structures (Wood 1949).

The shells are constructed in a modular manner by the addition of new chambers. The first-formed chamber is composed of a layer of organic material and either a single layer of mineral ("monolamellar") or more commonly two layers of mineral on either side of the organic layer ("bilamellar") (Smout 1954; Hansen and Reiss 1972). This organic layer is generally referred to as the "primary organic membrane." With the formation of each additional new chamber, a layer of organic material and a layer of mineral are added usually onto the entire specimen. The result is that earlier formed chamber walls are often composed of numerous mineral and organic layers (Fig. 4.6).

The actual process of chamber formation has to date been described for only a few species (reviewed by Hemleben et al. 1986). The basic process is quite different from the miliolids, in that mineralization takes place in a preformed structure at the final deposition site. In the case of the benthonic foraminifer *Heterostegina depressa,* chamber formation begins with pseudopodia flowing out of the test into a slowly enlarging space. The pseudopodia contain mainly electron-dense organic particles that concentrate into a diffuse layer at the site of the future chamber wall. The particles separate into two layers and the primary organic membrane forms between the two particle layers. Calcite crystals nucleate on both surfaces of the primary organic membrane, first in localized patches that rapidly (within about an hour) fuse to form a continuous wall. As this is happening, the pseudopodia withdraw into the older parts of the test. The mineralized wall thickens to its final dimensions over a period of 15–20 hours (Spindler and Röttger 1973).

In planktonic foraminifera the basic process is the same, but the manner in which the mould or "anlage" is formed differs. At the onset of chamber formation these foraminifera retract their long extended rhizopods and, at the same time, extrude protoplasm. The protoplasmic bulge organizes itself into an array of fanlike rhizopodia that forms the mould of the future chamber. The protoplasm then differentiates into two zones and the *anlage* is constructed at the border between the two. The primary organic membrane forms inside the *anlage* and the mineralization process then more or less follows that described for *Heterostegina.* For more details see Bé et al. (1979) and Hemleben et al. (1986).

Control over crystal growth for radial foraminifera, at least, is clearly demonstrated by the fact that the calcite crystallographic *c* axes are aligned more or less perpendicular to the wall surface. Insufficient information is available, however, to determine whether nucleation is by epitaxy on an organic template (Towe and Cifelli 1967) or by less specific substrates such as that described, for example, by Addadi et al. (1987) that can also induce *c* axis oriented calcite crystal growth. The observations on shell formation, however, clearly implicate the surface of the primary organic membrane as being the site for nucleation.

a

b

Figure 4.6. Scanning electron micrographs on the broken, fixed, and etched surface of the shell of *Heterostegina depressa* showing the different mineral and organic layers at two different magnifications (a, b). Preparation methods are described in Weiner and Erez (1984). Scale bars: 1.0 μm.

The structure of the primary organic membrane is not known, nor is its biochemical composition. Technically this is not easily accomplished. The membrane is insoluble after the mineral is dissolved in EDTA (Hemleben et al. 1977), but so are the quantitatively abundant organic linings on the inside and possibly on the outside of the shell as well. Thus, the primary organic membrane cannot be easily purified for separate analysis.

Whole foraminiferal shells do contain small amounts of protein (about 0.02 weight percent) and amino acid composition analyses confirm that they are acidic (King and Hare 1972). When *Heterostegina* shells are dissolved in EDTA at neutral pH, soluble acidic glycoproteins are obtained. These have been fractionated and partially characterized (Weiner and Erez 1984). Significantly they resemble in many respects the acidic glycoproteins found in other mineralized tissues. They could be coating the surfaces of the primary organic membrane and, thus, be directly involved in nucleation, and/or they could in vivo be in solution and be interacting with the growing crystals. Much more information is needed before we begin to understand their functions.

Meanwhile a few insights can be gained from examination of the ultrastructure of the mature test and the manner in which the crystals grow (see Weiner 1986). The crystal shapes and sizes are highly irregular except for the plane adjacent to the organic membrane. Furthermore, the boundaries between adjacent crystals are sinuous or sutured (Wood 1949; Stapleton 1973). It has also been observed that during growth large crystals dissolve at the expense of small ones (Angell 1979). This suggests that following nucleation on the organic membrane surface, crystals do not grow in compartments isolated from each other, but all share a common supersaturated aqueous medium. The presence of sinuous boundaries between crystals may imply that adjacent crystals are intergrown.

4.2.4 Concluding Comment

Foraminifera thus present many challenging and fascinating problems in biomineralization. Their diversity in terms of basic modes of mineralization (agglutination, vesicle-bound and extracellular framework) is almost unparalleled. The many "curiosities" [for example, the miliolidlike foraminifera that forms an opaline shell, the Spirillinacea that form whole tests of "single" crystals of calcite, or the wall of *Cibicides floridanus* that has a thick outer layer in which the calcite *c* axes are parallel to the surface (Bellemo 1976)] force us to reexamine our basic concepts in trying to accommodate these veritable "magicians" into the scheme of things. We can only encourage more people to study the foraminifera by predicting that important contributions to the field as a whole will result from their efforts.

4.3 Haptophyta (Coccolithophoridae)

The Haptophyta or Prymnesiophyceae are a group of mostly marine golden yellow algae. They derive their name from an unusual threadlike structure called the haptoneme that emanates like a flagellum from the cell. This is formed during the motile stage of their life cycle. Another characteristic feature of the members of this phylum is the presence of numerous, usually nonmineralized surface scales that cover the outside of the cell plasmalemma. These are present in both the motile and nonmotile stages in the lifecycle of the cell. One cell may produce a variety of scales, and, in a number of genera, the scales or periplasts have most unusual shapes (reviewed by Manton 1986). Their primary function, as proposed by Manton (1986), is to protect the plasmalemma.

A scale basically has a two-layered plate structure with the portion adjacent to the cell having a pattern of radial fibrils and the exterior part having spirally oriented fibrils (Manton and Leedale 1963). In most genera the scales are nonmineralized. In some, however, the upper plate is mineralized. This occurs occasionally in motile cells, but is more frequent in the nonmotile cells, where the mineralized structures are often relatively large and elaborate.

The Haptophyta that form these large mineralized structures are collectively called *Coccolithophoridae.* They derive their name from the term first used to describe the individual mineralized plates, namely *coccoliths* (Huxley 1858). Coccoliths were first identified by Ehrenberg (1836) in sedimentary rocks and their algal origin was only realized in 1898, when Murray and Blackman (1898) observed coccoliths in situ arranged on the outer surface of algal cells (called *coccospheres*). The mineral formed is calcite (Black 1963; Wilbur and Watabe 1963; Isenberg et al. 1963), with only one suspected exception (Table 4.1).

Coccolithophoridae abound in all the oceans of the world. In terms of calcium carbonate production, they are undoubtedly *the* major contributors to the ocean sediments. The most abundant coccolithophorid species is *Emiliania huxleyi.* In fact it could well be the most abundant calcium carbonate-secreting species on earth (Westbroek et al. 1984). The enormous extent to which coccoliths accumulate in the sediments can perhaps best be appreciated by pointing out that chalk deposits are formed primarily out of coccoliths.

Most coccoliths are organized along similar lines at least with respect to the arrangement of the calcite crystals—an important observation made by Black (1963). The major exceptions are those in which the whole surface of the scale is nonuniformly covered with very small rhombohedral-shaped crystals. They in fact do turn out to be formed by a different process (described in the following section). As Black (1963) points out, the vast majority of coccoliths contain relatively few crystals. They tend to nucleate on the scale rim and the growth directions are predominantly along the surfaces of the scale. In some cases the crystals express their natural faces that are, for the most part, the stable [104] faces of the cleavage rhombohedron, judging from Black's illustrations. Neighboring crystals are usually intergrown. In other cases, no natural surfaces are formed and the individual crystals adopt very "unnatural" shapes for calcite. These differing coccolith types certainly must represent increasing degrees of control exerted over the crystal formation process. Unfortunately, even in the few better studied cases, described in the following section, we still understand very little about these processes.

In this section we will describe the various known biomineralization processes in the Haptophyta, naturally paying particular attention to the most thoroughly investigated and important process of coccolith formation. The organization of the section basically follows the review by Green (1986), first describing intracellular coccolith calcification, but refers also to the less common extracellular holococcolith formation and finally the few cases of known noncoccolith-associated mineralization. Note that numerous terms are used to describe the two basic types of coccoliths. Parke and Adams (1960) used the following terminology. Those formed during the motile phase are called holococcoliths or crystalloliths, and those formed during the nonmotile phase are called heterococcoliths or placoliths. For simplicity, we will follow Green (1986) and call the latter "coccoliths" and the former "holococcoliths."

4.3.1 Intracellular Coccolith Formation

The most commonly formed type of coccolith is an elliptical object, a few microns in diameter, of quite amazing complexity (Fig. 4.7a). Viewed from the side it appears to be composed of two superimposed plates. In fact, the basic building blocks are not the plates but individual anvil-shaped segments that fit together side by side in a jigsawlike manner to form the entire structure. One of these coccoliths can comprise as many as 20 or more such segments. The shape of one individual segment of *E. huxleyi* is schematically illustrated in Fig. 4.8 following the description given by Watabe (1967). In a very elegant study, Watabe (1967) showed by electron diffraction that one entire segment behaves as if it is a single crystal. The *c* axis is parallel to the base of the segment. The observation has been confirmed by Mann et al. (1988) using high resolution TEM. The surfaces of the crystalline segments of *E. huxleyi* coccoliths do not show the natural faces of calcite crystals, but are essentially smooth (Fig. 4.7b). This combination of an apparently single crystal with smooth, curved surfaces is reminiscent of the calcitic echinoderm skeleton (Chapter 8). In this regard it would be of interest to discover whether the coccolith segments cleave along the natural [104] crystal planes or with conchoidal glassy cleavage, as occurs in echinoderms.

Echinoderm skeleton formation and coccolith formation share another important property: mineral formation occurs in vesicles located within the cell. The coccolith is then extruded to the cell surface to form interlocking clusters. This was first observed by Dixon (1900). The smooth surfaces of these coccolith segments presumably reflect their vesicular origin. The vesicle itself is derived from the cisternae of the Golgi body. Its shape in *E. huxleyi* is moulded by being juxtaposed to various cellular organelles, including the nuclear envelope and portions of the Golgi cisternae collectively called the reticular body (Wilbur and Watabe 1963; Westbroek et al. 1986). It appears that the folded membranes of the reticular body are mainly responsible for the detailed sculpturing of the coccolith vesicle (van Emberg et al. 1986), although the details of this process remain obscure.

The stages of coccolith formation inside the vesicle have been described in most detail for *Pleurochrysis carterae* (Pienaar 1969; Outka and Williams 1971) and *E. huxleyi* (Wilbur and Watabe 1963; reviewed by Westbroek et al. 1984; 1986). For brevity, we will refer here only to *E. huxleyi*. The initial step is the formation of an oval-shaped nonmineralized base plate. The first crystals nucleate on the rim of the base plate. Significantly they adopt the characteristic rhombohedral morphology of inorganically formed calcite crystals, implying that following nucleation crystal growth proceeds pseudoinorganically. Judging from the morphology of the crystals, it would appear likely that the calcite crystallographic *c* axes

a

b

Figure 4.7. Scanning electron micrographs of the coccoliths of *Emiliania huxleyi* from a sediment sample of unknown origin. (a) Different views, some of which show the superimposed platelike structure. (b) Higher magnification view showing that the surfaces have no discernible topography. Scale bars: 1.0 μm.

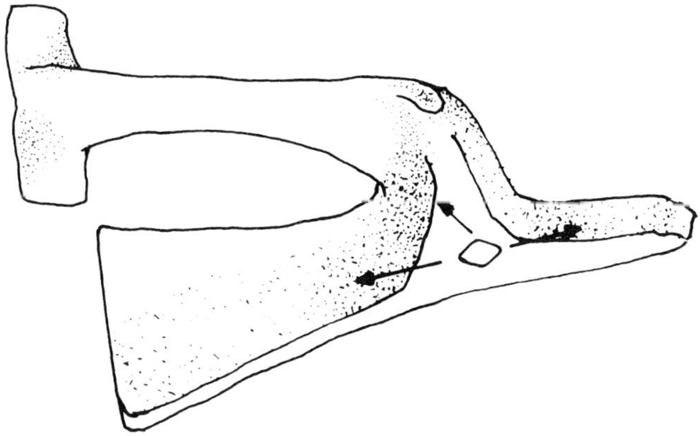

Figure 4.8. Schematic illustration of an individual unit of the coccolith of *Emiliania huxleyi* showing the site of initial crystal formation (rhombohedron) and the directions of growth. Redrawn from an illustration in Watabe (1967).

are more or less perpendicular to the nucleation surface of the base plate rim. Crystal growth proceeds in both directions along the *c* axis direction and parallel to the base plate surface to form the lower portions of the coccolith segment. It also proceeds upwards at an angle of roughly 45° from the nucleation site to form the upper element (Wilbur and Watabe 1963). The growth directions are indicated in Figure 4.8. During this growth process the membrane of the coccolith vesicle continuously expands slightly ahead of the mineralization front (Westbroek et al. 1986).

If a mature coccolith is carefully demineralized, the whole structure can still be seen in the form of an organic framework (illustrated by Watabe in Pautard 1970). Neither the composition nor the origin of this framework is known. However, when the coccolith is dissolved in EDTA after its surfaces are well cleaned with sodium hydroxide, a soluble acid polysaccharide is released (Westbroek et al. 1973). This macromolecule constitutes some 3% by weight of the coccolith and is presumably located somewhere inside the mineral phase. It has been extensively studied by de Jong, Westbroek, and their colleagues and the results are reviewed by De Vrind-de Jong et al. (1986). There is also a water-insoluble fraction that by weight is less than 0.1% of the coccolith and it is composed mostly of protein (de Vrind-de Jong et al. 1986). Note that *P. carterae* also contains similar acidic polysaccharides.

The polysaccharide from *E. huxleyi* has an estimated molecular weight of 60,000 by osmometry or 90,000 by light scattering. The complex structure of the repeating unit has been determined (Fichtinger-Schepman et al. 1981). Of particular interest from the mineralization point of view is the presence of two different acidic ligands: carboxylate groups associated with uronic acids and sulfate esters. With the addition of calcium, the polysaccharide changes conformation (Borman et al. 1987). An in vitro study of the effects of the polysaccharide and calcium on carbonate crystallization shows that the polysaccharides inhibit crystallization but that this inhibition is not affected by the removal of the sulfate groups. This observation highlights the importance of the carboxylate ligands in this process (Borman et al. 1982). Immunochemical localization of the polysaccharide shows that it is present in the coccolith vesicle, the Golgi cisternae, the reticular body, and the coccolith itself, provided the latter is first etched. These observations are consistent with the polysaccharide being produced in the Golgi and entering the vesicle via the reticular body (van Emburg et al. 1986). Its presence inside the mineral phase itself suggests that one function of the polysaccharide could be to modulate calcite crystal growth. This could occur by the polysaccharide interacting from solution with the growing crystal surfaces and eventually being overgrown and incorporated inside the mineral phase, in a manner analogous to that proposed for sea urchins (Berman et al. 1988; Chapter 8). Westbroek et al. (1984) envisage a more elaborate multifunctional role for the polysaccharide in which it is involved in the induction of crystals, their inhibition, and stabilization. Needless to say, the real functions that this interesting macromolecule performs are still obscure.

4.3.2 Extracellular Holococcolith Formation

Holococcoliths are characterized by many small rhombohedral-shaped calcite crystals. Only about 30 extant species are known to form holococcoliths, and Haq (1978) indicates that a survey of the literature shows that most, and perhaps all, these species produce holococcoliths only during the motile phase of their life cycle. The only species known to produce coccoliths during its nonmotile phase and holococcoliths during its motile phase is *Crystallolithus hyalinus* (Parke and Adams 1960). The large nonmotile cells were observed in culture to produce their coccoliths intracellularly and then extrude them onto the cell surface, in the familiar manner. The smaller motile cells produced quite different mineralized scales. These are covered with many small rhombohedral-shaped calcite crystals. Manton and Leedale (1963) noted that they never observe mineralized scales inside the cell and therefore suggested that these crystals may form extracellularly only after the unmineralized scale is extruded from inside the cell. This was confirmed by Rowson et al. (1986) who actually observed the scales being formed inside the cell, but always without any mineral. Furthermore they noted that during extracellular crystal formation the crystals were still isolated from the environment by an organic "skin," implying that it is still a biologically controlled process.

4.3.3 Non-Coccolith-Associated Mineralization

Species of the genus *Chrysotila* are observed to undergo a different type of mineralization process (Green and Course 1983). Vegetative or nonmotile cells found in coastal and inland habitats produce a thick lamellate mucilaginous sheath around themselves. Cells in culture are observed to induce crystals to form within the sheath. The crystals grow to form spherulitic clusters that eventually encapsulate the cell. The final form of the mineralized body is graphically described by Green and Course as a "disorganized blackberry" (p. 181). The type of crystals was not identified, although from their morphology the authors *suggest* that they are probably aragonite, and that they may well be produced as a byproduct of the cell's metabolism. In other words, a relatively poorly controlled process as compared to coccolith formation.

4.3.4 Silicification of Cysts

Almost nothing is known about this process of mineralization in Haptophyta, except that it has been confirmed to occur in the cysts of two species *Prymnesium parvum* and *Prymnesium patellifera* (Pienaar 1981; Green et al. 1982). The mineral is associated with the cyst wall.

Thus the Haptophyta, like so many other protoctists, prove to be diverse mineralizers. Mineralization involves both different minerals and varying degrees of control over the mineralization process. Even the coccoliths themselves are formed in two different ways. The facts that these organisms are so abundant in the world's oceans and that some species can be cultured in the laboratory should provide much impetus to improve our understanding of mineralization in the Haptophyta.

5

Cnidaria

The phylum Cnidaria or Coelenterates includes sea anemones, jellyfish, hydras, sea fans, and, of course, the corals. With few exceptions they are all marine organisms and most are inhabitants of shallow water. In spite of the great variation in shape, size, and mode of life, they all possess the same basic metazoan structural features (Fig. 5.1): an internal space for digestion (gastrovascular cavity or coelenteran), a mouth, and a circle of tentacles, which are really just an extension of the body wall. The body wall in turn is composed of three layers: an outer layer of epidermis, an inner layer of cells lining the gastrovascular cavity, and, sandwiched between them, a so-called mesoglea (Barnes 1980). All these features are present in both of the basic structural types: the sessile polyp and the free-swiming medusa (Fig. 5.1). During their life cycle, some cnidarians exhibit one or the other structural type whereas others pass through both.

Most Cnidaria have no mineralized deposits. The ones that, to date, are known to have mineralized deposits are listed in Table 5.1. They are found in both the free-swimming medusae and the sessile polyps. Not surprisingly, these have very different types of mineralized deposits. In the medusae they are located exclusively within the statocyst where they constitute an important part of the organism's gravity perception apparatus. Interestingly the statoconia of the Hydrozoa, examined to date for their major elemental compositions only, are all composed of amorphous Mg-Ca-phosphate, whereas those of the Scyphozoa and Cubozoa are composed of calcium sulfate (Table 5.1). Calcium sulfate minerals (presumably gypsum) are not commonly formed by organisms and the only other known occurrence is in the Gamophyta among the Protoctista (Table 2.1). Spangenberg (1976) and her colleagues have expertly documented this phenomenon in the Cnidaria. (For a more detailed discussion of mineralization and gravity perception see Chapter 11.) The predominant mineralized hard part associated with the sessile polyps is skeletal. These can take the form of skeletons composed of individual spicules, spicule aggregates, or massive skeletons. They are composed of aragonite, calcite, or both.

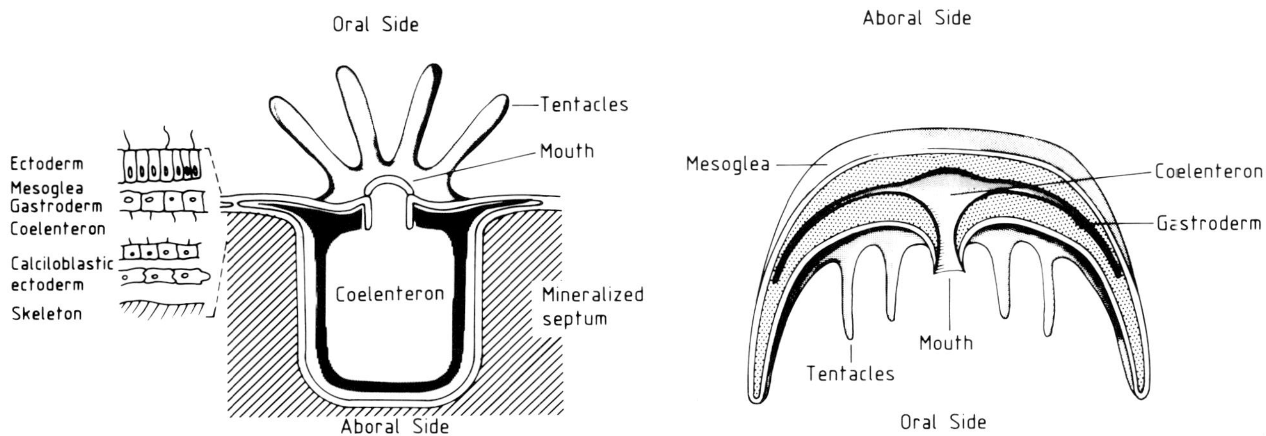

Figure 5.1. Schematic illustration of the polypoid body form (left-hand side) and medusoid body form (right-hand side) showing also the detailed layered structure of the epithelial cell layers in a coenosteal region of a colonial reef coral. Redrawn from figures of Barnes (1980) and Johnston (1980).

Table 5.1 The Diversity, Distribution, Localization, and Functions of Minerals Formed by the Cnidaria

Class	Lower taxa	Mineralization site	Mineral	Function	References
Hydrozoa	*Corymorpha*	Polyp statoliths	Indeterminate	Gravity receptor	Campbell (1972)
	Obelia	Medusae statoconia	Amorphous Mg-Ca-phosphate	Gravity receptor	Chapman (1985)
	Lovenella	Medusae statoconia	Amorphous Mg-Ca-phosphate	Gravity receptor	Chapman (1985)
	Phyalidium	Medusae statoconia	Amorphous Mg-Ca-phosphate	Gravity receptor	Chapman (1985)
	Aglantha	Medusae statoconia	Amorphous Mg-Ca-phosphate	Gravity receptor	Chapman (1985)
	Milliporina	Skeleton	Aragonite	Mechanical support	Boschma (1956)
	Stylaster	Skeleton	Aragonite	Mechanical support	Boschma (1956); Lowenstam (1964b)
	Distichopora	Skeleton	Aragonite	Mechanical support	Boschma (1956)
	Allopora	Skeleton	Aragonite and/or calcite	Mechanical support	Lowenstam (1964b)
	Errinopora	Skeleton	Calcite	Mechanical support	Lowenstam (1964b)
Scyphozoa	*Auralia*	Medusae statoconia	Gypsum	Gravity receptor	Spangenberg and Beck (1968)
	Chryosaora	Medusae statoconia	"$CaSO_4$"	Gravity receptor	Chapman (1985)
Cubozoa	*Chiropsalmus*	Medusae statoliths	"$CaSO_4$"	Gravity receptor	Chapman (1985)
Anthozoa	Octocorallia				
	Tubipora	Colonial skeleton	Mg-calcite	Mechanical support	Spiro (1971)
	Telestacea	Spicules (sclerites)	Mg-calcite	Mechanical support	Chave (1954)
	Alcyonaria	Spicules	Mg-calcite	Mechanical support	Chave (1954)
	Helioporida	Massive skeleton	Aragonite	Mechanical support	Chave (1954)
	Gorgoniacea				
	Scleraxonia	Axis spicules, cement cortex spicules	Mg-calcite	Mechanical support	Chave (1954)
	Holaxonia	Axis	Mg-calcite and/or aragonite	Mechanical support	Lowenstam (1964b)
		Base	Mg-calcite or aragonite	Mechanical support	Lowenstam (1964b)
		Cortex spicules	Mg-calcite	Mechanical support	Chave (1954)
	Pennatulacea	Axis: (coenenchyme) some mineralized mesoglea spicules	Calcite	Mechanical support	Ledger and Franc (1978); Dunkelberger and Watabe (1974)
		Peduncle (foot) spicules	Indeterminate		Lyke (1965)
	Zoantharia				
	Scleractinia	Solitary or colonial skeletons	Aragonite, trace of calcite	Mechanical support	Meigen (1903), Constantz and Meike (1988)

When both are present, the two polymorphs always occupy different anatomical sites (Lowenstam 1964b). This is the normal situation in biomineralization, when two or more minerals are present in the same structure. Very few exceptions are known. It is interesting to note that all the known minerals formed by Cnidaria are calcium minerals. This is probably more a reflection of the fact that we know very little about the true extent of mineralization in this phylum, rather than a representation of some physiological restriction of the members of the phylum. This, of course, still remains to be shown.

In this section we focus on cnidarial skeletal formation in the Anthozoa and deliberately contrast and compare spicule formation with massive skeletal formation processes. One objective is to suggest that these processes, at least among the Anthozoans, may not be unrelated.

From the point of view of skeletal formation the Anthozoa are by far the most important class of Cnidaria. The only other groups that form massive mineralized skeletons are the Hydrocorallina, with *Milliporina,* the fire coral, being a well known member (Table 5.1). The mineralized hard parts formed by the Anthozoa can also be conveniently classified into three types: spicules, aggregates of spicules, and massive or coherent structures. In some colonial genera, two of the three types may be present.

5.1 Spicules

Spicules are found in the Octocorallia and the Gorgoniacea and they are all composed of calcite. They are usually found in the axis as well as the cortex of the colonial skeletons, although they may also be present in the tentacles, pharynx, and upper part of the autozoids. Spicule shapes are species determined and within a species they generally differ according to the anatomical site at which they are formed, a first hint that spicule formation processes are well controlled.

The calcite crystals that constitute the spicules are arranged in different ways. Using polarized light microscopy to identify calcite crystallographic *c* axis orientations, Schmidt (1924) distinguished four different spicule types. Scale-shaped or rounded spicules with a spherulitic structure are formed by some gorgonians. Spindle to rod-shaped spicules with calcite *c* axes aligned with the spicule long axis are common among gorgonians and pennatulids. Another more complex spindle-shaped spicule found in the *Eunicea* has additional tubercles growing perpendicular to the spicule long axis. The calcite *c* axes of the tubercles are perpendicular to the spicule long axis! Finally a gorgonian species of *Briareum,* forms an entire spicule out of one single calcite crystal. Even more unusual is the fact that the *c* axis of this spicule is perpendicular to its long axis. Thus, there is a whole spectrum of spicule types formed by the Anthozoans. The formation of one that has been studied in some detail is described below.

Leptogorgia virgulata or the sea whip belongs to the Gorgoniacea. These are corals that have a plantlike growth form and a skeleton of predominantly collagen-containing organic material reinforced with mineral. In *Leptogorgia* the axial skeleton contains an amorphous calcium compound, probably calcium carbonate (Kingsley and Watabe 1982a). The mesoglea, which in these organisms is relatively thick, surrounds the axial skeleton and contains Mg-calcite spicules. They are elongate and polycrystalline, with their crystallographic *c* axes aligned with the long axis, and are decorated by many tubercles or "wartlike" projections (Kingsley and Watabe 1982b). Spicule formation is initiated in vacuoles within amoebocytic scleroblasts distributed throughout the mesoglea. Prior to crystal formation, a fibrous organic network is laid down inside the vacuole. Crystal formation is intimately associated with this organic material. TEM micrographs of decalcified spicule sections clearly reveal the reticulate nature of the organic network (Kingsley and Watabe 1982b). Spicule formation in *Leptogorgia* thus appears to be a true matrix-mediated process that takes place in a vacuole inside a cell. Further growth of the spicule reduces the cytoplasm to a thin envelope. The vacuole membrane fuses with the plasma membrane and eventually ruptures, at which stage the spicule extends into the extracellular space (Kingsley and Watabe 1982b).

The mature spicules contain 5.9% by weight organic material, most of which is protein (Kingsley and Watabe 1983). Gel electrophoresis reveals the presence of many proteins and the amino acid compositions of those that are still soluble after decalcification in 0.1 *N* HCl show that some 65 mol% of the proteins is aspartic acid. Acidic proteins, even as acidic as these, are known from other mineralized hard parts, including scleractinian corals (Mitterer 1978). A pulse–chase study using ^{3}H aspartic acid as a marker for the aspartic acid-rich proteins documents their pathway from synthesis in the rough endoplasmic reticulum, to transport to the Golgi complexes, followed by packaging into vesicles and eventually their appearance inside the spicule (Kingsley and Watabe 1984). A similar study using ^{45}Ca dissolved in seawater shows that calcium first accumulates in the axis and then part of it is transported into the spicules (Kingsley and Watabe 1985a). The uptake of calcium from the medium is actively mediated by Ca^{2+}-ATPase (Kingsley and Watabe 1985b). Immunochemical localization of carbonic anhydrase showed that this enzyme is concentrated in the scleroblasts of the axial skeleton, particularly on the membranes comprising the spicule-forming vacuole. Surprisingly, inhibition of the enzyme's activity by Diamox causes increased uptake of radioactive calcium, the opposite to that generally observed in invertebrate mineralizing systems (Kingsley and Watabe 1987). Although most of the growth of spicules in *Leptogorgia* is complete before rupture of the cell membranes, this is not always the case in other gorgonians. Spicule growth in *Pseudoplexaura flagellosa* is initiated inside a cellular vacuole but continues in the extracellular space. Pseudopodia from two or more cells cover the spicule surface, forming what is termed a *multicellular vesicle* inside which further growth occurs (Goldberg and Benayahu 1987).

The **Pennatulacea** or sea-pens like the gorgonians have spicules and a mineralized axial skeleton. Spicule formation is also a highly controlled, primarily intracellular process (Dunkelberger and Watabe 1974). In contrast, mineralization of the axial skeleton is quite different. In a notable study of the axial skeleton of *Veretillum cynomorium,* Ledger and Franc (1978) show that the initial calcite nodules nucleate extracellularly between the collagen fibrils. They grow, push aside the collagen fibrils, and eventually surround and encapsulate them. Significantly mineral does not form within the collagen fibril and collagen is not involved in the nucleation of the calcite. The crystals grow to relatively large sizes and are elongated along their *c* axes. The crystals radiate out in all directions from central points (spherulitic growth). So in addition to the nucleation event, crystal growth appears to be unimpeded except by spatial constraints.

5.2 Spicule Aggregates

Spicule aggregates are not as widespread as individual spicules or coherent massive skeletons, but do occur in a number of different Octocorallia. Unfortunately nothing is known about the processes involved in aggregated spicule formation. *Melithaea ochracea* is a gorgonian sea fan found in Indo-Pacific coral reefs. Its long anastomosing branches extend into the open water and its skeletal ultrastructure is, therefore, designed to withstand the considerable forces exerted by tidal currents. The major adaptation to this mode of life is the jointed nature of their branches resulting in their axial skeletons being constructed of alternating segments of dense fused spicule aggregates and collagen-rich segments. The latter spicules are loosely connected to each other (Muzik and Wainwright 1977). Figure 5.2a shows the skeletal texture of an unidentified *Melithaea* species. The individual spicules are 130–160 μm long and the cementing material is organic in nature. The spicules in the less dense segments are similar in shape, but smaller in size (70–80 μm) and are arranged in a polyhedral meshwork. Examination of broken spicules shows that they are composed of acicular lath-shaped crystals (Fig. 5.2b). Organic material connects these spicules to each other to form segments that provide flexibility to the structure. The whole axial skeleton is surrounded by a thin coenenchyme that contains individual nonaggregated spicules, which show differentiation in their morphology depending upon their function (Fig. 5.2c). Individual spicules are also found in the central core of the axis. Both these spicule types are quite different in shape from the aggregated form, vividly demonstrating the remarkable complexity that exists with respect to mineralized hard parts in these animals (Muzik and Wainwright 1977).

5.3 Fused Spicular Aggregates

The family Tubiporidae contains one representative, the scarlet-colored species *Tubipora musica,* appropriately named because of its resemblance to an organ pipe. The massive structure of vertical tubes and horizontal stolons (Bayer 1956) (platforms) is composed of tightly fused high magnesium calcite spicules cemented together by crystals. The latter are similar in shape to those of the spicules themselves. In heavily cemented areas the individual spicules are difficult to recognize in the scanning electron microscope (SEM). The individual spicules are rod-shaped or branching structures some hundreds of microns long. Cross sections through an individual spicule show concentric layers of crystals arranged parallel to the axis (Spiro 1971). Regrettably, we know nothing about the mode of formation of the skeleton of *Tubipora.*

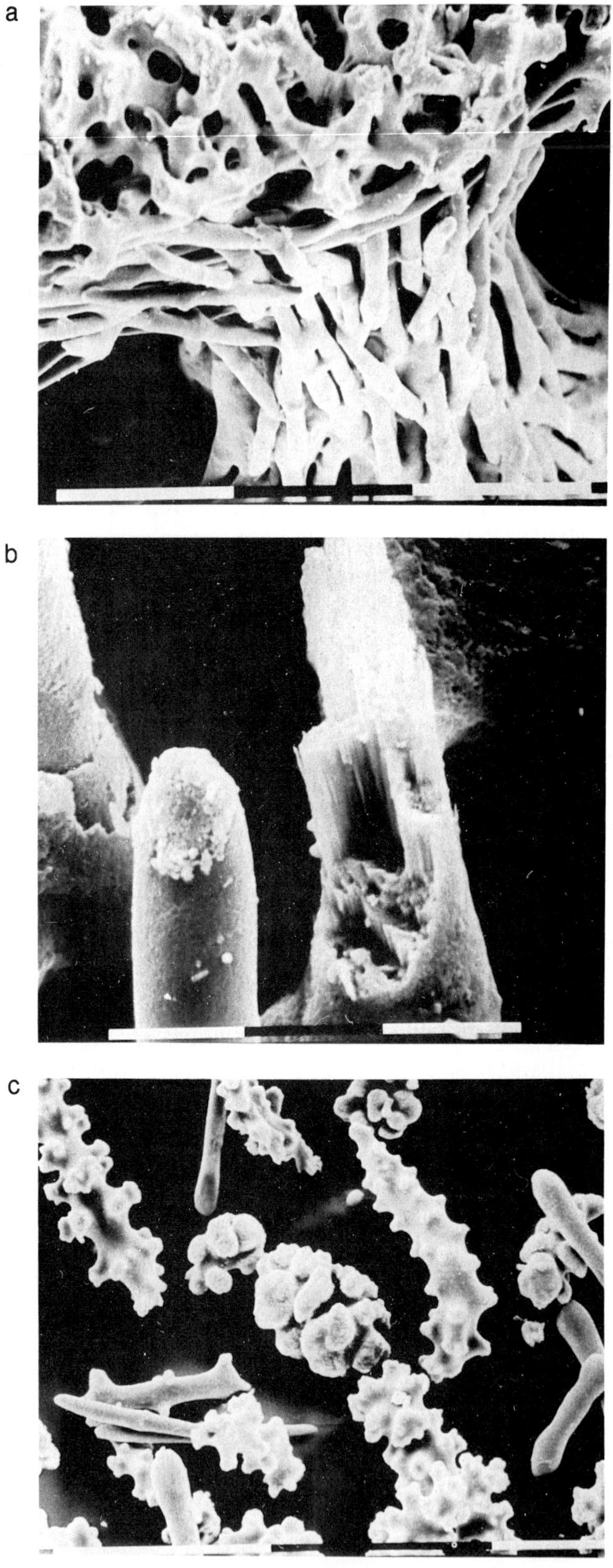
a
b
c

5.4 Massive Skeletons: The Scleractinian Corals

Scleractinian corals are distributed throughout all the oceans in waters with a salinity of 27 parts per thousand or more and to depths of almost 6000 m. The highest concentrations of scleractinian corals and the taxonomically most diverse assemblages are found in waters in which the temperature does not fall below 15°C at any time during the year. This zone is generally restricted to a circumequatorial belt around the globe. It is in these environments that scleractinian corals along with coralline algae grow to form huge reefs.

Some scleractinian taxa live as isolated individuals, whereas others form large colonies. Each individual animal or polyp forms its own exoskeleton. In the colonial forms the skeletons of neighbors fuse together or are connected by skeletal deposits called coenosteon, to form a massive or branching structure. The colonial mode of living is most common in coral reefs, whereas the solitary forms are most abundant in deep and/or cold water. Figure 5.1 schematically shows the morphology of an individual polyp and its surrounding exoskeleton.

One feature that characterizes reef-building colonial corals is that they usually possess endosymbiotic, unicellular dinoflagellate algae (also called zooxanthellae) within their endodermal tissues. The members of this "ecologic" group of corals are often referred to as *hermatypic* to distinguish them from the non-reef-building forms called *ahermatypic.* The gymnodinoid dinoflagellates in the genus *Symbiodinium* are the symbionts harbored by all reef corals and all other benthic, marine, symbiotic cnidarians (Trench and Blank 1987). It is widely believed that the zooxanthellae do influence mineralization because corals calcify more rapidly in the light than in the dark (Goreau 1959). Goreau's pioneering studies have stimulated enormous interest in this subject. These studies show just how complex a problem this really is and, although much has been learned about sources of carbon, internal pools, isotopic fractionations, diurnal rhythms, and rates of mineralization etc. (Pearse and Muscatine 1971; Goreau 1977; Chalker and Taylor 1978), the nature of the symbiotic relationship between the dinoflagellate and its host is still shrouded in mystery.

The cells that are adjacent to the forming mineral surface form the calicoblastic epidermis (Fig. 5.1) and from the very earliest studies it has been assumed that they are involved in the mineralization process (von Heider 1881). They do have high concentrations of intracellular calcium (Hayasi 1937) and are most certainly

←

Figure 5.2. (a) The spicule arrangement of an unidentified *Melithaea* species from Palau is similar to that described by Muzik and Wainwright (1977). Scale bar: 0.1 mm. (b) SEM micrographs of the dense, fused aggregates from the stiff internodes show that the individual spicules are themselves composed of acicular narrow lath-shaped crystals that are tightly packed and parallel to the long axis of the spicule. Crystal growth starts on the initially formed surface layer and spreads inward to fill the lumen of the spicules. Another feature not reported for *M. ochracea* is that in the internodes close to the border with the gastrodermal tubes, the solenia is lined by an anastomosing layer of fused spicules. Scale bar: 10 μm. (c) Variously shaped cortical spicules together with smooth spicules derived from the axis. Scale bar: 0.1 mm.

involved in directing mineralization. For a good overview of this subject, see Johnston (1980).

5.4.1 Larval Scleractinian Skeleton

Studies of larval skeletal formation have made possible the observation of different stages of mineralization during development following the settling of the planula on a substrate. Wainwright (1963) reported that the first stage of mineralization of *Pocillopora damicornis* is the deposition of mineral in the basal plate. Preliminary X-ray microdiffraction patterns showed that the mineral is calcite. The layer itself is some 1–2 μm thick. A more detailed study of the larval basal plate (Vandermeulen and Watabe 1973) showed that two different mineralized structures are present: flattened spherulitic platelets and smaller rod-like granules. The platelets are found in the central columella region and underlie the septal rudiments. The granules are between the platelets and at the growing edge of the basal plate. Significantly these granules do not coalesce to form larger structures and, although of irregular shape, they are all roughly similar in size. This suggests that the growth of each granule may take place within a space defined by an organic framework rather than in a large undivided space. The second stage of larval mineralization is the growth of larger aragonitic crystals similar to the bulk of the adult skeleton on top of the basal plate. Electron diffraction of the basal disk showed the unequivocal presence of calcite as well as aragonite. The spatial distribution of the two minerals was not determined (Vandermeulen and Watabe 1973), but the observation is consistent with the X-ray results of Wainwright (1963).

5.4.2 Adult Scleractinian Skeleton

The basic units of ultrastructure of the majority of scleractinian corals are spherulitic arrays or bunches of needlelike crystals of aragonite (Fig. 5.3A). This was firmly established by Ogilvie (1896) and the analogy to inorganically formed spherulitic structures was noted by Bryan and Hill (1941). Thin sections viewed in the optical microscope show that the loci from which the aragonite crystals grow have a darker appearance than the radiating crystal arrays (Fowler 1887; Ogilvie 1896). They form spots or lines or some combination thereof (Fig. 5.3B) and, in fact, these variations must be genetically controlled as they have proved most useful in the classification of stony corals (Wells 1956). The traditional term used to describe these dark patches from which the aragonite crystals radiate out is *centers of calcification* (Ogilvie 1896). When they are viewed in the scanning electron microscope they have a granular appearance quite distinct from the aragonite crystals (Fig. 5.3C) (Constantz 1986). The first definitive information on their composition was obtained by Constantz and Meike (1988) for the species *Mussa angulosa,* which has relatively large centers of calcification. Very thin samples prepared by ion beam thinning were examined in the TEM. Electron diffraction patterns of the centers of calcification show that they are composed of calcite crystals that are frequently twinned (Fig. 5.4). Constantz and Meike (1988) were not able to locate the centers of calcification in other species because they are small and constitute only a very minor portion of the total mass. It is, therefore, not yet known if other scleractinian corals also have calcite crystals at their centers of calcification.

X-Ray and electron diffraction studies of the spherulitic arrays of aragonite crystals show that they are elongated along their *c* axes and that the *a* and *b* axes of crystal assemblies are randomly oriented (Wainwright 1964; Constantz 1986). Examination of the aragonite crystal morphologies reveals that genera belonging to different families have characteristically shaped aragonite crystals (Constantz 1986) and that clearly this aspect of crystal formation is also under some biological control.

Scleractinian coral skeletons contain small amounts of organic material. Johnston (1980) observed some of this material in a study of the epidermal cell–skeleton interface of *Pocillopora damicornis.* SEM micrographs of specimens prepared by cryofracture as well as TEM micrographs of demineralized thin sections show that the layer adjacent to the cells is composed of a fine reticulate network of organic sheets. The layer itself is only a few microns thick and the voids delineated by the organic material have diameters in the submicron range. On the other side of this layer are sheets of organic material spaced approximately 4–6 μm apart and oriented more or less perpendicular to the layer (see text Figs. 14 and 15 in Johnston 1980). Johnston interprets these observations as an indication that the material comprising the reticulate network is lost with time after deposition of the mineral, although there is no independent evidence to confirm this.

Pocillipora damicornis is an unusual coral in that it is one of the few in which the major organic constituent of the skeleton is chitin (Wainwright 1963; Wilfert and Peters 1969). Chitin as determined by the rather insensitive chitosan test has been found only in *Pocillipora* among all the mineralized coral species (Wainwright 1963; Wilfert and Peters 1969). Lipids are an integral component of the skeletons of scleractinian corals (Young et al. 1971; Isa and Okazaki 1987). They constitute between 0.02 and 0.03% by weight of the skeleton. Unlike most of the lipids associated with the skeleton, the phospholipids are released from the skeleton only when the mineral is dissolved. The phospholipids are able to bind calcium (Isa and Okazaki 1987). Protein is always present in coral skeletons (Young 1971) and it too comprises about 0.03% by weight of the dry skeleton (Wilfert and Peters 1969; Young 1971; Mitterer 1978). The amino acid compositions of the proteins from carefully cleaned material show that the acidic amino acids, aspartic and glutamic acid, can constitute between 40 and 50 mol% of the protein (Mitterer 1978). Aspartic acid is much more abundant than glutamic acid and it is of interest to note that γ-carboxyglutamic acid (Gla) was found in total extracts of two scleractinian corals (Hamilton et al. 1982). Gla is a component of some proteins in vertebrate bones (Hauschka et al. 1975; Price et al. 1976). The significance, if any, of its presence in corals with respect to biomineralization should be carefully examined.

5.4.3 Processes of Scleractinian Coral Mineralization

The limited information available to date on both larval and adult scleractinian skeletons presents a confusing picture. Some observations point to a well-controlled process, whereas others suggest the opposite.

Bryan and Hill (1941) make an important analogy between the similarity of

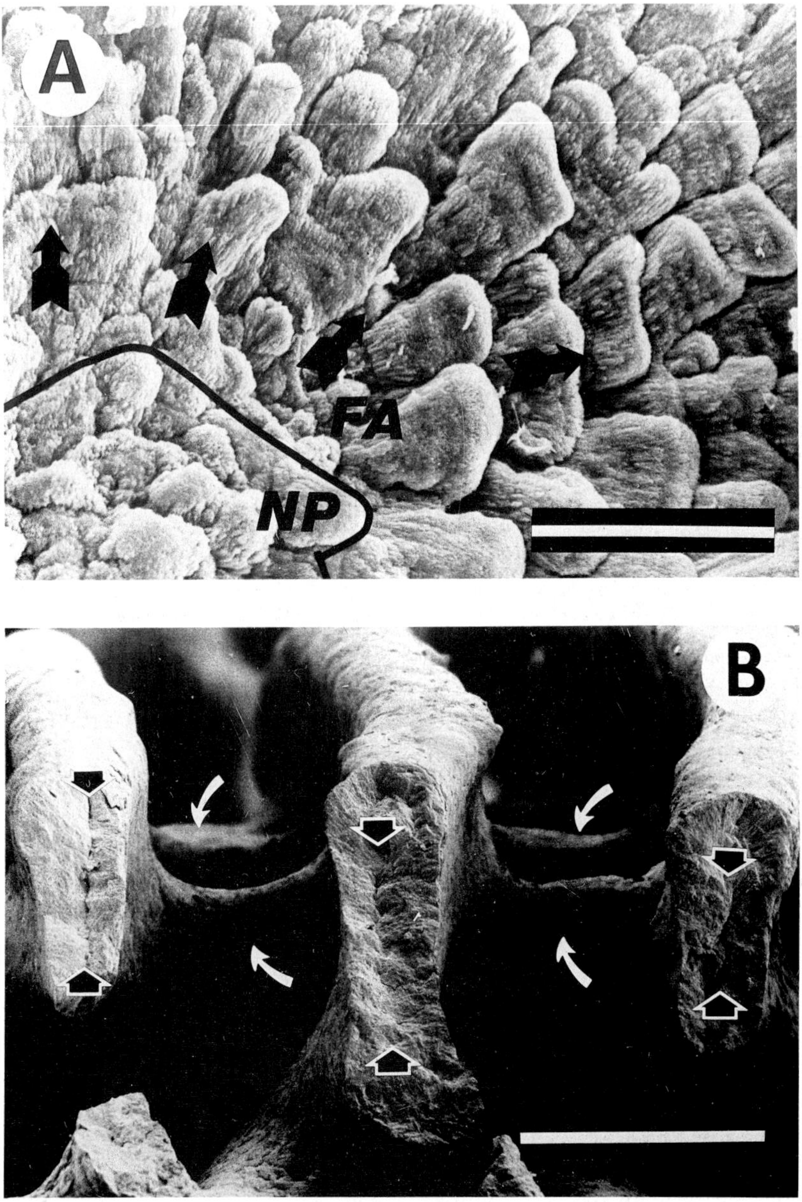

Figure 5.3. (A) Scanning electron micrograph of *Acropora palmata* from Barbados showing the fiber bundles (FA) radiating out of the region called the center of calcification (NP). Scale bar: 30 μm. (B) Scanning electron micrograph of *Acropora cervicornis* (US Virgin Islands). The fractured surfaces reveal sections through the septa of the axial corallite. Black arrows show the surface from which the aragonite fiber bundles radiate. The white arrows show the synapticulae that connect to the sides of the septa. Scale bar: 0.25 mm.

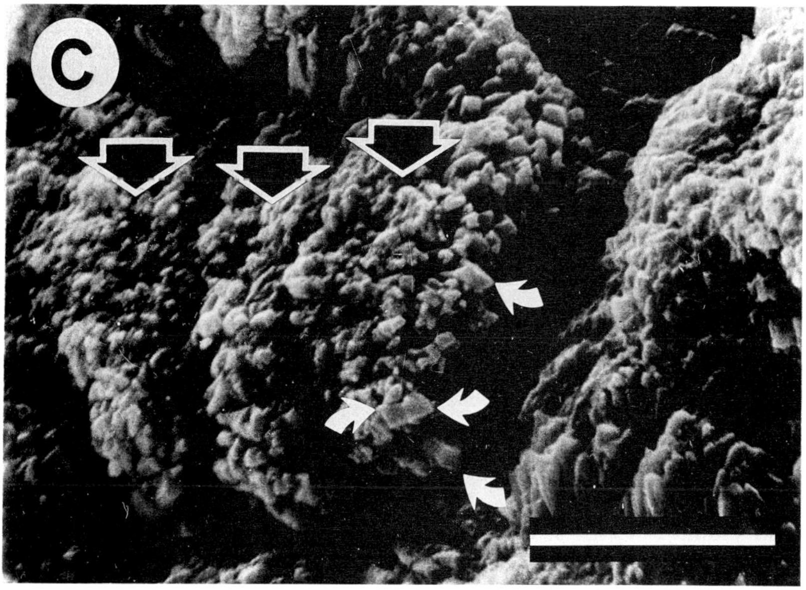

Figure 5.3. (*Continued*) (C) Scanning electron micrograph showing the center of calcification of *Acropora palmata*. Black arrows point to fine granular crystals and the white arrows to larger crystals with fairly well-defined morphologies. Scale bar: 10 μm. Courtesy of Dr. B. R. Constantz.

the aragonitic crystal arrays in corals and the spherulitic structures formed by inorganic processes. The latter are characteristic of precipitates from highly supersaturated solutions and obviously form very rapidly. It is of interest to note that spherulitic structures are found in other mineralized tissues (Bryan and Hill 1941) including avian egg shells, which also form very rapidly (Wilbur and Simkiss 1968). Thus, as the major portion of a coral skeleton is composed of spherulitic structures, it can be concluded that physicochemical processes predominate in its formation (Constantz 1986).

That the aragonite crystal morphologies vary between taxa implies that the microenvironments in which these crystals grow must also vary between taxa in a consistent and stable manner (B. R. Constantz, personal communication). This aspect of coral spherulitic growth is, therefore, certainly under control by the organism. How this is achieved is not known. Obvious possibilities include careful regulation of the ionic composition of the medium from which the crystals form or the presence of additives in solution that can alter crystal morphology (see Chapter 3).

The locations at which the spherulites form must be under strict biological control, judging from their taxonomic usefulness. The distribution and structure of these sites play key roles in the determination of the overall skeletal ultrastructure. The exciting discovery by Constantz and Meike (1988) that in *Mussa angulosa* the centers of calcification contain calcite crystals, although to date known only from one species, underscores the potential importance of these sites in directing spherulitic aragonite growth. In fact the involvement of calcitic structures in directing

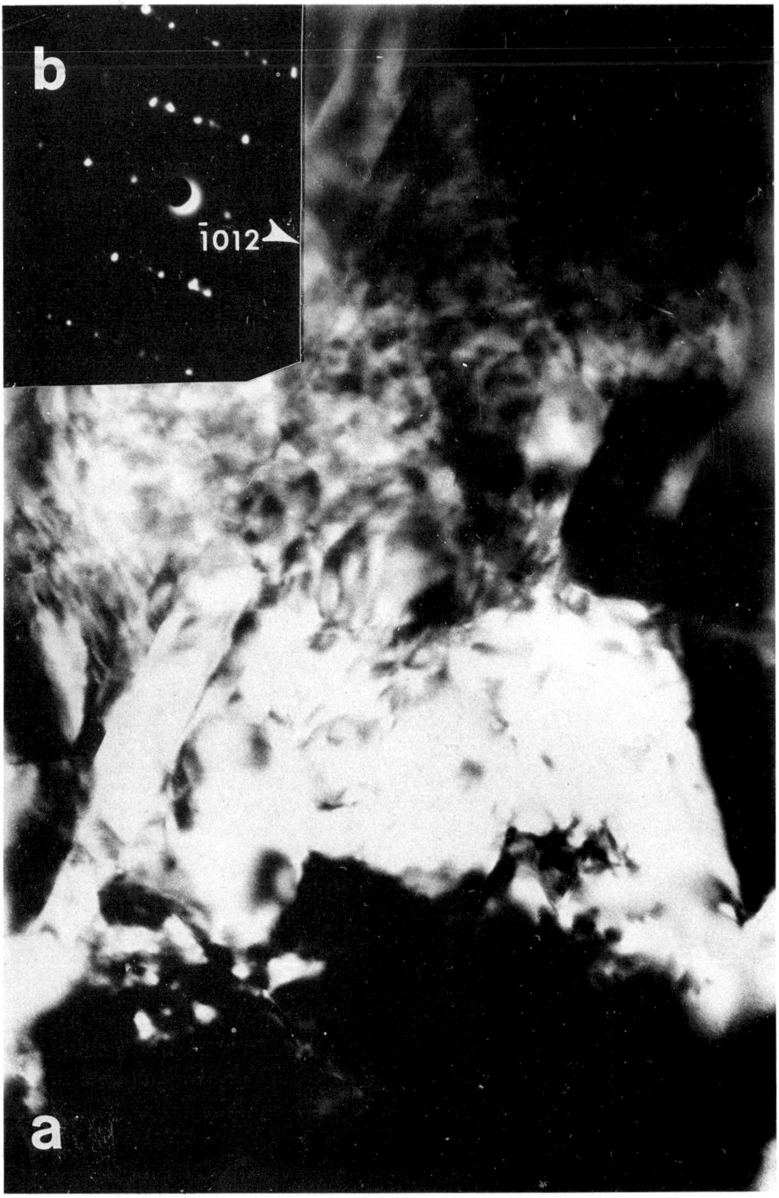

Figure 5.4. Transmission electron micrograph of an ion beam thinned section of *Mussa angulosa*. The image is produced in dark field using the $(10\bar{1}5)\alpha$ reflection of calcite. Note that the center of calcification is well illuminated when compared to the surrounding aragonitic fiber bundles. The inset (b) shows a twinned calcite diffraction pattern from the center of calcification. Courtesy of Drs. B. R. Constantz and A. Meike.

spherulitic growth in adults is consistent with observations in the larval skeleton, in which the first formed layer, upon which the aragonite crystals grow, is also calcitic. Furthermore, in considering the spectrum of mineralization processes in the whole Anthozoan class, calcitic structures feature very prominently (Table 5.1) and, most significantly, are formed under well-controlled conditions. In the better documented cases, the spicules form intracellularly in vacuoles, which even have an organic matrix framework (for example, *Leptogorgia*). It would, therefore, be worthwhile to discover whether the larval scleractinian calcite crystals as well as those in the adult skeleton really have an intracellular origin and form within an organic matrix. This is a subject that has been addressed in many studies (Hayasi 1937; Hayes and Goreau 1977; Johnston 1980; Kinchington 1980), but with still rather inconclusive results.

It is certainly premature to speculate about how the centers of calcification influence spherulitic growth. That they constitute only a few percent or less by weight of the adult skeleton makes them extremely difficult to study. The presence of acidic phospholipids and acidic proteins in the skeleton raises many other intriguing possibilities by which crystal formation *could* be controlled. Bearing all this in mind we conclude, as we do at the end of almost every chapter in this book, that much more remains to be discovered.

6

Mollusca

6.1 Introduction

Mollusks have a well-deserved reputation for being expert mineralizers based only on their much-admired shell-making abilities. Table 6.1 shows that the reputation is deserved 10-fold as shell formation is just one of many different processes that these animals perform in which biogenic minerals are utilized. The table lists no less than 21 different minerals and about 17 different functions! The list contains both amorphous minerals (amorphous fluorite, calcium carbonate, calcium phosphate, calcium pyrophosphate, and silica) and many crystalline ones, including rather uncommon ones such as weddelite, calcium fluorite, barite, magnetite, lepidocrocite, and goethite. Weddelite, for example, is a calcium oxalate mineral frequently formed pathologically in vertebrates. Certain gastropods use the rather soft weddelite nonpathologically to cap pestlelike objects (gizzard plates) in their stomachs (Lowenstam 1968), which they use for crushing shelled prey. One mollusk, the chambered *Nautilus,* forms no less than five different minerals. An individual tooth of a chiton contains three different mature minerals that are products of two other transient minerals. In addition to the more familiar functions of mineralized tissues, mollusks use biogenic minerals as buoyancy devices, trap doors, egg shells, and love darts. The varieties of crystal shapes, sizes, organizational arrays, and tissue sites present a picture of overwhelming diversity all within one phylum.

It is illustrative to compare the mollusks with the echinoderms. The echinoderms also use minerals for a wide variety of functions, but in contrast to the mollusks they use essentially the same "building material" for many different purposes. Thus, understanding how one echinoderm mineralized tissue forms provides insight into how most of the others form. This is not so with mollusks. It seems futile to expect that they too have adapted one basic process to form all their mineralized tissues. It seems just as futile to look for a different explanation for each type of mineralized product. The mollusks force us to seek a level of understanding of mineralization that identifies common approaches, strategies, and principles and, at the same time, appears to dispel any "dreams" about discovering *the* mechanism of mineralization.

The mollusk phylum contains seven different taxonomic classes. Five of these are shell bearing: the Cephalopoda, Gastropoda, Bivalvia, Scaphopoda, and Monoplacophora. One class, the Aplacophora, contains no shell at all. The body is entirely or partially covered by a rather dense layer of spicules or scales. The seventh class, the Polyplacophora, covers part of their body with shell-like plates whereas other parts are covered by spicules. In this chapter we will briefly comment on aspects of mineralization of Aplacophora, Monoplacophora, and Scaphopoda. The Polyplacophora will be discussed in more detail, with particular emphasis on tooth formation—a process that has provided the basis for many widely discussed "principles" in biomineralization. The final section will focus on shell formation and draws on information primarily from the three major mollusk classes, the Cephalopoda, Bivalvia, and Gastropoda.

6.2 Aplacophora

These are shell-less mollusks with a wormlike shape. Their identity as mollusks is firmly established from their embryological development. The only known mineralized hard parts are spicules that cover either the entire outer body surface or at least the dorsal side. In many species, the spicule shapes vary markedly in different regions of the body from elongated acicular needles to stubby paddle-shaped structures, some of which also have keels. They are generally a few hundred microns long (Schwabl 1963). All the spicules examined to date are aragonitic with their *c* axes aligned with the long axis of the spicule (H. A. Lowenstam, unpublished observation). Schmidt (1924) reports the presence of some calcite. He noted, however, that the material examined was poorly preserved. Insoluble organic material remains after the spicule is dissolved (von Graff 1876). The process of spicule formation is most unusual in that each spicule is first formed intracellularly, but soon protrudes from the epidermal cell as it grows (Woodland 1907). The cell enlarges, becomes multinucleated, and stalklike in shape. The one end has the form of a cup and surrounds the base of the enlarging spicule (Hoffman (1949). The other end of the cell penetrates the cuticle but is firmly embedded in the epithelial layer. Thus, the entire spicule is formed by one cell! This unusual circumstance should certainly be studied in great detail.

6.3 Monoplacophora

The monoplacophorans have a unique status among mollusks as their capshaped shells with numerous pairs of muscle insertion scars were for years known only from fossils ranging in age from the early Cambrian (about 550 million years) to the mid-Devonian (about 380 million years). In 1957 Lemche identified a living representative dredged from the deep-sea floor by the Danish Galathea Deep-sea Expedition. The soft part anatomy of *Neopilina* was characterized by numerous

Table 6.1 The Diversity, Distribution, Localization, and Functions of Minerals Formed by Mollusca

Taxon	Taxa reported	Mineralization site	Mineral	Function	References
Aplacophora	All	Cuticle spicules formed by epithelial cells	Aragonite and calcite?	Protection	Schmidt (1924)
Polyplacophora	All	Shell plates	Aragonite	Protection and muscle attachment	Cornish and Kendall (1888)
	Most	Dorsal girdle spicules or scales	Aragonite	Protection	Cornish and Kendall (1888)
	All	Ventral perinotum spicules	Aragonite	Unknown	Lowenstam (unpublished)
	Cryptochiton stelleri	Forming teeth of radula	Ferrihydrite	Precursor of magnetite	Towe and Lowenstam (1967)
		Mature teeth of radula	Magnetite, amorphous hydrous ferric phosphate, and trace of opal	Mechanical grinding	Lowenstam (1962a); Lowenstam and Rossman (1975)
	Chitonida (several species)	Forming teeth of radula	Ferrihydrite	Precursor to magnetite	Towe and Lowenstam (1967)
			ACP	Precursor to carbonate apatite	Lowenstam and Weiner (1985)
		Mature teeth of radula	Magnetite, lepidocrocite, francolite, or dahllite	Mechanical grinding	Lowenstam 1967); Lowenstam and Weiner (1985)
Monoplacophora	All	Shell	Aragonite	Protection	Schmidt (1959)
		Statoliths	Indeterminate	Gravity receptors	Lemche and Wingstrand (1959)
Scaphopoda	All	Shell	Aragonite	Protection	Bøggild(1930)
		Statoliths	Indeterminate	Gravity receptors	Barber (1968)

Gastropoda	All	Shell	Aragonite or aragonite and calcite	Protection	Bøggild (1930)
	Some land and freshwater snails	Shell	Vaterite	Precursor of mature form	Levetzow (1932)
	Viviparus	Shell	Vaterite, aragonite	Protection	Kessel (1933)
	Helix	Shell	Vaterite, aragonite	Protection	Stolkowski (1951)
	Archidoris; Anisodoris	Skin spicules	Amorphous monohydrocalcite and amorphous fluorite	Protection?	Lowenstam and McConnell (1968); Lowenstam (unpublished)
	Some land snails;	Operculum	Aragonite, calcite	"Trap door"	Kessel (1942)
		Epiphragm	Aragonite, calcite	Shell seal	Block (1971)
	Pomacea paludosa	Mantle cells (granules)	Amorphous $CaCO_3$, vaterite, calcite	Temporary storage	Watabe et al. (1976); Saleuddin (1970)
		Foot (granules)	Amorphous $CaCO_3$, vaterite	Temporary storage	Watabe et al. (1976)
	Pulmonates	Hepatopancreas (granules)	Calcite, amorphous phosphates	Unknown	Burton (1972)
	Helix aspersa	Hepatopancreas (granules)	Amorphous calcium pyrophosphate	Detoxification	Howard et al. (1981)
	Land snails (some)	Egg capsules	Calcite; calcite and vaterite	Calcium source?	Prenant (1927); Tompa (1976)
	Some freshwater snails	Egg capsules	Vaterite	Calcium source?	Hall and Taylor (1971)
	Helix pomatia	Love dart	Aragonite	Reproduction	Hunt (1979); Tompa (1980)
	Anguispira	Blood vessels	Amorphous $CaCO_3$	Unknown	Tompa and Watabe (1976a)
	Patella vulgata	Hemolymph	Ferrihydrite	Iron storage	St Pierre et al. (1986)
	Patellacea	Radula	Goethite and opal	Food gathering (scraping)	Lowenstam (1962a, 1971)
	Heteropoda	Statoconia	Calcite?	Gravity receptors	Tschachotin (1908); Barber (1968)

Table 6.1 (*Continued*)

Taxon	Taxa reported	Mineralization site	Mineral	Function	References
	Aplesia	Statoconia	Ca mineral	Gravity receptors	Wiederhold et al. (1986)
	Cephalaspidea				
	Scaphander div. sp.	Gizzard plates	ACP and fluorite	Crushers of shelled prey	Lowenstam and McConnell (1968)
	Scaphander	Gizzard plates	ACP and weddelite	Crushers of shelled prey	Lowenstam (1968)
	Philline div sp.	Gizzard plates	ACP and monohydrocalcite	Crushers of shelled prey	Lowenstam (1972b)
	Cylichna cylindracea	Gizzard plates	ACP and weddelite	Crushers of shelled prey	Lowenstam (1972b)
	Acteocina cf. *culcitella*	Gizzard plates	ACP	Crushers of shelled prey	Lowenstam (1972b)
Bivalvia	All	Shell	Aragonite and/or calcite	Protection	Prenant (1927); Bøggild (1930)
		Shell myostracum	Aragonite	Unknown	Stenzel (1963); Taylor et al. (1969)
	Mytiloida, Mytilacea, Anomalodesmata	Shell periostracum (spikes or granules)	Aragonite	Enhanced stability	Carter and Aller (1975)
	Lithophaga nigra	Shell periostracum (outer layer)	Dahllite	Protection against shell abrasion	Waller (1983)
	Lithophaga antillarum	Shell periostracum (outer layer)	Francolite	Protection against shell abrasion	Carter and Clark (1985)
	Some	Shell hinge ligament	Aragonite	Unknown	Wada (1961); Marsh and Sass (1981)
	Anodonta cyanea	Mantle tissue	Ca mineral	Temporary storage	Istin and Girard (1970)
	Rangia cuneata	Mantle innermost shell lamella	Barite, aragonite	Unknown	Marsh (1986)
	Neotrigonia margaritacea	Gill supports	ACP	Mechanical strengthening	Lowenstam (1972b)
	Ligumia subrastrata	Gills (spherules)	Calcium phosphate	Temporary storage	Silverman et al. (1983)

	Some	Burrow linings, tubes	Aragonite and/or calcite	Mechanical strengthening	Bøggild (1930); Carter (1980a)
	Mercenaria mercenaria; Argopecton, irradians	Kidneys	ACP	Excretory process	Doyle et al. (1978)
	Pecten inflexus	Statoconia and/or statolith	Indeterminate	Gravity receptor	Buddenbroek (1915)
	Anomalodesmata	Statoconia and/or statolith	Indeterminate	Gravity receptor	Morton (1985)
	Some	Pearls	Aragonite, calcite	Insulation of foreign particles	Schmidt (1924)
Cephalopoda	All	Shell (wall, septa)	Aragonite	Protection and/or bouyancy device	Bütschli (1908); Bøggild (1930)
Nautiloidea	*Nautilus* div. sp.	Shell callus	Aragonite	Drag reduction?	Lowenstam et al. (1984)
	Nautilus all sp.	Shell siphuncle	Aragonite	Osmotic pumping divice	Mutvei (1972); Grégoire (1973)
	Nautilus div. sp.	Mandible	Aragonite, calcite, brushite, ACP, weddelite	Biting	Lowenstam et al. (1984)
	Nautilus div. sp.	Statoconia	Aragonite, ACP	Gravity receptors	Lowenstam et al. (1984)
	Nautilus div. sp.	Uroliths	ACP	Calcium and phosphorus storage	McConnell and Ward (1978)
Coleoidea	All	Statoliths	Aragonite	Gravity receptor	Lowenstam et al. (1984)
	Spirula, Sepia	Shell (internal)	Aragonite	Bouyancy device	Bøggild (1930)
	Argonauta	Egg case	Calcite	Brood chamber	Kelly (1901)

metameric or repeat structures including parts of the nervous system, gills, kidneys, gonads, and major pedal retractor muscles. Lemche and Wingstrand (1959) thought these represented primitive features, although this has since been questioned (e.g., Stasek 1972; Trueman 1975). Since 1957 a number of other species belonging to the monoplacophorans have been discovered in the deep-sea (summarized in Lowenstam 1978) and one species has recently been located at shallow depths between 180 and 420 m (McLean 1979).

In three recent monoplacophorans the shell mineral is composed entirely of aragonite. The shells of these species contain three mineralized layers: a thick outer layer composed of a regular spherulitic prismatic structure, a middle nacreous layer, and a thin inner prismatic layer (Schmidt 1959; Erben et al. 1968; H. A. Lowenstam, unpublished). The outer surface is covered by a layer of organic material called the periostracum (Lemche and Wingstrand 1959). From an ultrastructural point of view (Erben et al. 1968) and from the biochemical properties of the shell macromolecules (Meenakshi et al. 1970; Poulicek and Jeuniaux 1981), the shell of the monoplacophorans is certainly not unique among mollusks.

6.4 Scaphopoda

Scaphopoda are marine mollusks that burrow into the sediments. They are found from shallow depths down to more than 7000 m (Lindberg 1985). Their shells are tooth or tusk shaped and are open at both ends. They position themselves with their heads down in the sediment. The shell is composed of aragonite (Bøggild 1930) and is differentiated into three layers: a thick middle layer (crossed-lamellar ultrastructure) and two thin surface layers (homogeneous or finely prismatic) (Bøggild 1930). The mantle is completely fused into a cylindrical cone. Almost nothing appears to be known about the shell formation processes of Scaphopoda.

6.5 Polyplacophora: Tooth Formation

The Polyplacophora or chitons are a class of marine mollusks that is strikingly different in appearance from all other shell-bearing mollusks in that their "shell" is made up of a series of eight overlapping aragonitic plates rather than one solid mass. Interestingly, this pseudometameric "trait" is also reflected in their soft part anatomy, i.e., musculature, gills, and nervous system. Another unusual characteristic of chitons is the development of a thick fleshy girdle surrounding and, in some forms, even covering the plates. The girdle is an adaptation of the mantle and is commonly "armored" by mineralized aragonitic spicules or, more rarely, by aragonitic scales as well (Schmidt 1924). Aragonitic spicules are, in addition, present subcutaneously on the underside of the girdle rim (H. A. Lowenstam, unpublished). The mode of formation of the dorsal mantle spicules is similar to that of aplacophorans. The initial mineralization stages are thought to take place within a vesicle inside an epithelial cell. (Some doubt about its intracellular location still exists.) With further growth, the spicule protrudes through the cell membrane. Continued growth occurs only at one end of the spicule where the epithelial cell has formed a cup-shaped structure enveloping the spicule base (Haas 1976). In contrast, the structure of the shell plates has many properties in common with the shells formed by the shell-bearing mollusks, although some obvious differences are present. One such difference is that the plates have a most unusual layer between the outer organic protective covering (periostracum) and the inner shell layers. The so-called *tegmentum* is composed of spherulitic structures aligned *parallel* to the plate surface—a very unusual circumstance for mollusk shells. This layer also houses sensing organs (esthetes), which in some species are light sensitive (Haas 1976). Beedham and Truemann (1967) have described the adult polyplacophoran shell, mantle, and cuticular spicules and the anatomical relations between them. Their study shows that the Polyplacophora must have diverged from the main molluscan stock at an early stage in the evolution of the phylum. Polyplacophora thus appear to occupy an intermediate niche between the Aplacophorans and the shell-bearing mollusks.

The elaborately mineralized teeth of the chitons are formed in an assembly linelike fashion offering the opportunity to study the different stages of mineralization. They have thus revealed some of our most important insights into the understanding of basic processes of controlled biological mineralization. The following section, therefore, discusses the processes of chiton tooth formation.

The teeth are used in surface water species for scraping rocky substrates to remove the surficial as well as the endolithic organisms living within the rock (Lowenstam 1962a). Most chitons live in the intertidal and subtidal zones together with patellacean limpets. On tropical limestone coasts the constant scraping of the chitons has actually produced massive undercuts, which eventually cause recession of the coastline itself (Lowenstam 1974). Although this grazing behavior of the common chitons was well known, it was only in 1962 that Lowenstam, recognizing that their major lateral teeth must be harder than the substrate, discovered that they were capped with the hard iron oxide mineral, magnetite. This was the first discovery of biologically formed magnetite, a biogenic mineral now known to be widely distributed among organisms (Table 2.1 and summarized in Lowenstam and Kirschvink 1985).

The magnetite-capped teeth are also present in some rather rare carnivorous chitons that do not scrape the substrate, as well as in deep-sea chitons that are deposit feeders and ingest the sediment to extract its nutritional content. The latter group has lost the massive buccal musculature content normally present in chitons (Menzies et al. 1973). Presumably both groups evolved from the chitons living on the rocky substrate.

The chiton teeth are arranged in numerous transverse rows along a long ribbonlike structure known as the radula. The location of the radula inside the organism is shown in Figure 6.1. Each row comprises one central tooth and eight pairs of flanking teeth. Only the second pair of flanking teeth, the major lateral teeth, are mineralized. One end of the radula extends into the mouth parts and at any given time only the teeth in the anterior region (in *Cryptochiton* the last 8 to 10 rows) are used for scraping. These teeth become abraded or fractured and are then discarded at an average rate of a row every 12 to 48 hours depending upon the species (Nesson 1968; Kirschvink and Lowenstam 1979). To maintain an "equilibrium" new teeth are formed at the other end (by the odontoblast cells) at the same rate. Thus,

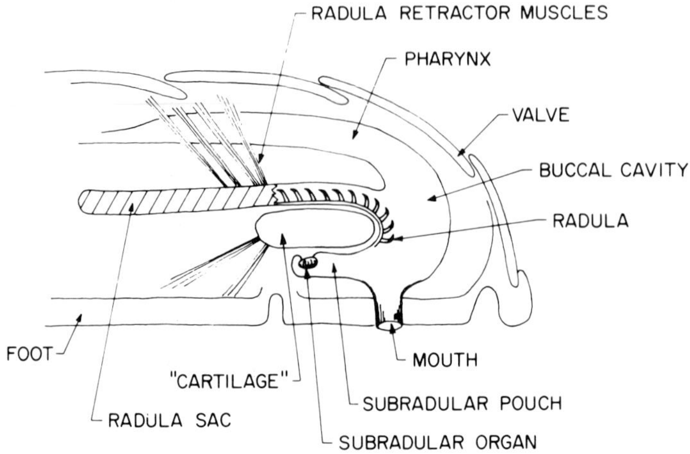

Figure 6.1. Diagrammatic longitudinal section through the anterior of a chiton showing the location of the radula apparatus and the orientation of the radular teeth while in the radular sac. Reproduced from Kirschvink and Lowenstam (1979), by courtesy of Elsevier Science Publishers.

the entire structure is an assembly line or conveyor belt for tooth formation and each row contains teeth in a slightly more advanced stage of formation than the previous row, until mineralization is complete. The chiton radula is, therefore, one of the most accessible systems for studying the dynamics of biomineralization and it is no surprise that often the only reliable source of "hard data" for some of the most widely discussed topics in this field is the radula.

The patellacean gastropods (limpets) and the chitons occupy a similar ecological niche in shallow water environments. The limpets also have mineral-capped teeth and some species, in addition, have mineralized tooth bases (Lowenstam 1962b, 1971). Their overall tooth structure differs quite significantly from the chitons. For more details on the radular teeth of limpets see Runham et al. (1969), Lowenstam (1962b, 1971), and Mann et al. (1986b). It should be noted that replacement of worn teeth is by no means unique to limpets and chitons. As a matter of fact all snails as well as the sharks exhibit a rather similar phenomenon. Rodents and sea urchins (Chapters 8 and 9) use a slightly different strategy. Rodent incisors and the lateral teeth in sea urchins are continuously abraded at one end and are continuously formed at the other.

An individual mature chiton tooth is composed of a mineralized cap and a stylus or stalklike flexible attachment to the radular substrate. The caps vary in size from a fraction of a millimeter to about 2 mm in the giant *Cryptochiton.* A major organic component of the tooth is α-chitin (S. Weiner, unpublished), presumably associated with an assemblage of proteins. Because of their small size, almost nothing is known about the biochemistry of the teeth. Each mature tooth of the common **Chitonida** contains no less than three minerals, all occupying different locations (Lowenstam 1967) (Fig. 6.2). Magnetite ($Fe^{2+}Fe_2^{3+}O_4$) occupies the tooth margin, the area used for scraping, as well as the entire posterior surface. Lepido-

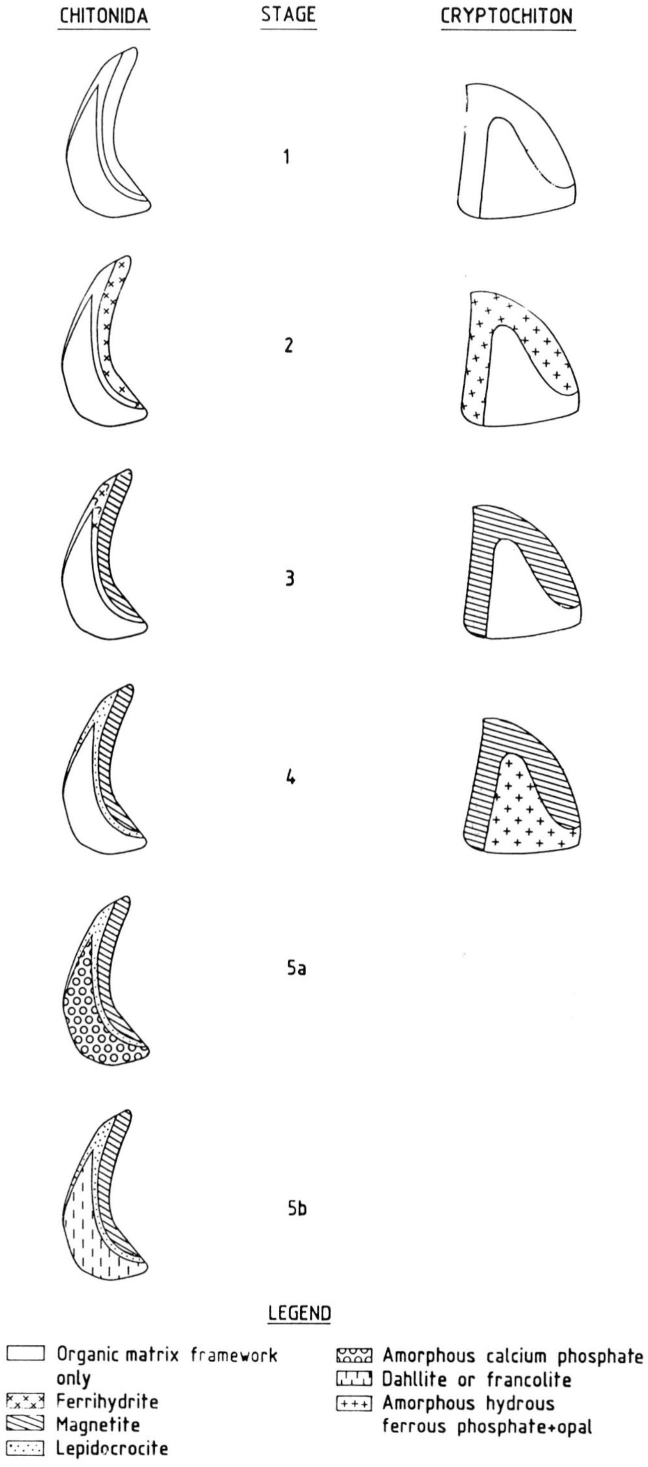

Figure 6.2. Schematic representation of the five stages of tooth formation in the Chitonida and in *Cryptochiton.*

crocite [γ-FeO(OH)] forms a thin layer that lines the inner surface of the magnetite. Dahllite, the carbonate apatite mineral [$Ca_5(PO_4CO_3)_3(OH)$] (or in some species francolite, the fluorinated form), fills in the central core and most of the anterior tooth surface. In *Cryptochiton* the inner core is filled with amorphous hydrous ferric phosphate that also contains traces of amorphous silica (opal), (Lowenstam 1972b; Lowenstam and Rossman 1975) instead of the lepidocrocite and dahllite components of the Chitonida.

Row by row studies of tooth formation reveal all the stages of mineralization in great detail. During the entire process the tooth is completely surrounded by a layer of closely adhering superior epithelial cells (Nesson and Lowenstam 1985). These cells orchestrate the entire formation process. Tracer experiments using a radioactive iron isotope (^{59}Fe) show that the source of iron is ferritin present in the bloodstream (Nesson 1968). This is reduced and solubilized, and then transported to the superior epithelial cells where it is again temporarily stored in the form of ferritin prior to mineralization of the tooth itself (Nesson and Lowenstam 1985). The process of tooth formation can be conveniently divided into the following five stages, which are also illustrated in Figure 6.2.

Stage 1. Construction of the organic framework takes place before any mineral is introduced into the structure. X-Ray diffraction patterns of individual teeth from the first five or so rows of chitons show progressively sharper reflections of the α-chitin pattern indicating that the polymers are undergoing self-assembly (S. Weiner, unpublished observation). During stage 1 the entire tooth achieves its mature morphological shape prior to the introduction of any mineral. This does not necessarily mean that every one of the tooth macromolecules is in place by this time.

Stage 2. Ferrihydrite is the first mineral deposited within the organic matrix framework (Towe and Lowenstam 1967). The ferrihydrite is not translocated from the superior epithelial cells, but is transported to the tooth organ in the reduced soluble form even though it is derived from intracellular ferritin in the form of ferrihydrite (Nesson 1968). The deposition of ferrihydrite in the tooth takes place over two to five rows and these are easily recognized by their light brown to reddish color (Towe and Lowenstam 1967).

Stage 3. Magnetite formation starts in the last light brown colored tooth usually within the ferrihydrite deposits (Kirschvink and Lowenstam 1979). Mass conversion of ferrihydrite to magnetite takes place in the next tooth row (the first black tooth) (Towe and Lowenstam 1967) and continues to build up for quite some time after that. As there is always a thin layer of ferrihydrite on the forming magnetite surface, it would seem that all the magnetite forms from a ferrihydrite precursor (Kirschvink and Lowenstam 1979). It has, however, also been proposed that the bulk of the magnetite forms de novo from solution (M. Nesson, personal communication to H. A. Lowenstam).

Stage 4. In the Chitonida *lepidocrocite* is laid down in the form of a narrow strip on the inside surface of the magnetite layer. Lepidocrocite forms even as the magnetite continues to build up on the posterior tooth surface. It is likely, although it still remains to be demonstrated, that ferrihydrite is also the precursor of lepidocrocite. In *Cryptochiton,* amorphous hydrous ferric phosphate with a trace of opal begins to be deposited during this stage and continues through stage 5 as well (Lowenstam 1972b; Kirschvink and Lowenstam 1979).

Stage 5. Amorphous calcium phosphate (ACP) is the last mineral phase deposited in the remainder of the tooth of the **Chitonida** (Lowenstam and Weiner 1985). In the SEM this has the appearance characteristic of many amorphous minerals, namely, rounded spherically shaped objects. After roughly 13 tooth rows the ACP begins to transform into crystalline dahllite or francolite and significantly the *c* axes of the crystals have a preferred orientation. They are more or less aligned with the organic matrix sheets inside the mineralized layer suggesting that the matrix surface somehow directs the transformation process. The transformation probably proceeds by dissolution and reprecipitation as the magnesium content of the product is much lower than in the amorphous precursor (H. A. Lowenstam, unpublished observation). This may also explain why the fluorine starts entering into the calcium phosphate phase of *Clavarizona hirtosa* at a late stage (Kim et al. 1986), probably after the transformation has taken place. The degree of crystallinity increases continuously until tooth mineralization is complete. At this stage the curtain of tightly adhering superior epithelial cells detaches from the tooth caps and the now mature teeth are free and ready for use until discarded.

These observations of tooth formation in the chiton establish unequivocally a few important principles of biomineralization that may well have general relevance. (1) A preformed mould or framework is constructed prior to the initiation of mineralization. In this respect, tooth formation is clearly a "matrix-mediated" process (Lowenstam 1981). (2) It is established for two quite different minerals that the intially deposited phase is mineralogically different from the mature product. Chiton teeth provide the only documentation to date for the transformation of ACP to dahllite, although this process is widely discussed (or hotly debated!) in the field of bone mineralization (Chapter 9). (3) A mature product formed from an amorphous precursor can have oriented crystallographic axes.

Aside from magnetite, which is always present in mature chiton teeth, the occurrence of amorphous ferric phosphate has so far been found only in cold to temperate water species. In warm water species lepidocrocite and carbonate apatite minerals substitute for the amorphous compound. Moreover, in warm waters the magnetite fraction of the tooth minerals is notably reduced. Thus, it appears that variations in environmental temperatures correlate with the observed differences in tooth mineralogy (Lowenstam 1974). The reason for this is not known, but is presumed to be in the realm of preferred selection of certain genes over others, which are somehow influenced by the prevailing environmental temperature (see Chapter 11 for more examples and discussion).

6.6 Cephalopoda, Bivalvia, and Gastropoda: Shell Formation

6.6.1 The Mantle

The mantle is the organ responsible for shell formation. It is a thin sheet of tissue lining the inner surface of the shell and is composed of two epithelia separated by connective tissue. A detailed schematic illustration by Waller (1980) of the anatomy of the mantle and the shell in the ventral region of a mature bivalve, *Arca zebra,* is shown in Figure 6.3. It is representative of most bivalves. [For comparison

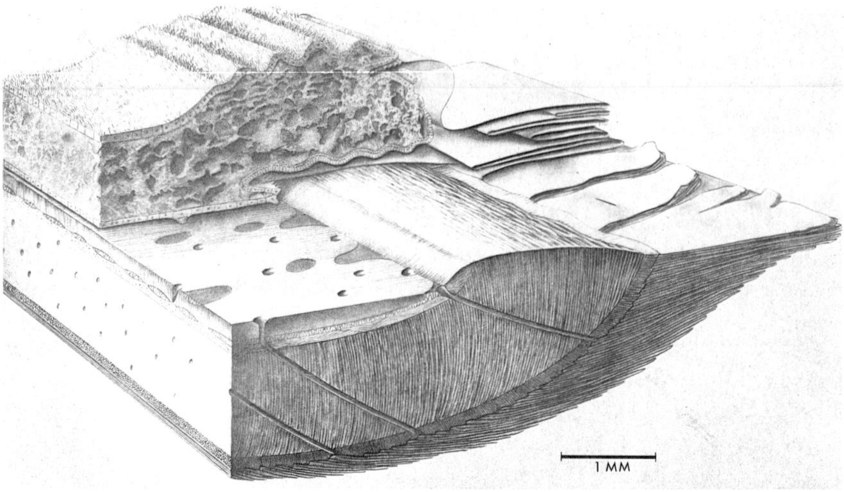

Figure 6.3. Schematic block diagram showing the relationship between the mantle (top) and shell (bottom) in the ventral region of the bivalve *Arca zebra*. For details see Waller (1980). Reproduced from Waller (1980), by courtesy of the Smithsonian Institution Press.

with a freshwater bivalve, refer to the schematic illustration of mantle and shell in Saleuddin and Petit (1983).] The most conspicuous difference is the ultrastructural characteristics of the mineralized shell layers. These differ from one taxonomic group to another (see below). However, the detailed anatomy of the mantle in bivalves differs significantly from gastropods. For example, the mantle edge in bivalves usually has three folds, whereas gastropods only have two (Zylstra et al. 1978). The mantle enlarges by cells undergoing mitotic division at numerous growth centers spread throughout the layer of epithelial cells (Kniprath 1978). The cells of the outer mantle epithelium (Fig. 6.3) that abut the inner shell surface are the ones most directly involved in mineralization. They synthesize and secrete the array of macromolecules that self-assembles outside the cell and through these macromolecules direct crystal formation (see below). The outer membrane epithelial cells receive their calcium and bicarbonate from the hemolymph (Schoffeniels 1951). It has been proposed that in bivalves, the epithelial cells located within the pallial line (central zone of the shell) allow calcium to passively diffuse into the mineralizing site through channels formed between cells (Neff 1972), and that outside the pallial line, in the marginal zone, calcium must pass through the cells to reach the mineralization site (Crenshaw 1980). It is not clear, however, how generally applicable this view of calcium transport through the mantle is.

The mantle edge is the most active zone of shell deposition (Wilbur and Jodrey 1952). The cells of the outer mantle epithelium edge zone are ultrastructurally quite different from their counterparts in the central zone. They are elongated and contain numerous mitochondria, well-developed endosplasmic reticula, and Golgi apparatuses, whereas those of the central zone are more cuboidal with less conspicuous organelle development (see Crenshaw 1980). At the mantle edge the shell is not only thickened, as occurs throughout the whole inner surface, but is also enlarged to accommodate the growing animal. Appropriately then, it has been discovered that in the freshwater snail, *Lymnae stagnalis,* a growth hormone that regulates growth of all organs is also responsible for controlling enlargement of the shell (Geraerts 1976). It acts directly on what is essentially the first step in mineralization during the shell enlargement process, namely the elaboration of the first organic substrate (periostracum) on which mineralization ensues (Dogterom and Jentjens 1980). The growth hormone is also responsible for maintaining a high concentration of calcium (Dogterom et al. 1979) and a calcium-binding protein in the mantle edge (Dogterom and Doderer 1981). It does not influence matrix formation and calcification in the inner shell surface. This body of work represents the best established link, to date, between the mineralization process itself and the hormones that regulate it in the invertebrate world. Other indications in mollusks that sophisticated endocrinal control of calcium homeostasis occurs and hence, directly or indirectly, control of shell formation as well is the demonstrated presence of an active vitamin D metabolism in land snails (Weiner et al. 1979) and in the marine bivalve *Mytilus edulis* (Lehtovaara and Kosinen 1985).

6.6.2 The Periostracum

The mantle edge of all mollusks is folded so as to produce one or more grooves. The groove closest to the shell surface is the site of periostracum formation. The periostracum is a primarily proteinaceous layer, which in the mature shell covers the outside surface. It is formed by a group of specialized cells at the base of the groove. The cytological aspects of periostracum formation are reviewed by Saleuddin and Petit (1983) and the biochemical aspects by Waite (1983). Whereas the details differ from case to case, the generally recognized stages of periostracum formation are the synthesis of the precursor macromolecules in the Golgi cisternae where they fold into structurally ordered units. These separate into secretory vesicles that are then transported out of the cell and the contents presumably self-assemble into the large continuous sheet that emanates from the groove. In some cases other cells along the groove secrete additional material to thicken the sheet (Saleuddin and Petit 1983). The periostracum is described in some detail as it is a well-studied example of polymerized macromolecules used for space delineation (Chapter 3).

Quinone-tanned protein (Brown 1952; Beedham 1958) containing DOPA (3, 4-dihydroxyphenylalanine) (Degens et al. 1967b) is the major constituent of periostracum, and chitin is a minor one. Chitin, however, is not always present, at least at detectable levels (Goffinet and Jeuniaux 1979; Peters 1972). Periostracum is not easily solubilized and biochemical studies of its constituents are difficult to perform. To date, the only detailed study of an isolated periostracum protein is by Waite et al. (1979). The protein, called periostracin, was extracted from the margin of the bivalve *Mytilus edulis* with formic acid and is actually a precursor form of the mature end product. It has some very unusual properties. It is basic, hydrophobic, water insoluble, and has an apparent molecular weight of 22,000. The unpurified form spontaneously degrades and polymerizes. Glycine is the most abundant amino acid (55 mol%) and DOPA is present at a concentration of 22 residues per 1,000. DOPA is thought to form by posttranslational hydroxylation of the tyrosyl residues. The DOPA concentration decreases with sclerotization of the periostracum, strongly suggesting that it is an integral part of the cross-linking process. Waite (1983) proposed the secretion of a pre-DOPA protein that is activated to the aggregating form by proteolysis. The aggregated mass is acted upon by the enzyme *O*-diphenoloxidase that converts DOPA residues to quinones and/or semiquinones that spontaneously polymerize. This enzyme is present in periostracum (Hillman 1961). It is interesting to note that sclerotization of insect cuticle occurs in quite a different way. Instead of using DOPA-containing proteins, insects mix *O*-diphenol *N*-acetyldopamine with oxidases, cuticular proteins, and chitin to form a highly cross-linked macromolecular complex (Waite 1983). In *Mytilus* the amino acid composition of the mature periostracum differs from that of periostracin in having much less DOPA and cystine and about twice as much aspartic acid. The DOPA and cystine are consumed during the cross-linking reactions. The excess aspartic acid suggests that mature periostracum also has adhering shell matrix proteins (Waite et al. 1980; Waite 1983), some of which are characteristically rich in aspartic acid (Weiner 1979).

TEM examinations of the periostracum generally show little or no ultrastructural order at the molecular level, although subdivision into various layers is often apparent (Kniprath 1972). One notable exception is the periostracum of the gastropod *Buccinum undatum* (Hunt and Oates 1970, 1978). The stained periostracal fragments released by sonication show well-defined ribbonlike units with distinctive longitudinal banding patterns. It has been proposed that the repeating units are dumbbell shaped and some 310Å long. The ribbons aggregate into sheets and, interestingly, these sheets form a three-dimensional helicoidal structure characteristic of insect cuticle (Bouligand 1965; see Chapter 7).

Mature periostraca in the mollusks vary greatly with respect to thickness, ultrastructure, color, and texture. The periostracum coats the outer shell providing it with some protection from the environment, although in the older portions of the shell it is often lost by abrasion (see review by Clark 1976). Another important function of the periostracum at the mantle edge is to isolate the tissue from the environment (space delineation). This is achieved by the production of a continuous sheet of periostracum that emanates from the mantle groove and curls around to the outside of the shell. The continuity is not disrupted by mantle movements, as the periostracum is still flexible for some time after formation (Clark 1976).

Periostracum fulfills an additional role in mineralization, namely as a substrate upon which mineralization is initiated (Iwata 1980; Clark 1974a). Another hypothetical role proposed for the periostracum in mineralization is based on its demonstrated ability to act as a semiconductor that differentially concentrates ions at the inner surface. This may help or cause nucleation of the initially formed crystals (Digby 1968). A far more active role for the periostracum in shell formation has been proposed by Petit et al. (1980) (also reviewed in Saleuddin and Petit 1983) for the freshwater clam *Amblema*. The middle periostracal layer in particular is observed to be intimately involved in organizing the formation of the aragonitic prisms. The fact that shell formation at the margin of *Amblema* apparently differs

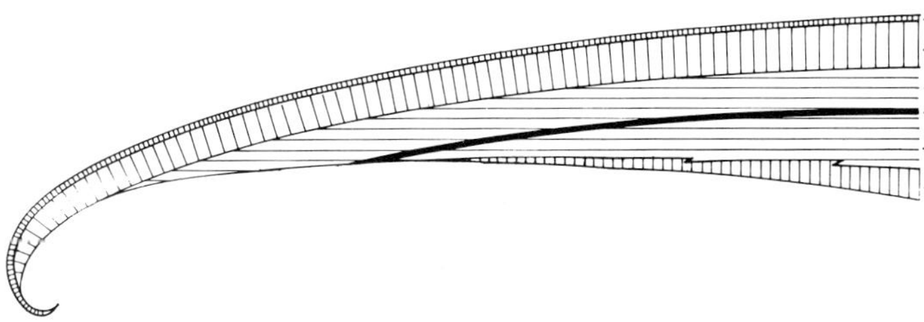

Figure 6.4. Schematic transverse section through an idealized posterior portion of a bivalve shell.

from other mollusks may be related to its freshwater habitat and its need for efficient isolation from the environment.

6.6.3 *The Shell*

The mollusk shell is a characteristically layered structure (Fig. 6.4) composed of calcium carbonate in the form of calcite, aragonite, or both (Table 6.1), with small amounts of organic material [usually between 0.01 and 5% by weight (Hare and Abelson 1965]. In shells in which both polymorphs of calcium carbonate are present they are always spatially separated into different layers. The layers are usually differentiated by their ultrastructural motifs. The basic ultrastructural types recognized today [see, for example, the guide to bivalve shell microstructures by Carter (1980b)] were originally described by Bøggild in 1930. Figure 6.5 shows a series of SEM micrographs illustrating some common ultrastructural arrangements identified by Bøggild (1930). Bøggild also documented the distributions of calcite and aragonite between taxa and, where appropriate, their locations within individual shells. The spatial distribution of individual layers in a given shell can be fairly complex because of interfingering of adjacent layers and the presence of additional thin shell layers called myostracum, which are formed at the site of attachment of muscles to the shell (Fig. 6.4). These myostracal layers are always aragonitic and have a prismatic ultrastructure (Table 6.1). As the shell grows, the muscle attachment areas advance and leave behind a "trail" of myostracum. This is then covered by inner shell layers.

The distribution of organic material within the shells of the large majority of mollusks is primarily as an envelope surrounding each individual crystal. There is also reason to believe that some of the organic macromolecules may be located within crystals as well (Crenshaw 1972a; see Chapters 3 and 8). In the process of crystal formation, however, it is generally thought that an extracellular framework of organic material is first formed and only then do the crystals grow within the voids of this framework. The best illustrated evidence documenting this is the forming nacreous layer (Nakahara 1979, 1983), although whether one or more than one organic sheet is in place prior to a new round of crystal formation is still disputed (Watabe 1984). Actually by far the most convincing and unequivocal documentation of a preformed matrix in mollusk mineralization is the earlier noted

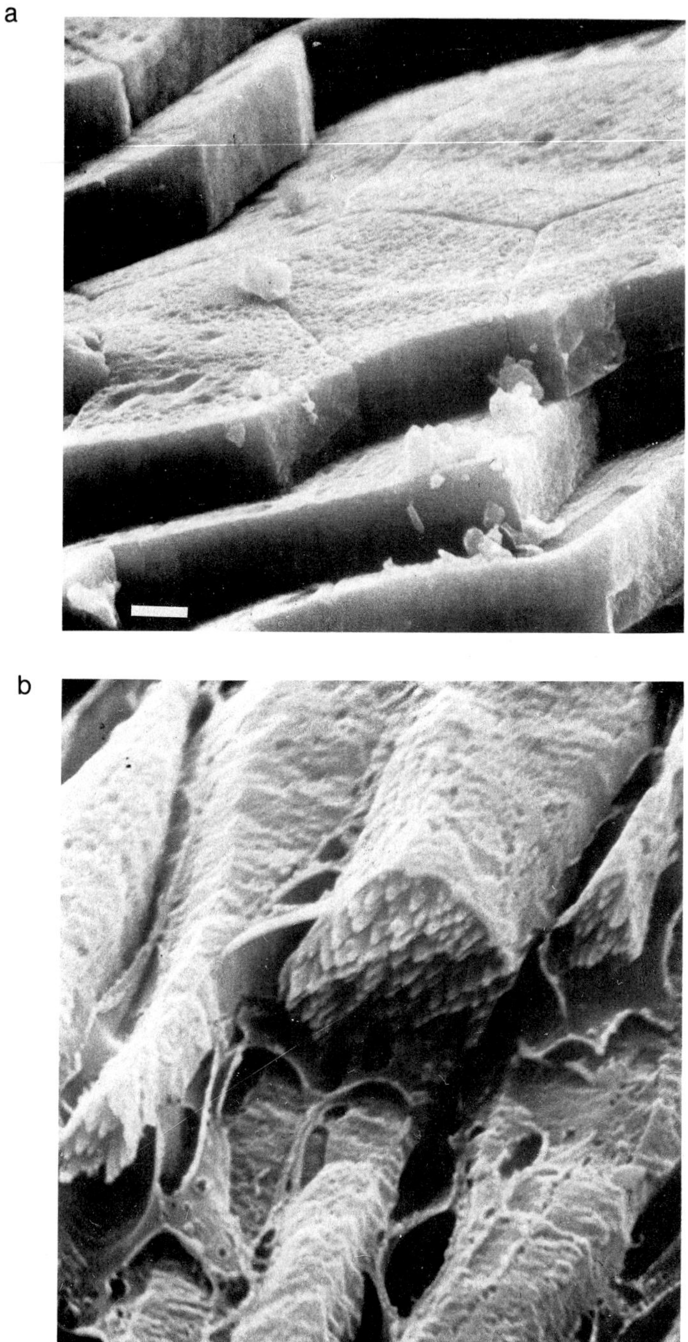

Figure 6.5. Scanning electron micrographs of broken surfaces of mollusk shells. (a) Nacreous ultrastructure composed of flat tablet-shaped crystals of aragonite *(Mytilus californianus)*. Scale bar: 1.0 μm. (b) Prismatic structure showing calcite prisms each enveloped in a sheath of organic matrix (*Mytilus californianus*). The specimen was fixed, etched, and critically point dried. Scale bar: 1.0 μm.

c

Figure 6.5. (*Continued*) (c) Crossed-lamellar structure made up of bundles of thin elongated crystals of aragonite oriented in different directions *(Chama gryphoides)*. Scale bar: 10 μm.

process of tooth formation in the Polyplacophora (chitons) and in certain gastropods (see next section).

Mollusks form an almost unbelievably diverse array of shell ultrastructural types. A very helpful recent catalog lists about 50 types and subtypes (Carter and Clark 1985). This diversity is a valuable resource for improving our understanding of basic processes of biomineralization. Many of the available descriptions of these ultrastructural types do not, however, provide the information that is often most relevant to the understanding of crystal formation. Even straightforward measurements of the dimensions of the individual crystals can provide important clues about the types of control exerted over crystal growth. For example, the mature nacreous layer is composed of tablet-shaped crystals of aragonite (Fig. 6.5) whose thicknesses (height) are remarkably uniform in a given species but vary from species to species (Watabe 1965). In contrast the lengths, widths, and shapes of the tablets vary considerably even within a single layer. This strongly suggests that the processes that control crystal thicknesses are different from those that control the other dimensions. For a detailed discussion of these aspects of nacreous layers, see Weiner (1986).

The relative orientations of the crystallographic axes of each of the crystals with respect to its neighbors may provide information about the spatial organization of the nucleation sites themselves. Again the relatively well-studied nacreous layer is a good example. The aragonite crystals of gastropod nacre show no preferred alignment of their *a* and *b* axes with respect to each other over large areas (square millimeters) using X-ray diffraction (Wise 1970) or even over small areas (few square microns) using electron diffraction (Weiner et al. 1983b). The *c* axes are all well aligned. This means that adjacent crystal tablets are randomly rotated about their *c* axes with respect to their neighbors. The cephalopod *Nautilus repertus* and the bivalve *Mytilus californianus* have moderately well-aligned *a* and *b* axes, whereas the nacreous crystals of the bivalves *Pinctada radiata* and *Neotrigonia margaritacea* are perfectly aligned over areas as large as a few square millimeters (Weiner and Traub 1980, 1981). As the *ab* crystallographic plane is parallel to the plane of crystal nucleation from the organic matrix in mollusk nacre and as we also know that the *a* and *b* crystallographic axes are aligned with the bulk of the matrix macromolecules off which they nucleate [see below; Weiner and Traub 1980; Weiner et al. 1983b)] we can infer that the nucleation sites on the surface of an organic matrix sheet are not aligned in gastropods, are poorly aligned in *Mytilus and Nautilus,* and are superbly aligned in *Pinctada and Neotrigonia.*

The crystal faces that are expressed can provide important information about the controls exerted over crystal growth. A basis for beginning to interpret this information in terms of acidic matrix protein–crystal interactions has been developed by Addadi and Weiner (1985). Crystal faces with a particular stereochemical property are able to interact far more effectively with the acidic proteins than those without it. This is discussed in more detail in Chapter 3. The implication is that if faces of this type are expressed it is more reasonable to invoke an active role for acidic matrix proteins in controlling crystal formation. Furthermore, it is of interest to ascertain whether or not the same faces are expressed in all the crystals that make up a particular ultrastructural type. By determining the crystallographic identity of the faces it is also possible to differentiate between thermodynamically more or less stable faces. This, in turn, raises interesting questions about the processes responsible for their stabilization. A notable study along these lines is that of Runnegar (1984) who identified the expressed calcite crystal faces of the foliated layers of a variety of bivalves. The foliated layers are composed of thin (about 250 nm) sheets of subparallel laths, which are usually inclined at less than 10° to the surface of the shell. Using X-ray and electron diffraction as well as secondary overgrowth of calcite, Runnegar could differentiate between three types of foliated layers based on expressed crystal faces. In the most common type of crystal face of the lath surface is the (104) face, which is the stable face of the calcite cleavage rhombohedron. This face includes both calcium and carbonate ions. This is in striking contrast to the upper and lower surfaces of the aragonitic nacreous tablets (Schmidt 1924) or the crystals of certain articulate brachiopods (Runnegar 1984) in which the (001) face is expressed. The (001) face is composed entirely of either calcium or carbonate ions. The most surprising of Runnegar's observations is that the face of the foliated lath surfaces of the bivalve anomiids (jingle shells) is very unstable, and, as Runnegar (1984) points out, may well require a highly specific interaction with an organic surface to account for its presence.

The crossed-lamellar structure is commonly found in mollusks and is usually composed of aragonite. There are numerous varieties of this ultrastructural type

(Carter and Clark 1985), but its basic characteristic is that it is composed of two sets of elongated crystals, each of which is tilted at some angle to the surface of the shell (Fig. 6.5). For comparison, the foliated layer has one set of tilted crystals. Crossed-lamellar ultrastructure has been studied in some detail (Uozumi et al. 1972), but little is known abut the orientations of its crystallographic axes (see Haas 1972) or the identity of expressed faces, not to mention the processes by which these alternating sets of crystals are oriented. This organizational motif is reminiscent of vertebrate tooth enamel that contains two major sets of oriented crystals as well as a third less abundant set (see Chapter 9). In enamel, the relative orientation of the crystal sets is determined, in part, by the specific morphology of the ends of the epithelial cells that are active in mineralization (Watson 1960; Nanci and Warshawsky 1984). It would be most instructive, therefore, to examine mollusk mantle epithelial cells to determine whether any such relation exists.

The organic material in the shell is a true "matrix" in that it acts primarily as a substrate on which and/or a mould in which the crystals grow (Wada 1961; Bevelander and Nakahara 1969b). The first really "modern" studies of mollusk organic matrix to include both TEM examined ultrastructure and biochemical analyses were performed by Grégoire et al. (1955). The matrix is composed of many different macromolecules, only some of which dissolve after the mineral has been removed by chelation with EDTA (Meenakshi et al. 1971), a methodology first introduced by Grégoire in 1949. The major portion of this soluble fraction is sulfated glycoproteins (Simkiss 1965; Crenshaw 1972a). The protein moiety is characteristically rich in aspartic acid (Weiner 1979) and a common amino acid sequence comprises stretches of aspartic acid residues usually separated either by glycine or serine (Weiner and Hood 1975). One purified aspartic acid-rich glycoprotein (it contains about 30 mol% Asp) from the bivalve *Mytilus californianus* has an apparent molecular weight of approximately 15,000 and contains as many as 70 or so sulfates per 100 amino acids (Addadi et al. 1987). These proteins bind calcium ions (Crenshaw 1972a; Krampitz et al. 1983a; 1983b) and, as a consequence, a major portion of their polypeptide chains adopts the β-sheet conformation (Worms and Weiner 1986). The proteins of the insoluble fraction tend to be hydrophobic (Crenshaw 1982) and are rich in glycine or glycine and alanine (Meenakshi et al. 1971; Grégoire 1972). Chitin in its β-form is also an important constituent of the insoluble fraction (Jeuniaux 1963; Goffinet and Jeuniaux 1969; Weiner and Traub 1980). For details see reviews of matrix biochemistry by Crenshaw (1982), Krampitz et al. (1983a), and Weiner et al. (1983a).

TEM photographs of stained sections cut through individual matrix sheets from the nacreous layers of gastropods (Nakahara 1979, 1983) show that they are composed of up to five different sublayers. Figure 6.6 shows a schematic block diagram of such a matrix layer. Nakahara (1983) identified the thin core layer as chitin and showed that the surface layers are rich in the aspartic acid-rich glycoproteins (Nakahara et al. 1982). X-ray and electron diffraction patterns of the whole insoluble fraction revealed that the layers on both sides of the chitin are in the antiparallel β-sheet conformation and that the mean orientations of the chitin fibrils and the protein polypeptide chains are more or less mutually perpendicular (Weiner and Traub 1980; Weiner et al. 1983b). The insoluble complex is further stabilized by quinone-tanning (Gordon and Carriker 1980; Samata et al. 1980). All organic

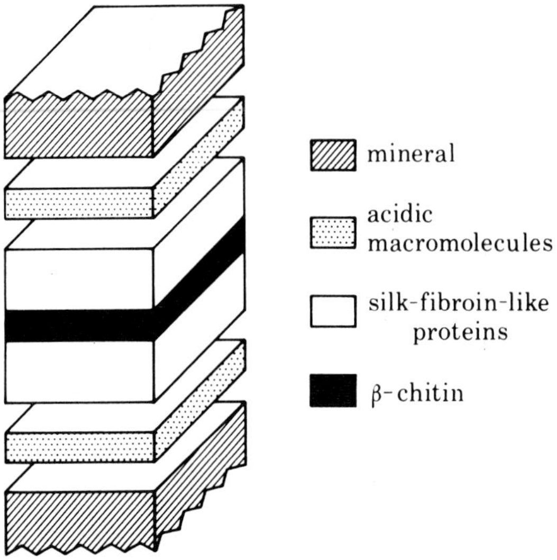

Figure 6.6. Schematic illustration of a composite section of one individual matrix sheet bounded on both sides by mineral. Reproduced from Weiner and Traub (1984), by courtesy of the Royal Society, London.

matrices do not appear to contain the full complement of five layers. The large and robust shell of the gastropod *Strombus gigas* has only the electron-dense acidic layers (Bevelander and Nakahara 1980; Weiner 1979). In the nacreous layer of the gastropod *Turbo,* the vertical matrix sheets do not contain the core complex (Nakahara 1979). Both these observations emphasize the importance of the acidic matrix constituents in mineralization and suggest that the "insoluble" core complex may not be an essential component. We emphasize, however, that the chitin layer alone is omnipresent (Poulicek et al. 1986), implying that it too plays a fundamental role in the shell formation process.

A very definite spatial relation at the molecular level exists between the constituents of the insoluble complex and the associated aragonite crystals in the nacreous layers of mollusks, from three different taxonomic classes (Weiner and Traub 1980; Weiner et al. 1983b). The aragonite *a* axes are aligned with the chitin fibrils and the *b* axes with the protein polypeptide chains of the core complex (Fig. 3.2). This strongly suggests that crystal nucleation occurs by epitaxy (Weiner and Traub 1984), although incontrovertible proof of an epitaxial nucleation process requires the demonstration of a matching of the lattice dimensions of the matrix surface at the nucleation site and the ions of the *ab* plane of aragonite.

To date, our understanding of the molecular organization of the matrix components and crystals of the nacreous layer probably surpasses any other mineralized tissue. The main reason for this is that the simple geometry of the nacreous layer greatly facilitates these types of structural investigations. This tissue should, therefore, be more fully exploited to address questions such as the stages in self-assembly of the organic matrix, the precise distribution of individual macromolecules on the matrix surface, the size and structure of the nucleation site, the structure of the crystal surface (as opposed to the crystal bulk) overlying the nucleation site, and the real possibility that some matrix components are located within the crystal itself.

6.6.4 The Zone between the Mantle and the Shell

In terms of understanding mantle–shell interactions, this zone is obviously of great importance. The disjointed information currently available emphasizes its importance, but does not provide a cohesive picture. Analyses of the fluid (known as extrapallial fluid) (Wilbur 1964) in this zone from marine bivalves shows that its ionic composition differs from both the blood and the environmental water (Crenshaw 1972b). The difference is enormous in freshwater bivalves, in which the extrapallial fluid is highly concentrated (Wada and Fujinuki 1968). Changes in the O_2 tension, pH, and calcium and succinic acid concentrations of the extrapallial fluid have been monitored continuously over periods of hours (Crenshaw and Neff 1969). Significant variations occur that correlate with the opening and closing of the valves. One most unusual attribute of this mantle-shell zone found only in some mollusks is the presence of a thin organic film that tightly adheres to the shell surface. This has been observed in the Arcoid bivalves outside the pallial line (Fig. 6.3) (Waller 1980) and in *Mercenaria mercenaria* and *Rangia cuneata* within the pallial line (Neff 1972; Marsh and Sass 1983). Various particles are associated with the film. In the estuarine clam *R. cuneata* the types of particles present vary according to the season collected (Marsh 1986). These include some very interesting phosphoprotein- and calcium-containing particles that are also found free in the fluid and between mantle cells. When these are absent, the film contains a different type of particle rich in barium sulfate (Marsh 1986). The functional significance of the film or the particles is not known.

Much can still be learned about this compartment that would shed important light on biomineralization processes in general. In the future this may well include information such as the programmed transport through the extrapallial fluid of organic matrix macromolecules synthesized in the epithelial cells.

6.6.5 Shell Dissolution and Remodeling

It is a common misconception to regard mollusk shells as "inert" structures whose sole function is to protect the animals internal soft parts. It is well known that calcium ions can move from the shell to the blood and tissues. This has been shown to occur in marine, freshwater, and terrestrial mollusks (Collip 1921; Dugal 1939; Wagge 1952; Greenaway 1971). The shell thus acts as a reservoir of calcium ions in a manner analogous to vertebrate bone, even though it is located external to the tissues. The mechanism by which this occurs is not known. It may be related to another well-known phenomenon in mollusks, namely their demonstrated ability to respire anaerobically (see review by Lutz and Rhoads 1980). The end product of this process is succinic acid (Simpson and Awapara 1966) and it has often been postulated that the acid is buffered by calcium carbonate from the shell. This has been proven in the marine bivalve *Mercenaria mercenaria* (Crenshaw and Neff 1969). Measurements of O_2 tension in the space between the mantle outer epithelial cells and the shell showed that the clam respired anaerobically soon after the valves closed. Using ^{45}Ca as a tracer, Crenshaw and Neff also showed that the succinic acid produced was indeed neutralized by previously deposited shell material.

Evidence for massive shell dissolution and remodeling primarily in gastropods is also available. A number of gastropods, for example, have complex structures within and around their apertures. Prior to periods of growth, these have to be resorbed. In many species of *Conus,* the Cypreids, and *Nerita,* the earlier formed shell walls are noticeably thinned in the mature animals as a result of remodeling. This is common among the terrestrial gastropods as well. A quite different form of remodeling takes place in some species of gastropods belonging to the genus *Turbo.* In these snails the earlier shell whorls are filled in with additional shell material.

The mollusks offer many unusually advantageous opportunities to study basic processes of biologically controlled mineralization. The simple geometry of the nacreous layer together with the fact that the nacreous crystals are relatively large have made detailed probing of the molecular organization of this tissue possible. The assembly line manner in which chiton and limpet teeth are formed has also provided an invaluable "break" allowing investigators to probe fundamental processes that in other tissues are almost impossible to sort out. The tremendous diversity of mineralization processes in this phylum, coupled with the fact that some mollusks, as far as we can ascertain, are able to exert very sophisticated controls over mineralization processes, will no doubt continue to make the study of this phylum a most productive enterprise.

7

Arthropoda

7.1 Introduction

The arthropods are distinguished by having segmented bodies and appendages, as well as hardened external skeletons. Growth is achieved by shedding the exoskeleton and then regenerating a new and larger one. The hardening of the exoskeleton usually occurs by chemical cross-linking (sclerotization) of the macromolecular constituents, mostly proteins and the polysaccharide, α-chitin. The major exception is the class Crustacea. The members of this group harden their skeleton not only by sclerotization, but also by the addition of inorganic minerals. After each molting, the new exoskeleton is remineralized. The result is that many Crustacea, particularly those that live in freshwater or on land where the availability of calcium is limited, have evolved novel and diverse temporary storage sites for mineral (reviewed by Greenaway 1985). From the perspective of biomineralization processes, this adaptation is certainly one of the "highlights" of the Arthropod phylum. Interestingly one taxonomic order within the Crustacea, the Cirripedia or barnacles, does not moult their heavily mineralized cuticles, even though their "organic" exoskeleton does go through periodic molting cycles (Darwin 1854).

Table 7.1 lists many of the known reports of biomineralization processes in the Arthropoda. The table is already impressively long. However, as this phylum is by far the largest in the animal kingdom, we have no doubt whatsoever that the true extent of mineralization processes in the Arthropoda is far from having been ascertained. In the insects alone nearly half a million species have been described, and our list comprises just a few documented cases of insects that mineralize. Interestingly, the list of minerals formed by insects includes a number of so-called "organic minerals," for example, uric acid, crystalline wax, and long chain paraffins. We strongly suspect that many more "organic minerals" have yet to be discovered among the insects.

Table 7.1 The Diversity, Distribution, Localization, and Functions of Minerals Formed by the Arthropoda

Taxon[a]	Taxa reported	Mineralization site	Mineral	Function	Reference
Crustacea					
Malacostraca					
Decapoda	Most	Cuticle	Calcite and/or amorphous $CaCO_3$ with traces of ACP	Mechanical strengthening	Vinogradov (1953); Chave (1954)
	Pseudosquilla bigelowi	Carapace	Amorphous calcium carbonate and phosphate	Mechanical strengthening	Lowenstam (1972b)
	Carcinus maenas (crab)	Cuticle, regenerated	Aragonite, amorphous $CaCO_3$	Mechanical strengthening	Dilliman and Roer (1980)
	Orconectes virilis (crayfish)	Gastroliths	Amorphous $CaCO_3$	Temporary storage	Travis (1963)
	Carcinus maenas (crab)	Midgut granules	Calcium phosphate	Temporary storage	Paul and Sharpe (1916)
				Detoxification system	Hopkin and Nott (1979)
	Callinectes sapidus	Mitochondria in midgut gland	Calcium phosphate	Temporary storage	Chen et al. (1974)
	Holthuisana transversa (crab)	Hemocoel (blood)	Amorphous $CaCO_3$ and phosphate	Temporary storage	Sparkes and Greenaway (1984)
		Eyes	Uric acid	Unknown	Kleinholz (1959)
Isopoda	*Lirceus brachyurus*	Carapace (integument)	Calcite	Mechanical strengthening	Hawkes and Schraer (1973)
	Limnaria lignorum	Midgut	Fe-containing protein	Detoxification?	Fahrenbach (1959)
	Some	Eyes	Calcite, Vaterite?	Lenses	Dudich (1931)
Amphipoda	*Orchestia cavimana*	Midgut caeca	Ca mineral	Temporary storage	Meyran et al. (1984)
	Gammarus setosus	Organ of Bellonci	Calcium mineral	Magnetoreceptor (?)	Steele and Oshel (1985)

	Leucothoe spinicarpa	Skin (cuticle?)	$CaCO_3$ crystals	Unknown	Reid (1943)
	Some	Eyes	Calcite	Lenses	Dudich (1931)
Mysidacea	*Praunus flexuosus; Neomysis intiger; N. rayi*	Statocyst	Fluorite	Gravity perception	Lowenstam and McConnell (1968)
	Diamysis bahirensis	Statocyst	Vaterite	Gravity perception	Ariani et al. (1981)
Tanaidacea	Some	Eyes	Calcite	Lenses	Dudich (1931)
Ostracoda	All	Cuticle	Calcite	Mechanical strengthening	Müller (1884)
	Various freshwater species	Pigment granules	Poorly crystalline Ca phosphate (ACP?)	Temporary storage	Rosenfeld (1979)
Cirripedia	All barnacles	Cuticle	Calcite	Mechanical strengthening	Meigen (1903)
	Some	Base plate	Aragonite	Unknown	Lowenstam (1964b)
	Balanus balanoides	Midgut (granules)	Zinc phosphate mineral	Detoxification?	Walker et al. (1975)
Copepoda	Some	Mandibular blades	Opal	Mechanical grinding	Sullivan et al. (1975)
Uniramia					
Insecta					
Orthoptera	*Periplaneta americana, Blattella germanica* (cockroach)	Oothecae	Weddelite, whewhellite, calcium oxalate, indeterminate	Mechanical strengthening?	Stay et al. (1960); Hackman and Goldberg (1960)
	Mantids	Oothecae	Calcium citrate	Citrate storage?	Parker and Rudall (1955)
Coleoptera (Scarab beetles)	*Plusiotis* s.l.	Cuticle	Uric acid	Light reflector	Caveney (1971)
Lepidoptera	*Bombyx mori* (silkworm)	Malpighian tubes	Whewellite	Temporary storage?	Teigler and Arnott (1972)
	Epipyrops anomala	Cuticle surface	Paraffin hydrocarbon	Unknown	Marshall et al. (1974)
Homoptera	*Prociphilus tesselatus* (aphid)	Cuticle surface	Wax (long chain) crystals	Mechanical strengthening	Dorset and Ghiradella (1983)
	Cercopid larvae	Midgut (granules)	Ca, Mg, Fe carbonates and phosphates	Temporary storage	Gouranton (1968)
Diptera	*Musca autumnalis*	Puparium	Calcium phosphate mineral	Mechanical strengthening	Darlington et al. (1983)

Table 7.1 (*Continued*)

Taxon[a]	Taxa reported	Mineralization site	Mineral	Function	Reference
		Malpighian tubes	Amorphous calcium phosphate	Temporary storage	Grodowitz and Broce (1983)
Hymenoptera	Bees	Abdomen	Magnetite (inferred)	Navigation	Gould et al. (1978)
Diplopoda	*Thyroglutus malayus, Julus* (millipedes)	Cuticle	Unknown	Unknown	Shrivastava (1970)
Collembola	*Tomocerus minor*	Midgut (granules)	Ca, K, Mg phosphates	Temporary storage and detoxification	Humbert (1978)
Chelicerata					
Arachnida	Spider	Endocuticle	Uric acid, calcium mineral	Unknown	Millot (1949); Kovoor (1978)
Trilobitomorpha	All	Carapaces	Calcite	Mechanical strengthening	Sorby (1879); Teigler and Towe (1975)
	Asaphus raniceps	Eyes	Calcite	Lens	Clarkson (1973) Towe (1973)

[a]Taxonomy according to Barnes (1980).

One fascinating case of mineralization in the arthropods occurs in the now extinct Trilobita, which evolved some 570 million years ago. Some genera used oriented calcite crystals as lenses for their compound eyes (Clarkson 1973). Calcite crystals are highly birefringent and to avoid a double image they oriented their crystals such that the crystallographic *c* axis was parallel to incoming light (Clarkson and Levi-Setti 1975; Towe 1973). Addadi and Weiner (1985) pointed out that this can be achieved by nucleation of rather simple substrates comprising acidic proteins in the β-sheet conformation and sulfated polysaccharides. For more information, see Addadi et al. (1987).

The first part of this chapter describes the basic structure of the "normal" arthropod cuticle as well as the mineralized crustacean cuticle and the second part describes the temporary mineral storage sites of the Crustacea. Since little is known about mineralization processes in the insects, the discussion is restricted almost entirely to the Crustacea.

7.2 Arthropod Cuticle

The commonly observed ultrastructure of arthropod endocuticle (Fig. 7.1a) is dominated by a well-ordered network of fibrillar structures, as seen in the TEM (reviewed in Filshie, 1982). The basic building units of the cuticle are so-called microfibrils. These are composed of α-chitin [poly-β-(1,4)-*N*-acetyl-D-glucosamine] and proteins, with diameters between 2.5 and 5.0 nm. The molecular structure of the chitin–protein complex of the cuticle is not well understood. Based on a fiber X-ray diffraction study of the ovipositor of an ichneumon fly, Blackwell and Weih (1980) proposed that the chitin forms the fibril core and is surrounded by a sheath of six globular proteins. Four such complexes, hexagonally arranged, constitute the microfibril. In a more recent study of the ichneumon fly, Atkins (1985) has shown that the central core of chitin is itself composed of 19 chains. The protein not only surrounds the core, but strongly adheres to it as a result of a lattice match between the chitin and protein interacting surfaces. It is of interest to note that both these models are quite different from that proposed for the chitin–protein complex in mollusk shells (Nakahara 1983; Weiner and Traub 1984), in which the chitin forms a thin layer sandwiched between thicker protein layers (Chapter 6). In mollusks, however, the chitin is in the β-form and the proteins are structural (as opposed to globular) in nature as they are primarily in the β-sheet conformation.

TEM micrographs of the endocuticle show that fibers are arranged into lamellae each of which has a characteristic arcuate pattern [see photographs in Neville (1975) and Filshie (1982)]. The widely accepted interpretation of these patterns is that of Bouligand (1965). Figure 7.1b schematically illustrates the proposed model in which fibers are aligned in horizontal planes. The fiber orientation in successive planes is rotated by a constant angle. A 180° rotation of the fiber direction constitutes a single lamella. This so-called helicoidal structure has been compared to the cholesteric structure of liquid crystals (Bouligand 1972). An alternative interpretation of the endocuticle structure was proposed by Mutvei (1974).

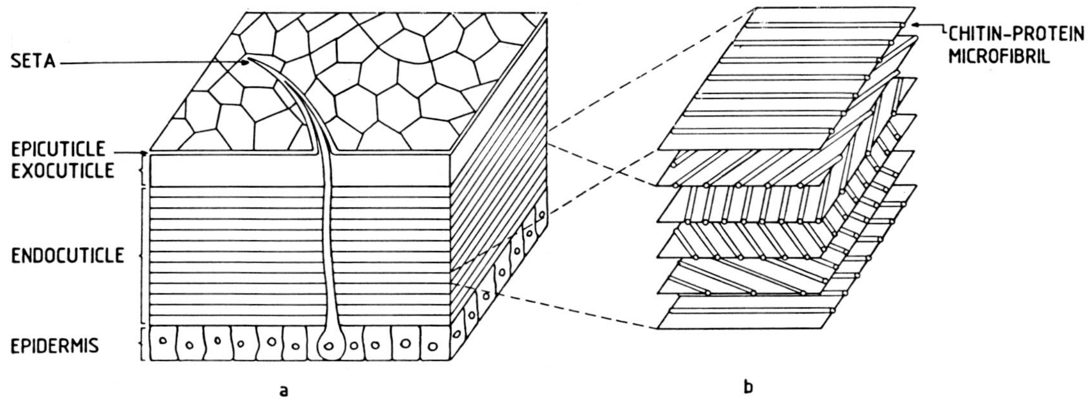

Figure 7.1. (a) Schematic illustration of the arthropod integument, showing the lamellar structure of the endocuticle, and (b) the arrangement of the chitin–protein microfibrils in a single lamella in the form of a helicoidal or cholesteric structure as proposed by Bouligand (1965).

7.3 The Mineralized Crustacean Cuticle

The basic structure of the mineralized crustacean cuticle is similar to that of the nonmineralized cuticle in other arthropods. Presumably the helicoidally arranged chitin–protein complexes constitute the "framework" in which mineralization occurs. The situation is reminiscent of vertebrate type I collagen that mineralizes in bone but remains unmineralized in skin (Chapter 9). In both cases it can be assumed that additional factors are responsible for the mineralization process, although in bone and cuticle their identity is not yet known.

The minerals present in the decapod crustacean cuticle are calcium carbonate, either in the amorphous form, the crystalline form as calcite, or a mixture of both (Vinogradov 1953), and amorphous calcium phosphate (ACP), usually in trace amounts. In some cases (see Malacostraca Table 7.1) appreciable quantities are present. Prenant (1927) found that amorphous calcium carbonate is present when the concentration of phosphate is relatively high (a ratio of P_2O_5/CO_2 greater than 0.104) and calcite is present when the concentration is low. The correlation was based on analyses of 13 different crustaceans [recalculated and reported in Vinogradov (1953) but incorrectly referenced to Prenant (1928b) instead of Prenant (1927)]. This correlation may arise as a result of phosphate ions acting as inhibitors of calcite nucleation. This most interesting phenomenon should be studied rigorously as it may be quite generally applicable. The amounts of mineral in a cuticle vary from species to species (reviewed in Greenaway 1985) and considerable variations occur even within the exoskeleton of an individual (Huner et al. 1978).

Determination of the relationship between the mineral and the chitin–protein framework is essential for a basic understanding of the mineralization processes themselves. One of the most thorough studies to date is that of Hegdahl and his associates (1977a,b,c) on the carapace of the crab *Cancer pagurus* in which the mineral is calcite. All three layers, the epicuticle, exocuticle, and endocuticle, are mineralized, but their ultrastructural organizations differ. The outer epicuticle (Hegdahl et al. 1977c) is sparsely mineralized and the distribution of mineral is patchy and confined to pore canals aligned more or less perpendicular to the cuticle surfaces. The crystals occur as individuals or aggregates, have a large range of sizes (up to 3000 Å), and no uniform shape, all properties characteristic of a poorly controlled process. The exocuticle (Hegdahl et al. 1977b) constitutes about one-fifth the entire integument. The distribution of mineral is also very uneven. There appears to be three different sites for mineral deposition: within pore canals, between chitin–protein fibrils, and within fibrils. In sparsely mineralized areas the crystals are small, elongated, and plate shaped, and their lengths, at least at the ultrastructural level but not necessarily at the molecular level, are aligned with the fibril axes. They tend to form long chains that are clearly influenced in their distribution by the helicoidal fibril ultrastructure. Pore canal mineralization is similar to that found in the epicuticle. In heavily mineralized areas the crystals from all three locations tend to grow into large dense aggregates. The endocuticle (Hegadahl et al. 1977a) is the major mineralized layer of *C. pagurus*. It is dominated by the helicoidal arrangement of the chitin–protein fibrils, and the distribution of elongated plate-shaped crystals is dictated by the structure, as in the exocuticle. Mineralized pore canals are also present. From the above observations two quite different mineralization

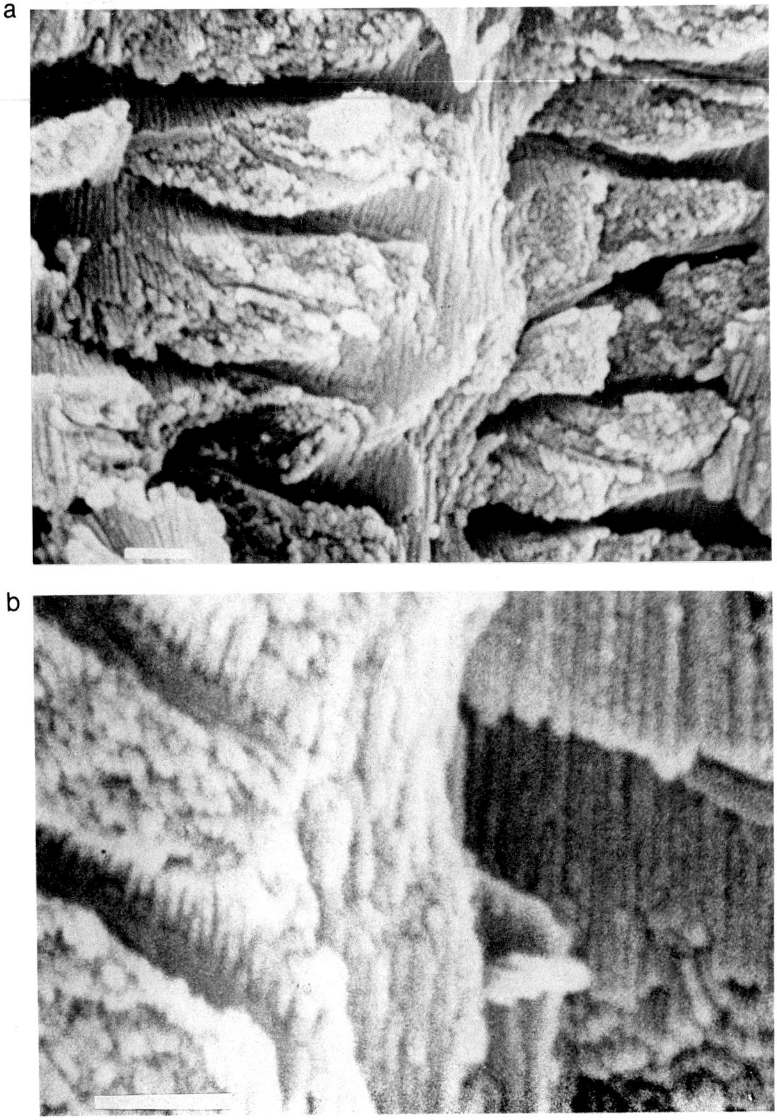

Figure 7.2. (a) Scanning electron micrograph of the dorsobranchial endocuticular carapace of *Carcinas maenas*. Cuticle was rendered anorganic by treatment with sodium hypochlorite. Mineral is organized into lamellae of rod-shaped aggregates, which correspond to the architecture of the organic matrix. Scale bar: 1 μm. (b) Higher magnification of (a) showing individual spherulitic elements that comprise rod-shaped crystal aggregates. Scale bar: 0.5 μm.

processes seem to occur: mineralization is controlled in some way by the chitin–protein framework, but is poorly controlled in the pore canals. TEM studies by Travis (1963) and Yano (1980) support the concept of framework directed mineralization. A most informative SEM study by Roer and Dilliman (1984) of the crabs *Carcinus maenas* and *Menippe mercenaria* provides a more graphic view of the

c

Figure 7.2. (*Continued*) (c) Scanning electron micrograph of the dorsobranchial carapace of *Menippe mercenaria.* This sample is from the lamellar region of the endocuticle rendered anorganic by treatment with sodium hypochlorite. The rod-shaped aggregates are arranged in lamellae corresponding to the pattern of the organic matrix. Scale bar: 1 μm. Reproduced from Roer and Dilliman (1984) by courtesy of the American Zoological Society.

endocuticle ultrastructure. Calcite crystals are seen to faithfully track the chitin–protein fibrils to form "long rod-shaped elements" (Fig. 7.2). Each element is itself composed of spherules. The rod-shaped elements are well aligned in the plane parallel to the cuticle surface. In fact, these SEM views of the mineralized cuticle in many ways illustrate the helicoidal organization of the fibrils better than the non-mineralized cuticle.

In ostracods the cuticle mineral is also calcite (Table 7.1) and the chitin–protein fibrils are arranged in a helicoidal structure (Bate and East 1972). A careful X-ray diffraction study (Jorgensen 1970) shows that the *c* axes of the calcite crystals are well aligned perpendicular to the cuticle (valve) surface. Speculation, at least for the ostracods, suggests that the crystals may grow into the space between sheets of chitin–protein. The most likely site for nucleation would be the surfaces of the sheets as they are parallel to the cuticle surfaces. Addadi and Weiner (1985) noted that nucleation of calcite is most likely to occur in the plane perpendicular to the *c* axis. This view of crustacean mineralization is consistent with Yano's (1980) observation that the crystal clusters (rod-shaped elements) are located between the "parabolic fibrils of the interlaminar sheets" (p. 193). Much work still remains to be done, however, to determine the exact spatial relationship between mineral and matrix at the molecular level.

The most obvious candidates among the macromolecules for initiating and regulating mineralization are some of the proteins present in the cuticle. With this in mind, Degens and co-workers (1967a) performed an elegant experiment in which amino acid composition analyses of different parts of the exoskeletons of portunid crabs were made. A factor analysis of the data showed a good correlation between certain amino acids and cuticle calcium content, suggesting that some mineral-associated proteins may be present. A thorough analysis of the mineral-associated proteins themselves has not been carried out. The problem is formidable as the cuticles of arthropods are known to contain a plethora of proteins (Vranckx and Durliat 1980) that includes some of those present in the hemolymph (blood) as well (Durliat et al. 1980). An immunoelectrophoretic study of the cuticular proteins of the crayfish *Astacus leptodactylus* showed the presence of three quantitatively abundant proteins in the calcified carapace but not in the noncalcified carapace (Durliat and Vranckx 1986). Additional evidence that proteins involved in mineralization are present is that inhibition of the enzyme carbonic anhydrase using "Diamox" (acetazolimide) reduces the growth of the calcite crystals in the crab *(Carcinus maenas)* cuticle by a factor of two (Giraud 1977). Characterization of these proteins may well improve our understanding of the processes involved in carapace mineralization.

7.4 Molting and Mineralization in the Crustacea

In the molt cycle (ecdysis), the new cuticle is formed without mineral before the old one is broken down and discarded. Once in place the new cuticle is rapidly mineralized. Thus, in addition to the other myriad of biochemical events that regulate this behavior, there is a sophisticated system for dissolving some of the mineral out of the old cuticle, transporting it, storing part of it temporarily, and then remobilizing it for mineralizing the next cuticle. Greenaway (1985) has expertly reviewed many of these aspects of the crustacean molt cycle.

Crustaceans live in marine, brackish, freshwater, and terrestrial environments. The availability of calcium in these environments varies tremendously. In the marine environment calcium is essentially limitless. In freshwater the calcium concentrations are very low and on land dissolved calcium in drinking water can constitute only a secondary source of calcium. Thus, the need to temporarily store at least some of the cuticle mineral during molting is greatest in the terrestrial and freshwater crustaceans. The stored material may represent up to 75% of the calcium in the old cuticles (Greenaway 1985). Temporary storage in marine species does occur to a limited extent, usually less than 10% of the calcium in the old cuticle. This material is utilized primarily to minimize the lag period between molting and hardening the cuticle by having a readily available supply of calcium.

Food, which in this context can include the old molted cuticle (exuviae), is another major source of calcium, particularly for terrestrial and freshwater species. However, food alone cannot solve a logistic problem that arises from the fact that after molting the cuticle is soft and the animals are barely able to walk, let alone feed. Thus, they are dependent initially on internal storage of some mineral to at least harden their mouth parts.

The storage sites, the amounts of mineral stored, and the nature of the mineralization processes involved vary greatly within the Crustacea (Table 7.1) and may well have evolved independently. The following are the main temporary storage sites in Crustacea.

1. *Gastroliths* are mineralized bodies formed in the lining of the cardiac stomach of some decapods. They are formed during the stages leading up to the molt (Travis 1963). Gastrolith formation has been studied in greatest detail in the freshwater crayfish *Orconectes virilis* by Travis (1963). From the point of view of ultrastructure the *O. virilis* gastroliths are similar to the cuticle itself. TEM micrographs show the characteristic helicoidal structure of the chitin–protein fibrils. The mineral phase differs from the cuticle in one major respect—it is amorphous calcium carbonate (by X-ray and electron diffraction), whereas the exocuticle of *O. virilis* contains poorly crystalized calcite as well as crystalline calcite (Travis 1963). TEM micrographs of the gastrolith mineral show spheroidal electron-dense particles, characteristic of amorphous phases. The use of an amorphous phase in gastroliths as opposed to a crystalline phase may facilitate rapid redissolution of the mineral after molting is complete. During the formation of the gastrolith the organic framework is formed first and then the mineral is introduced. Some of the mineralized particles appear to align themselves with the fibers (Travis 1963), but this does not necessarily mean that the fibers actively direct the formation of the mineral.
2. *Midgut gland or hepatopancreas* acts as a temporary storage site for mineral in many marine and some freshwater decapoda (reviewed in Greenaway 1985). Significantly the mineral formed is a calcium phosphate mineral (Table 7.1) rather than calcium carbonate. A TEM study of the hepatopancreas cells in a crab, *Callinectes sapidus,* shows that at least some of the crystals are located within the mitochondria. They are needle shaped and hence clearly crystalline in nature, show no preferred orientation, and their lengths are approximately 300 ± 150 Å (Chen et al. 1974). Mineral deposits of quite a different kind are also found within a specific cell type of the hepatopancreas (Becker et al. 1974; Hopkin and Nott 1979). These are large (0.5–3.7 μm diameter) concentrically layered granules. X-ray microanalysis shows that the major elements are calcium and phosphate. Significantly some of the granules also contain large concentrations of lead, suggesting that they might function as part of a metal detoxification system (Hopkin and Nott 1979) rather than as a temporary storage site. Interestingly, the hepatopancreas of gastropods contains cells with similar concentrically layered granules composed of amorphous calcium pyrophosphate, which may also be involved in detoxification (Howard et al. 1981). Thus, the various intracellular mineralized bodies of the hepatopancreas of crustaceans (and mollusks for that matter) have at least two quite different functions—temporary storage and detoxification.

   One of the most intriguing examples of a crustacean temporary storage system is in the midgut caeca of the amphipod *Orchestia cavimana.* Concretions composed of a calcium mineral are formed during the period prior to molting with material derived from resorbed cuticle mineral. An organic matrix is first elaborated and then mineralized. In contrast to other temporary storage granules, these are formed entirely in the extracellular space. The concretions are resorbed immediately after exuviation and the mineral is transported through the epithelium along an extracellular pathway. During this transport process it is again temporarily stored as mineralized spherules. Unfortunately direct mineralogical identification of the various precipitates is not yet available (Graf and Meyran 1985; Meyran et al. 1984, 1986).

A rather similar phenomenon occurs in one of the few insects known to harden its puparium with mineral. The larvae of the face fly *Musca autumnalis* accumulate spherical granules composed of amorphous calcium phosphate in the anterior Malphigian tubules (Darlington et al. 1983). When the fully developed larvae begin pupariation, the granules are resorbed. This coincides with the mineralization of the puparium and it was proposed that the granules act as temporary storage sites for mineral (Grodowitz and Broce 1983). For more details on the granule structure and their associated acidic proteins, see Grodowitz et al. (1987).

3. *Hemocoel.* In general the hemocoel (blood) does not act as a calcium storage site during premolt (Greenaway 1985). One well-documented exception is the freshwater/land crab *Holthuisana transversa* (Sparkes and Greenaway 1984) in which an enormous increase in calcium occurs just prior to molting and the hemolymph turns cloudy due to the presence of microspherules (about 200 nm diameter) of calcium carbonate.

7.5 Concluding Comment

Crustaceans must be the champions of mineral mobilization and deposition in the animal kingdom. Clearly the cyclic nature of the molt that triggers all these mineral changes offers a unique opportunity to improve our understanding of the hormonal regulation of mineralization in invertebrates. β-Ecdysone appears to be the key hormone regulating the molt cycle and, hence, indirectly the calcium movements as well. (See section on control of calcium balance in Greenaway 1985.) In all likelihood other hormones, perhaps even those that are active in the vertebrate world, also play a part in crustacean calcium metabolism. In this context it is worth noting that mammalian parathyroid hormone does effect calcium mobilization from the exoskeleton of *Orconectes virilis* (Greenaway 1985) and that vitamin D metabolites are active in the calcium metabolism of terrestrial gastropods (Weiner et al. 1979) and sea urchins (Hobb et al. 1987).

8

Echinodermata

The Echinodermata are certainly one of the most unusual and interesting phyla from the biomineralization point of view. They all live in the marine environment. The five major taxonomic classes (Asteroidea or sea stars, Ophiuroidea or brittle stars, Echinoidea or sea urchins, Crinoidea or sea lilies, and Holothuroidea or sea cucumbers) have quite different anatomical shapes and are characterized by five-fold symmetry. Each group forms mineralized hard parts. In the Echinoidea the skeletal elements are fused together to form a rigid test, whereas in the Asteroidea, Ophiuroidea and Crinoidea the skeletal elements or ossicles are articulated with one another. In the Holothuroidea the skeleton is usually reduced to microscopic ossicles or spicules, and, in some cases, mineralized granules as well. The hard parts of echinoderms vary enormously in shape and function and include not only the diverse skeletal elements, but also spines and teeth. Remarkably, with very few exceptions, the mineralized hard parts are formed from the same mineral, magnesium-bearing calcite [usually 5–15% as magnesium carbonate (Chave 1952, 1954; Raup 1966)], which has some unique and interesting properties.

The ultrastructure of many of the macroscopic skeletal hard parts has a characteristic spongy or fenestrate structure (called the stereom) and is riddled with labyrinthine cavities (collectively called the stereom space) (Fig. 8.1). In echinoid spines the stereom spaces are secondarily filled in to form areas of solid mineral (Fig. 8.2). The surfaces of the mineral phase are very smooth, even when examined a high magnification in the SEM (Figs. 8.1 and 8.2) (Towe 1967; Millonig 1970). Furthermore, the broken surfaces show no characteristic ultrastructural motif, which is observed in almost all other mineralized tissues in which the individual crystals are enveloped by layers of organic material. The fracture surfaces of echinoderm calcite actually have a conchoidal cleavage (Towe 1967), which is characteristic of glassy or amorphous materials (Fig. 8.1). It is, therefore, most surprising that when individual skeletal plates, spines, spicules, ossicles, and even whole teeth are examined in polarized light or by X-ray diffraction, they behave as if they are

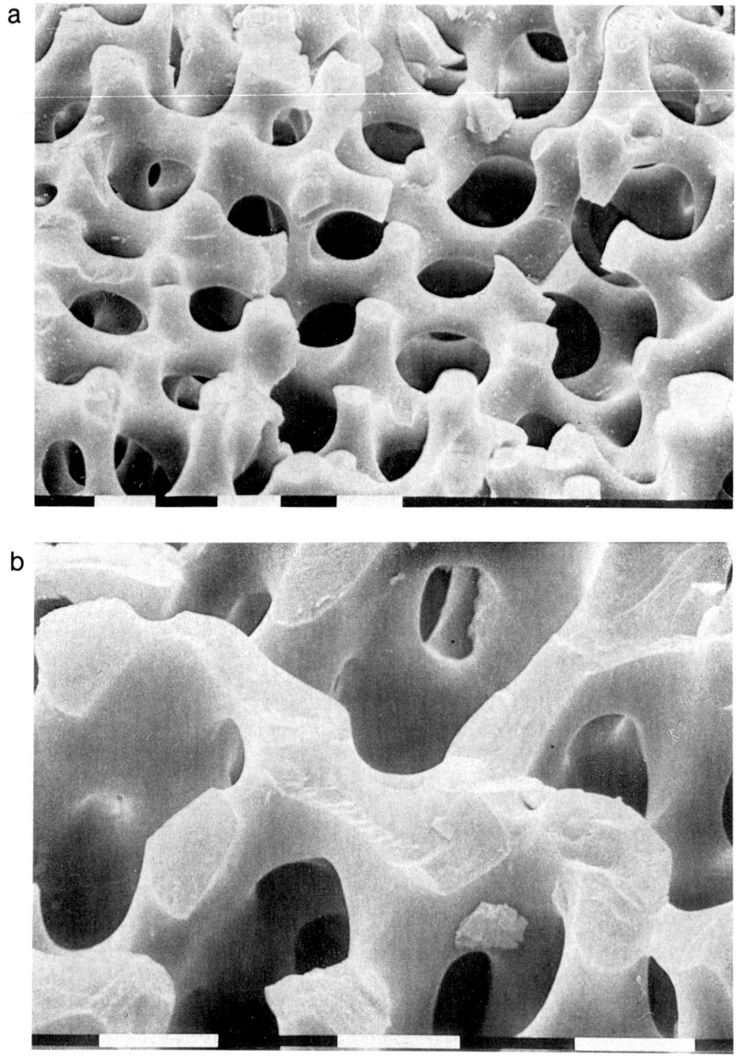

Figure 8.1. Scanning electron micrographs of broken pieces of the test of the sea urchin *Paracentrotus lividus* after treatment with sodium hypochlorite. (a) The stereom space and the highly curved surfaces. Scale bar: 10 μm. (b) High magnification view showing the "amorphous glassy" fracture surfaces of the calcite. Scale bar: 10 μm.

single crystals! (Towe 1967; Donnay and Pawon 1969). This must represent one of the most intriguing puzzles in biomineralization. It has been under discussion for more than 60 years (Prenant 1926; Schmidt 1932; West 1937) and will be addressed in some detail in this chapter.

Table 8.1 lists the minerals known to be formed by the Echinodermata. The list is surprisingly short for a phylum in which mineralized hard parts are a prominent feature of most of its members. The protodolomite formed in the core of echinoid teeth is really just Mg-calcite in which the magnesium content is so high

a

b

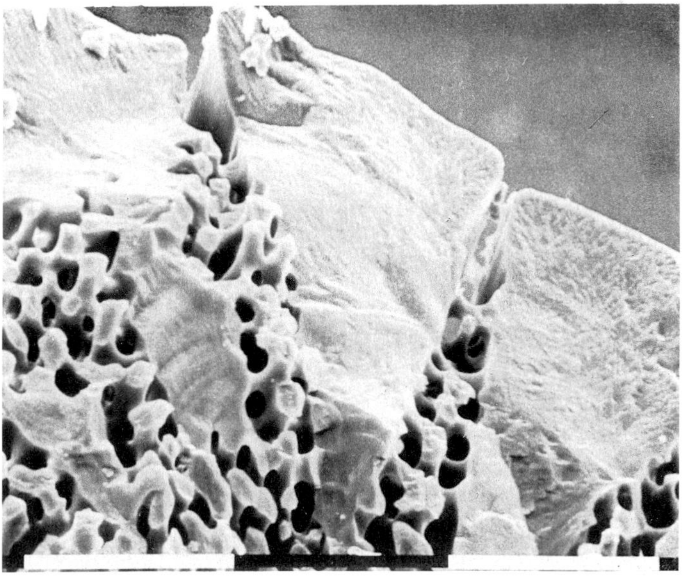

Figure 8.2. Scanning electron micrographs of broken spines of the sea urchin *Paracentrotus lividus* after treatment with sodium hypochlorite, showing both the stereom spaces and the secondarily infilled segments. Scales bars for a and b: 0.1 mm.

that it resembles dolomite. Structurally it is still calcite (Schroeder et al. 1969). The best documented case of an echinoderm forming a noncalcitic mineral occurs in the Holothuroidea. One group of holothurians, the Molpadiidea, form skin granules composed of two amorphous mineral phases—amorphous hydrated ferric phosphate and opal (Lowenstam and Rossman 1975). During ontogeny, the juveniles do form Mg-calcite spicules, but with development these are replaced by the

Table 8.1 The Diversity, Distribution, Localization, and Functions of Minerals Formed by the Echinodermata

Taxon	Skeletal element	Mineral	Function	Reference
Echinoidea	Plates, ossicles, spines, spicules	Mg-calcite	Mechanical strength and protection	Becher (1914)
	Teeth	Mg-calcite	Mechanical grinding	Becher (1914)
		Protodolomite	Mechanical grinding	Schroeder et al. (1969)
	Intracellular crystalloids	Fe-rich body	Mechanical grinding	Bachmann et al. (1980)
Asteroidea	Ossicles	Mg-calcite	Mechanical strength and protection	Becher (1914)
Ophiuroidea	Ossicles	Mg-calcite	Mechanical strength and protection	Becher (1914)
Holothuroidea	Ossicles (spicules)	Mg-calcite	Mechanical strength and protection	Becher (1914)
	Mesodermal granules	Amorphous hydrated ferric phosphate and opal	Mechanical strength and support	Lowenstam and Rossman (1975)
	Crystals	Weddelite	Excretionary product?	Lowenstam (unpublished)
Crinoidea	Skeletal elements	Mg-calcite	Mechanical strength and protection	Clarke and Wheeler (1922)

amorphous mineral. The granules and spicules are illustrated and described in more detail in Chapter 3.

The echinoids or sea urchins are the best studied of all echinoderms and this chapter will, therefore, be restricted to a discussion of their mineralization processes. Within the echinoids, one particular mineralization process is the focus of much investigation—the formation of the larval skeleton. This is because the developmental biology of sea urchin larvae is relatively well understood (Davidson 1986) and the larvae can be conveniently grown in synchronous culture, and are, therefore, most amenable to study by molecular biological techniques.

One of the most prominent morphogenetic events during embryo development is the formation of its spicular skeleton. Skeleton formation in the larva is therefore an attractive system to study gene expression, protein synthesis, macromolecular assembly, and, of course, mineralization itself. Furthermore, even the stem cells, whose descendants are responsible for spicule formation, can be isolated and cultured separately to produce "normal spicules" (Okazaki 1975; Kitajima and Okazaki 1980). For these reasons and as a result of the recent interest of developmental biologists in this aspect of larval development, spicule formation should soon be one of the best understood biomineralization processes.

Adult sea urchins also offer some unique experimental opportunities. The teeth of sea urchins are both large and continuously forming (Hyman 1955). This means that at any given time all stages of mineralization are present in one tooth. Experience has shown that whenever structures such as these are investigated in detail, they provide important basic insights into biomineralization processes. Good examples are the continuously forming rat incisor (Chapter 9) and the "conveyor-belt" type process of tooth formation in the chitons (mollusca) (Chapter 6). Sea urchin spines are capable of regenerating themselves after fracture (Carpenter 1870; Ebert 1967). This phenomenon offers yet another possibility for research, namely comparing the normal process of mineralization with that of "wound healing." Thus the Echinodermata and the Echinoidea in particular are an extremely interesting and important group in the field of biomineralization.

8.1 Spicule Formation in Sea Urchin Larvae

The lineage of cells that produces the calcite spicule in the larva can be traced back to a quartet of small cells (micromeres) that form after just four cell divisions. They are located at one end of the blastocoel (cavity of cells) and are the source of all the primary mesenchyme cells. After a series of additional cell divisions the round primary mesenchyme cells migrate into the blastocoel and occupy well-defined locations (Okazaki 1960; reviewed by Wilt et al. 1985 and Wilt 1987). The cells destined to form spicules produce long processes or pseudopodia. They fuse with the pseudopodia of their neighbors to form a "cable" or syncytium (Okazaki 1960; Wolpert and Gustafson 1961). A membrane-bound vacuole, which forms within the syncytium, is the site of spicule formation (Gibbins et al. 1969; Bevelander and Nakahara 1960). Figure 8.3 is a schematic illustration of the syncytium.

The geometry of spicule formation is very much a function of the shape of the vacuole, which in turn is directly influenced by the size and orientation of the syn-

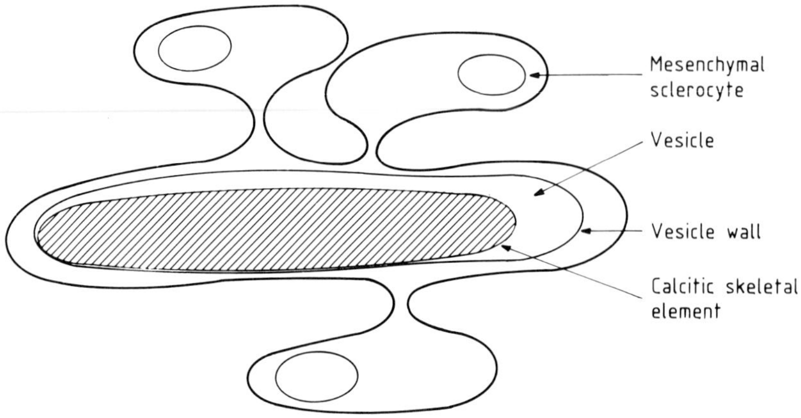

Figure 8.3. Schematic illustration of the syncytium and the vesicle inside the syncytium in which calcite deposition takes place. Redrawn in part from an illustration in Märkel et al. (1986).

cytium. Microtubules are intimately involved in the spatial organization of the whole system. The syncytium is always connected to many cell bodies through thin (0.5-μm-diameter) stalks (Gibbons et al. 1969). Observations of spicule formation in progress (Okazaki 1960; Bevelander and Nakahara 1960) show that a number of different mineral deposits first form, but subsequently all but one redissolve. With additional growth this initially formed granule adopts the triradiate form characteristic of the early spicule. Note that the spicule grows intracellularly throughout its development (Okazaki et al. 1980). Further growth is dictated by the shape of the syncytium. Induced changes that alter the shape of the syncytium result in gross changes in spicule morphology (Okazaki 1962).

Considering the above, it is surprising then that the morphology of the early formed triradiate spicule also faithfully reflects the crystallography of the minerals; each of the radii is perfectly aligned with one of the negative directions of the calcite *a* crystallographic axes. With additional growth, the "body rod" makes a right angle turn out of the plane of the triradiate spicule and its long axis is then perfectly aligned with another calcite crystallographic axis, the *c* axis (Okazaki et al. 1980). This is schematically illustrated in Figure 8.4 and compared to an inorganic calcite crystal. Thus, the shape of the spicule at this stage is dictated in some way by the crystallographic properties of calcite. Even more puzzling is the observation that the orientation of the triradiate crystal is initially quite random relative to the syncytium. Once formed it is rotated into the approximately, but not perfectly correct orientation for further growth into the three arms of the syncytium (Wolpert and Gustafson 1961). Understanding crystal growth in these spicules requires an elaboration of the delicate interplay between the roles of the biological structures and the crystal growth tendencies of calcite, a fundamental theme that is common to many biologically controled mineralization processes.

Does crystal growth occur within a preformed organic framework? The membrane-bound vacuole that surrounds the spicule itself constitutes such a framework

(Okazaki 1960) in the sense that it determines the final shape of the spicule. Cross sections of spicules examined in the TEM reveal concentrically layered structures with two more organic layers inside the spicule (Benson et al. 1983). These apparently further subdivide the spicule into internal compartments. Exactly how the mineral phase forms inside these compartments, whether a skeletal element is nucleated once or many times, what controls the growth of the crystal(s), and what stops their growth are still not understood.

A very promising start, at least, has been made on understanding spicule formation. An assemblage of N-linked glycoproteins is now known to be present *within* the spicule (Benson et al. 1986). These constitute about 0.1% by weight of the spicule, and their amino acid compositions resemble those found in other mineralized tissues. They bind calcium with dissociation constants ranging from 10^{-3} to 10^{-4} *M*. There are at least 10 different proteins. Antibodies raised against one of the quantitatively abundant proteins (about 50,000 molecular weight) were used to isolate and characterize a cDNA clone of the gene coding for this protein from a sea urchin embryo gene library (Benson et al. 1987). The structure of the gene (Sucov et al. 1987) reveals a typical N-terminal signal peptide and N-linked glycosylation site near the C-terminus. From the biomineralization point of view a most intriguing observation is that about 45% of the length of the protein is composed of consecutive repetitions of a 13 amino acid long segment. There is also a proline-rich domain and a very basic C-terminal region. The repeating domain suggests that the protein may interact in some as yet unknown way with the crystal. Curiously the proline-rich domain is, at the amino acid sequence level, somewhat homologous with human salivary proteins (Kauffman et al. 1982), which are known to function as inhibitors of calcium phosphate nucleation (Gron and Hay 1976). These intriguing results provide, for the first time, information on the primary sequence of a protein that is undoubtedly intimately associated with the mineral phase. The precise location of this protein or the locations of the other proteins present inside the spicule are not yet known nor, of course, are their functions.

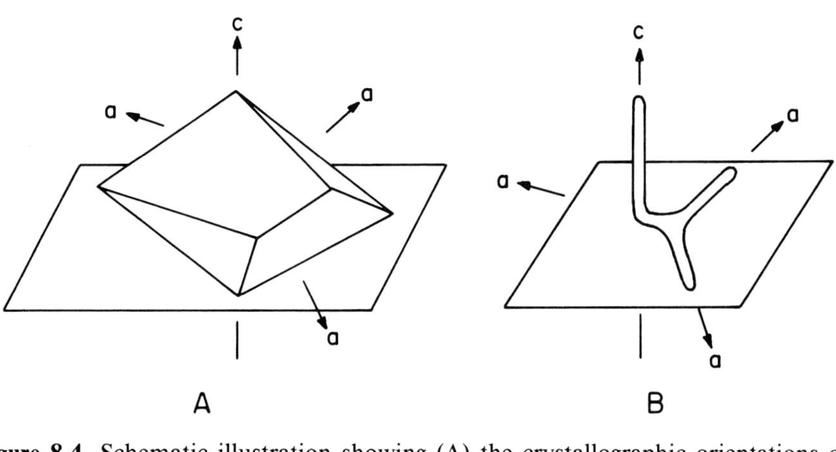

Figure 8.4. Schematic illustration showing (A) the crystallographic orientations of a synthetic calcite crystal and (B) a sea urchin larval spicule in the same orientation. The sea urchin larval spicule diffracts as a single calcite crystal.

Mintz et al. (1981) developed a simple in vitro culture system to study spicule formation that consists of 90–95% spicule-producing mesenchymal cells. These cells consistently produce spicules that elongate rapidly on the third day of culture and the system is most amenable for studying mineralization. One of the first applications of the culture system was the demonstration that a particular monoclonal antibody introduced into the culture on day 2 severely affects spicule formation on day 3 (Carson et al. 1985). The antigen to which the monoclonal antibody binds is a 130,000-Da cell-surface protein. Radioactive calcium uptake experiments suggest that this protein is involved in calcium uptake rather than in the mineral deposition process (Grant et al. 1985). Furthermore, the expression of this protein corresponds temporally to the increase in calcium uptake when spicule assembly is initiated (Farach et al. 1987).

The results of these "pioneering" studies are in themselves exciting and thought provoking. They vividly demonstrate the potential of immunochemical and molecular biological techniques to studying mineralization. Fortunately larval spicule formation does not appear fundamentally different from adult skeletal formation (see following) and some information can be transferred from one to the other.

8.2 Mineralization in Adult Sea Urchins

The adult skeleton is composed of Mg-calcite with ultrastructural, fracture, and mechanical properties similar to those of the sea urchin larval spicules (Emlet 1982). In fact a few of the early formed adult test plates (genital plates) actually use the larval spicule as a "nucleus" for further growth (Raup 1966; Emlet 1985). This comes about by projections of calcite growing out of the spicule, branching, and then rejoining others to eventually form the characteristic fenestrate (stereom) structure of the adult plate (Fig. 8.1). The formation of the adult skeleton, like that of the larva, is an intracellular process controlled by mesodermal tissue. In fact, the entire skeleton, at least during development, is always covered by a thin epithelium (Pilkington 1969), an observation that is beautifully illustrated for spines by Märkel and Röser (1983a). The cells responsible for mineralization initiate crystal formation intracellularly within vacuoles (Kniprath 1974) inside a syncytium (Fig. 8.3). For a detailed ultrastructural description, see Märkel et al. (1986). With growth of the skeletal element the vacuoles enlarge and eventually fuse with other elements to form the characteristic fenestrated structure. This intracellular process of mineralization is in essence the same as that reported for larval spicules and spine and tooth formation as well (Shimuzu and Yamada 1980; Kniprath 1974). Note that a contrary opinion, namely that crystal formation is an extracellular process, has been expressed (Heathfield and Travis 1975a,b).

In the adult skeletal elements that stereom spaces within the structure are occupied by a variety of cell types. *Epithelial cells* form the outer covering, although in some mature adult spines the outer epithelium is lost (Märkel and Röser 1983a). The characteristic cell type is the *sclerocyte,* some of which are directly involved in mineralization (Pilkington 1969; Märkel and Röser 1983a). Sclerocytes are highly convoluted in shape with many extended processes that envelope the mineralized skeleton. *Phagocytes* are the most abundant cell type in the stereom space (Bertheussen and Seljelid 1978). They are thought to be involved in the calcite resorption process (Märkel and Röser 1983b) that is known to occur in certain places in echinoid skeletons (Loven 1892) and spines (Märkel and Röser 1983a).

The growth of each of the sea urchin skeletal elements (plates, spines, and teeth) is a marvel of finely tuned differential growth processes. Very little is known about the responsible cellular or hormonal entities, although it has been shown that the sea urchin *Psammechinus miliaris* does produce vitamin D and is able to convert some of it to 25-hydroxyvitamin D_3 (Hobbs et al. 1987). The actual patterns of growth can be reconstructed in the formed adult test plates from the growth lines of individual test plates, or by periodic injections of radioactive calcium or fluorescent tetracycline into the living animal (Dafni and Erez 1982; Märkel 1981). New plates are formed in the upper (apical) hemisphere of the test. With additional growth of the test, each individual plate continuously moves down its column from the upper to the lower test hemispheres. The plates always grow by the addition of new material to their peripheries. The young plates in the upper hemisphere of the test grow faster on their upper edges as compared to their lower edges, and the reverse is true for the older plates in the lower hemisphere [see Märkel (1981) and the references therein].

The crystals that constitute each individual plate are aligned such that the whole plate behaves optically or when subjected to X-ray radiation as if it is a single crystal (Raup 1959). The *c* axes of individual plates are oriented either perpendicular to or near-tangential to the plane of the plate, depending upon the taxonomic affinity of the sea urchin (Raup 1966). Only a few species change *c* axis orientation with development (Raup 1960). Interestingly the sea urchins with near-tangentially oriented *c* axes are, with few exceptions, all "light negative." In other words they prefer shade or actually cover themselves with assorted materials to reduce the impinging solar radiation (Raup 1960). The near-tangential orientation of their crystal *c* axes, however, only exasperates their situation in that more light passes through these plates as compared to those with *c* axes perpendicular to the plane of the plate (Raup 1960).

Entire spines also behave optically as though they are single crystals. The *c* axes are invariably aligned with the long axis of the spine (Raup 1966). The spines (and the tests) are able to regenerate if accidentally broken. The stages of the regeneration process have been documented by SEM (Davies et al. 1972). It is interesting to note that the initial growth after fracture is predominantly in the distal direction, to form extended "spiculelike" projections in the direction of the crystallographic *c* axes. These are then thickened and side projections form that eventually fuse with neighbors to produce the fenestrate structure characteristic of echinoid mineralization.

The masticatory apparatus (Aristotle's lantern) of a sea urchin contains five elongated teeth. Only the hard tips of the teeth protrude into the buccal cavity. The constant wearing away of the oral ends is compensated by continuous growth at the soft aboral ends (Hyman 1955). Thus, at any given time a single sea urchin tooth contains all growth stages, making it a most attractive model for studying basic mineralization processes. Unlike continuously forming vertebrate teeth, the sea urchin tooth continues to grow even at its oral tip and indeed contains cells throughout its entire length. Pulse labeling studies with radioactive thymidine (Holland 1965) show in *S. purpuratus* that cells are generated at the soft aboral end and that subsequently little cell division occurs. The cells must therefore also migrate along specific pathways during tooth growth (Holland 1965; Veis et al. 1986). In the case of *S. purpuratus,* the tooth is completely renewed in 60 to 90 days.

The ultrastructure of the tooth is very complex and differs significantly from the other adult skeletal elements in that it does not contain the characteristic fenestrated structure (Fig. 8.5). It is composed of calcite with magnesium concentrations that vary significantly in different zones. The hard core of the tooth of *Lytechnus variegatus* actually contains 42–43.5% $MgCO_3$ and this is referred to as protodolomite (Schroeder et al. 1969) although its lattice structure is still that of a high Mg-calcite. Certain flanking regions around the core contain 28–29% $MgCO_3$, whereas in other regions there is only about 10% $MgCO_3$. The ultrastructure of the tooth is very complex and is related to the mechanical functions it performs (Märkel and Gorny 1973). Teeth differ significantly from one taxon to another (Märkel 1970a). In spite of their complexity, the teeth also behave optically as if they were constructed of just a few single crystals (Schroeder et al. 1969; Märkel 1970b)—a truly amazing phenomenon when the complex ultrastructure is considered (Fig. 8.5). Observations of initial crystal growth within the syncytium (Kniprath 1974) as well as the properties of some of the macromolecules present in the teeth (see following) confirm that the basic mineralization processes of tooth formation are similar to those of other sea urchin skeletal elements.

8.3 The Nature of the Mineral Phase

The paradox of the echinoderm calcitic mineral phase, namely that macroscopic skeletal elements diffract in X-ray beams as single crystals, but fracture as amorphous glassy materials, has been recognized for many years. Discussions in the literature on the nature of this material revolve around whether it is really a single crystal (Nissen 1969; Donnay and Pawson 1969) multicrystalline (O'Neill 1981; Blake et al. 1984) or some combination thereof (Towe 1967), but no explanation for the phenomenon has been proposed.

The adult skeletal test plates and teeth of the sea urchin *Paracentrotus lividus* are known to have associated with them an array of acidic glycoproteins, which in some ways are similar to organic matrix macromolecules in many other mineralized tissues (Weiner 1985). Earlier reports of the absence of an organic phase or the presence of collagen in the mineral phase were in error. More detailed biochemical analyses of the acidic macromolecules in the test of the sea urchin *Lytechinus variegatus* show that they contain both phosphate and significant quantities of carbohydrate (Swift et al. 1986; Veis et al. 1986). Furthermore, these acidic glycoproteins are capable in vitro of inhibiting calcium carbonate precipitation in supersaturated solutions. Surprisingly, an antibody to the principal phosphoprotein present in rat incisor has significant cross-reactivity with the sea urchin tooth proteins (Veis et al. 1986).

a

b

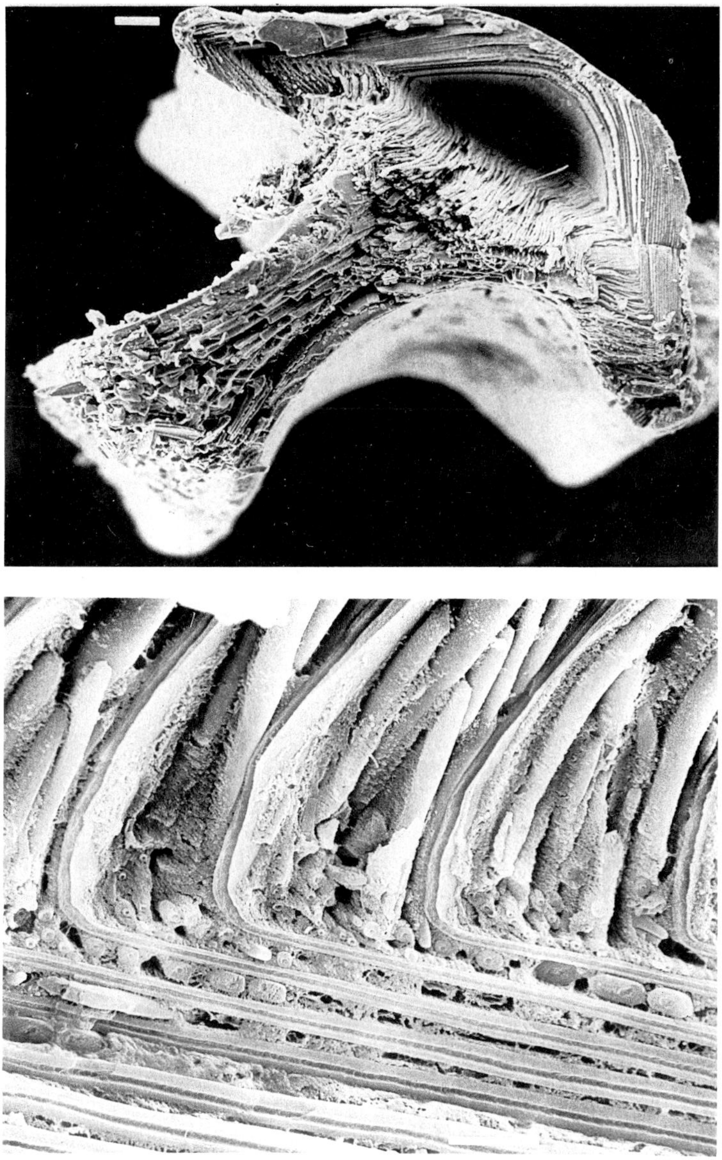

Figure 8.5. Scanning electron micrograph of (a) a broken section through a tooth of the sea urchin *Paracentrotus lividus.* Scale bar: 100 μm. (b) A higher magnification view of a portion of the tooth showing the absence of the stereom space structure. Scale bar: 10 μm.

Many of these proteins must be intimately associated with the mineral itself, as the test can be immersed in the highly aggressive oxidizing agent sodium hypochlorite, which readily breaks down almost all organic matter, but does not affect the acidic glycoproteins (Berman et al. 1988). Benson et al. (1986) made a similar observation for the sea urchin larval spicules. These observations imply that the acidic glycoproteins are protected from the oxidant by the mineral phase and therefore must be intimately associated with it. Possible sites for at least some of these macromolecules within the mineral phase and how their presence may explain the "sea urchin paradox" have been proposed by Berman et al. (1988). The proposal is based on the results of in vitro experiments in which synthetic calcite crystals are grown in a solution containing small amounts of sea urchin acidic glycoproteins. The sea urchin proteins interact with a specific set of calcite crystal faces that are more or less parallel to the crystallographic *c* axis. They adsorb onto these faces and the crystal continues to grow resulting in the proteins being occluded inside the crystal on these planes. When these synthetic crystals are cleaved they fracture with conchoidal cleavage and not along the normal cleavage planes of inorganic calcite. Thus, the presence of proteins on crystal planes that are oblique to the natural cleavage planes could explain how sea urchin calcite crystals also fracture with conchoidal cleavage. Significantly control experiments show that analogous acidic glycoproteins from mollusk shells behave quite differently. Berman et al. (1988) noted that the crystals with their occluded urchin proteins represent a "new" type of composite material that has the hardness and order of crystals, but not their brittleness.

9

Chordata

In the field of biomineralization the phylum Chordata is the most intensively studied, as one of its subphyla, the Craniata, includes our own species. The Craniata are often referred to as the *vertebrates,* a term that alludes to the importance of the endoskeleton in defining the essential character of these animals. The phylum Chordata also contains three other subphyla, only one of which has members that form mineralized hard parts. They belong to the Urochordata or tunicates. In fact, mineralization is confined to several families of a single class of urochordates, the Ascidiacea.

Table 9.1 is a compilation of the known biogenic minerals formed by members of the Chordata, together with the sites at which they form and their presumed functions. The table includes no less than 17 different minerals, which should dispel any notion that mineralization in the chordates is synonymous with "calcium phosphate" deposition. It is, of course, true that the mineralized *skeletal* hard parts of most of the Craniata or vertebrates contain a calcium phosphate mineral, usually in the carbonated form called dahllite. However, the vertebrates also form four different carbonate minerals that are most commonly found in the vestibulary apparatus (see Chapter 10). They form three different iron minerals, which includes magnetite found in the navigation system of various vertebrate genera. The Ascidiacea also form a diverse array of minerals. Interestingly, however, their diversity is essentially confined to one class, the Pyuridae, which form no less than six different minerals, including two phosphate minerals.

In this chapter we first describe biomineralization processes in the Ascidiacea followed by detailed discussions of mineralization processes in the Chordata or vertebrates. For convenience, the section on vertebrate mineralization is divided according to the major mineralized tissues: bone (dentin), cartilage, and tooth enamel. Mineralization in the vestibulary apparatus is discussed in Chapter 10.

Table 9.1 The Diversity, Distribution, Localization, and Functions of Minerals Formed by the Chordata

Taxon	Lower taxa	Mineralization site	Mineral[a]	Function	References
Urochordata					
Ascidiacea					
Didemnidae	Most species	Spicules	Aragonite	Protection and mechanical support	Schmidt (1924); Lowenstam (1963)
Polycitoridae	*Cystodytes* div. sp.	Spicules	Aragonite	Mechanical support	Monniot (1970)
Pyuridae	*Pyura australis*	Spicules	Dahllite and calcite	Mechanical support	Lowenstam (1988)
	Pyura sacciformis	Spicules	Dahllite and amorphous calcium carbonate	Mechanical support	Lowenstam (1988)
	Pyura stolonifera	Spicules	ACP and amorphous calcium carbonate	Mechanical support	Lowenstam (1988)
	Culeolus murrayi	Spicules	ACP	Mechanical support	Lowenstam (1988)
	Culeolus sp.	Spicules	ACP	Mechanical support	Lowenstam (1988)
	Pyura bradleyi	Spicules	ACP, amorphous calcium carbonate, and amorphous fluorite	Mechanical support	Lowenstam (1988)
	Pyura cataphracta	Spicules	Calcite, ACP, and trace of amorphous calcium carbonate	Mechanical support	Lowenstam (1988)
	Pyura littoralis	Spicules	Amorphous calcium carbonate and calcite	Mechanical support	Lowenstam (1988)
	Bathypera ovoida	Spicules	Calcite	Protection and mechanical support	Lowenstam (1988)
	Bathypera splendens	Spicules	Calcite	Protection and mechanical support	Lowenstam (1988)
	Bathypera sp. indef.	Spicules	Calcite	Protection and mechanical support	Lowenstam (1988)
	Herdmania momus	Spicules	Vaterite	Protection, mechanical support, and predator deterrant	Lowenstam and Abbott (1975)
Styelidae	*Polycarpa* (3 species)	Spicules	Undetermined	Mechanical support	In Monniot and Buge (1971)

	Kükenthalia borealis	Spicules	Undetermined	Mechanical support	
Molgulidae	*Molgula manhattensis*	Spheroids and crystals	Weddellite, calcite, and urate	Excretion products?	Saffo and Lowenstam (1978)
Corellidae	Most	Renal concretions	Undetermined	Excretion products?	In Berril (1950)
Ascididae	Most	Renal concretions	Undetermined	Excretion products?	In Berril (1950)
Cionidae	*Ciona intestinalis*	Pyloric gland concretion	Undetermined	Excretion product?	Millar (1953)
Craniata					
	All	Bone in endoskeleton	Dahllite	Mechanical support and ion reservoir	de Jong (1926)
	All, except Aves	Dentin in teeth	Dahllite	Mechanical support	Gross (1926)
	All	Bone, dentin, enamel	Octacalcium phosphate (OCP)	Precursor to dahllite	Brown (1966); Nelson et al. (1986)
	All	Cartilage	Dahllite	Mechanical support?	de Jong (1926); Gross (1926)
	All, except Aves	Cementum	Dahllite	Mechanical support	Gross (1926)
	Most	Vesicles	HAP (possibly dahllite)	Ion storage and/or nucleation center	Bonucci (1967); Anderson (1967)
	All	Ferritin micelle	Ferrihydrite	Iron storage	Towe and Bradley (1967); Harrison et al. (1967)
Agnatha (jawless fish)	Cyclostomata	Vestibulary apparatus	ACP	Balance and gravity perception	Carlström (1963); Lowenstam (unpublished)
Chondrichthyes (cartilagenous fish or elasmobranchs)	All	Skeletal cartilage in vertebral column and jaws	Dahllite	Mechanical support	Kyle (1927)
		Dermal bone	Dahllite	Protection?	Moss (1970)
	Raia clavata	Dermal scales, outer layer enameloid	Francolite?	Protection	Poole (1956)
	All	Tooth enameloid	Francolite	Mechanical grinding	Trautz et al. (1952); Glas (1962)

Table 9.1 (*Continued*)

Taxon	Lower taxa	Mineralization site	Mineral[a]	Function	References
		Tooth base—bone	Dahllite	Mechanical support	Moss (1970)
		Vestibulary appartus	Aragonite, calcite, monohydrocalcite, ACP	Balance and gravity perception	Carlström (1963); Lowenstam and Fitch (1981)
Osteichthyes (boney fish)	All?	Dermal scales	Dahllite	Mechanical protection and ion reservoir	Yamada and Watabe (1979)
		Tooth enameloid	Francolite, dahllite	Mechanical grinding	Suga et al. (1980)
	Anguilla anguilla (eel)	Intestinal mucus	Calcite	Calcium removal	Humbert et al. (1986)
	Tuna	Navigation system	Magnetite	Magnetic field perception	Walker et al. (1985)
	All	Vestibulary apparatus	Aragonite, calcite, vaterite	Balance and gravity preception	Carlström (1963); Lowenstam and Fitch (1981)
Amphibia	Gymnophiona	Dermal scales	Ca-PO_4 mineral	Protection	Zylberberg et al. (1980)
Reptilia	Some	Osteoderms or dermal ossicles	Dahllite?	Protection	Moss (1969); Levrat-Calviae and Zylberberg (1986)
	Turtles	Eggshells	Aragonite	Protection	Young (1950); Packard (1980)
	Crocodiles, geckos	Eggshells	Calcite	Protection	Krampitz et al. (1972); Erben (1970)
	Turtles	Navigation system	Magnetite	Magnetic field perception	Perry et al. (1985)
		Vestibulary apparatus	Aragonite, calcite	Balance and gravity perception	Carlström (1963); Marmo et al. (1983)

		Enamel	Dahllite	Mechanical grinding	Gross (1926); de Jong (1926)
Aves	All	Eggshells	Calcite	Protection	Cain and Heyn (1964)
	Pigeon	Navigation system	Magnetite	Magnetic field perception	Walcott et al. (1979)
		Vestibulary apparatus	Calcite	Balance and gravity perception	Carlström (1963)
	Female birds	Medullary bone	HAP	Ion reservoir	Kyes and Potter (1934); Ascenzi et al. (1963)
	Columba livia (pigeon)	Feathers and beak (keratin)	HAP	Mechanical strengthening	Blakey et al. (1963); Pautard (1963)
		Vestibulary apparatus	Calcite	Balance and gravity perception	Carlström (1963)
Mammalia	All	Enamel	Dahllite	Mechanical grinding	Gross (1926); de Jong (1926)
	Whales, dolphins	Tympanic bulla	Dahllite	Sound reception	In Oelschläger (1986)
	Most	Mitochondria in cartilage cells	ACP	Ion storage	Betts et al. (1975)
	Cow	Milk	ACP	Ion storage	McGann et al. (1983)
		Hair, horn, hoof, claw	HAP or Ca-PO_4 mineral	Mechanical strengthening	Pautard (1963); Blakey et al. (1963)
	Whale	Baleen (α-keratin)	HAP	Mechanical strengthening	Pautard (1963); St Aubin et al. (1984)
	Shrew	Teeth	Goethite	Mechanical grinding	Akerston et al. (1984)
	Human	Urine	Weddelite, whewellite, and struvite	Excretionary product	Prien and Frondel (1947)

[a]Abbreviations: ACP, amorphous calcium phosphate; HAP, hydyroxyapatite (probably dahllite)

9.1 Ascidiacea

All ascidians (tunicates) are marine organisms. Most of the adults are sessile and are firmly attached to the substrate. They are solitary or colonial and bear no resemblance to the more prominent members of their phylum, the Craniata. Ironically, they superficially resemble some of the simplest animals, the sponges. It is the free-swimming tadpole-like larva of ascidians that shares a common plan of embryonic development with all the other Chordata. They have a notochord, a dorsal tubular nerve chord, a postanal tail, pharyngial clefts, an endostyle that is homologous to the vertebrate thyroid gland, and the ability to synthesize thyroxine (Barrington 1965). All these traits firmly associate these organisms with the Chordata.

The name *tunicate* is derived from the development of a unique anatomical feature, the tunic, which forms an external covering over the epithelial cells. These in turn envelope the body organs. The tunic, which may be very thick, is rather pliable in some species whereas in others it is stiff and has physical properties reminiscent of cartilage. The tunic itself is composed mainly of the polysaccharide cellulose. This was originally thought not to be identical with cellulose and the term *tunicin* was used (Barrington 1974). The tunic also contains various glycoproteins and other polysaccharides, including in one species, *Halocynthia roretzi,* unusual sulfated glycosaminoglycans that are closely related to chitin sulfate. The latter is known to occur naturally only in the tunicates (Anno et al. 1974). Amoeboid cells are present within the tunic. In some species blood cells migrate into the tunic from the body mesenchyme, whereas in other species the tunic is supplied by blood vessels.

Our current understanding of the extent of biomineralization in the ascidians is far from complete and most of the studies are largely descriptive in nature. However, from the little we do know, the Ascidiacea constitute a fascinating group, not only because of their unique taxonomic status, but also because of their unusual biomineralization processes. Biogenic minerals are, as yet, known to occur only in the adults of certain taxonomic families, and even in these may be limited to certain genera or species within a genus. We suspect, however, that mineral formation is more widely distributed among adult ascidians and may even occur in the free-swimming larval tadpole stage (see the end of this section).

Table 9.1 lists the known mineralization sites of ascidians, the minerals formed, and their functions. The most common mineralization products are spicules. All spicules are composed of calcium minerals and although in some cases they may be closely packed (e.g., species of *Bathypera* and of Didemnidae), they are never cemented together. The spicules are usually located in the tunic, mantle, branchial basket, gonads, and siphon. In some species they are present at one anatomical site whereas in others they are found at several locations. They may also differ in morphology between the different sites. Spicules in the tunic are thought to provide protection and those in the body to provide mechanical support.

All the spicules formed by the Didemnidae and Polycitoridae are aragonitic and each species has only one spicule type. In contrast, the Pyuridae form an array of different shaped spicules, none of which is aragonitic. In some pyurid species, spicules with varying morphologies and composed of different minerals are found at up to four different anatomical locations. *Pyura bradleyi,* for example, contains antler-shaped, needle-shaped, and various flattened spicules. The antler-shaped spicules are composed of amorphous calcium phosphate in the core and amorphous calcium carbonate in the periphery. The needle-shaped spicules also contain amorphous calcium phosphate and calcium carbonate, but in addition have an amorphous fluoride compound (Lowenstam 1988). Nothing is known about the flattened spicules. Spicules formed by other pyurids are composed of vaterite (e.g., *Herdmania momus*), dahllite, and calcite (e.g., *Pyura australis*) (Table 9.1). Thus, in terms of their biomineralization products, the Pyuridae are a most unusual group. The evolutionary significance of these observations, including the fact that some pyurids form phosphate minerals, cannot at this stage be ascertained. *If*, however, the well-known hypothesis of Garstang (1928), namely that the vertebrates originated from the larval stages of ascidians by neoteny, has any basis, then some primitive pyurid may well have been the ancestor to the vertebrates (Lowenstam 1988).

A few species listed in Table 9.1 form what are thought to be excretion products in the form of spheroids and isolated crystals. Their mineralogy is not known, except in one case in which weddellite, calcite, and urate have been identified (Saffo and Lowenstam 1978).

In general little is known about spicule formation and the postformational fate of the spicules. Some data are, however, available for three different species. In *Trididemnum cereum* spicules composed of aragonite are formed in the tunic of the thorax within cells of the so-called paired lateral organs (Michelson 1919; Lafarque 1977; Lafarque and Kniprath 1978). Spicules form inside vesicles with a double layered membrane. During development the spicules migrate to their ultimate sites and in the process also increase in size. In adults, the spicules are fairly uniformly distributed throughout the tunic. Prenant (1925) and Monniot (1970) have formed synthetic aragonite crystal aggregates that are very similar to those of the tunicate and they suggest that little or no control is exercised over the process. Although almost nothing is known about the actual processes involved in biological spicule formation, the fact that it forms inside a vesicle suggests that some type of control is exerted (Kniprath and Lafarque 1980).

The pyurid ascidian *Herdmania momus* is a cosmopolitan, circumequatorial species, which is known to extend into temperate but still relatively warm waters only in Tasmania. Spicule formation in this species has been studied by Lambert and Lambert (1987). The spicules are found at a number of sites, but are all composed of vaterite and have a most unusual microarchitecture (Lowenstam and Abbott 1975) shown in Figure 9.1. The spicules from the tunic and body are similar, except that the former have a mace-shaped spiny base that anchors the mature spicule into the tunic. The upper portion of the spicule projects out of the tunic surface. The tunic spicules are much smaller than the body spicules (120–160 μm as compared to 1.5–2.55 mm). The vaterite crystals of both types are enclosed in a tightly adhering organic matrix envelope. The organic matrix comprises both EDTA-soluble and -insoluble components, but these have not been characterized.

The tunic spicules nucleate and develop within compartments tightly attached to the inner wall of the tunic blood vessels (Lambert and Lambert 1987). These compartments may actually be large cells. Nothing is known about the manner in which the crystals form. The spicules achieve their maximum size about 5 days

Figure 9.1. Scanning electron micrograph of (a) a vaterite spicule of the ascidian *Herdmania momus* (Tasmania). Scale bar: 20 μm. (b) Spicule at higher magnification. Scale bar: 2.5 μm.

after nucleation, when they exit the compartment, penetrate the blood vessel wall, and gradually migrate to the tunic surface. The migration of spicules appears to be a passive process resulting from forces generated by contractions of the body and tunic blood vessels.

The body spicules are principally located in the mantle, branchial basket, and siphons. They form in a sheath, which Lambert and Lambert (1987) describe as a "multilayered structure" (p. 151). Cells associated with spicule formation are located within the sheath and are referred to as sclerocytes. They appear to surround each of the fringing spines of a spicule. Forming spicules are arranged in an orderly manner within the sheath, with all their long axes oriented in one direction. Although the details of spicule mineralization in *Herdmania* are not known, the incredible spicule architecture and the facts that it is composed of vaterite that, for some undetermined reason is stable in seawater for a year (Lowenstam and Abbott 1975) and has the most unusual mode of formation within a layered sheath make this organism a fascinating one for further investigation.

The compound ascidian *Cystodytes* (suborder Aplousibranchia) contains disk-shaped spicules made up of acicular crystals that radiate from the center. X-ray powder diffraction patterns of six species analyzed to date show that the spicules are composed of aragonite (Lowenstam 1988). The spicules are located in the abdominal region of the innermost layer of the tunic, where they form in a spicular sack (e.g., van Name 1945). An unusual feature of *Cystodytes* is the presence of large, closely packed bladder cells located within the tunic just above the spicules. In *C. lobatus* these cells contain sulfuric and possibly also hydrochloric acid and have a pH of 1.3! (Abbott and Newberry 1980). When the animal is disturbed the acid is released and the spicules rapidly dissolve, apparently to allow the animal to contract efficiently. A light microscopic description of spicule growth in *C. lobatus* (Lambert 1979) shows that they form on the inside of the abdomen at the junction between the abdomen and the thorax. The spicules increase rapidly in size and number and eventually completely cover the abdomen. Another unusual phenomenon is that during budding of the zooids, the spicules are partitioned among the buds. New spicules are rapidly formed as the buds grow. All the spicules, including the inherited ones, continue to grow, with the result that the largest spicules in a bud may have passed through several generations.

The information available on spicule formation is fragmentary and, for the most part, descriptive. It is sufficient, however, to show that spicule mineralization processes in the ascidians are intriguing and should be investigated in detail. Even less is known about other sites of mineralization. One study of the renal concretions of *Molgula manhattensis* shows that they are made up of spheroids and isolated crystals. The spheroids are composed of calcite and the isolated crystals of weddellite (Saffo and Lowenstam 1978). As fungal symbionts occupy the renal sac, it is not known whether the minerals are precipitated by the ascidian or the symbiont. The extent of renal concretions as well as other mineralization products in adult ascidians is not known. Furthermore, nothing is known about mineralization in the larval tadpoles, even though it is well established that ascidian tadpoles have otoliths that have gravity receptor functions. The otolith is generally reported to be heavily pigmented, probably by melanin (Minganthi 1957). Indications that at least some of the otoliths may be mineralized are found in *Styella partita* in which the otolith is insoluble in acid fixatives (Grave 1944) and in *Botryllus schlosseri* in which it is brittle and fragments under a dissection knife (Grave and Riley 1935). It would be worthwhile to investigate this subject further.

9.2 Craniata (Vertebrates)

In terms of fundamentally different ultrastructural motifs, the vertebrates form few mineralized materials for constructing their skeletal and dental hard parts. Bone and dentin basically have very similar structures and for our purposes can be regarded as essentially the same material. Mineralized cartilage is quite different in ultrastructural organization from bone and dentin, even though its major constituents are also collagen fibrils and apatite crystals. Tooth enamel is distinct from bone, dentin and cartilage. Its apatite crystals are much larger than those of bone, dentin, and cartilage, and its organic phase does not include collagen. A fascinating "intermediate" form bridging enamel and bone or dentin is enameloid, a tissue with enamellike apatite crystals associated with collagen fibrils. All these constitute the major skeletal and dental building materials used by the vertebrates. It is convenient to discuss vertebrate mineralization in terms of these four materials and therefore the remainder of this chapter is divided accordingly into sections on bone, cartilage, and enamel together with enameloid.

Table 9.1 reveals many additional mineralized products formed by vertebrates. Some of these are discussed in other parts of the book (e.g., ferritin, the vestibulary apparatus, and navigation systems in Chapter 10), whereas others are overlooked. One particular omission that we would like to draw attention to is the process of eggshell formation. Reptiles and birds form their eggshells out of carbonate minerals. These processes have been studied in considerable detail (see Simkiss 1968; Silyn-Roberts and Sharp 1986) and have contributed significantly to our understanding of various aspects of mineralization.

9.3 Bone

The most widely used mineralized material among the vertebrates is bone. Its two most basic functions are first to provide structural support for the body and second to function as an ion reservoir (Neuman and Neuman 1958). Bones act as ion reservoirs in all but a few taxa. In one unusual case, the medullary bones of egg-laying birds function only as an ion reservoir (Kyes and Potter 1934; reviewed in Schraer and Hunter 1985).

The structure of bone is most easily understood by clearly differentiating between different levels of organization, as has recently been enunciated by Currey (1984). Following Currey, the basic level of organization is the molecular one that describes the crystals, the organic framework (mostly collagen fibrils), and the relationship between the framework and the crystals. This will be the major topic discussed in this section. The next level of organization describes the longer (a few to tens of microns) range order of the collagen fibrils and their associated crystals. In mammalian bone at least, Currey (1984) recognizes three fairly distinct forms at

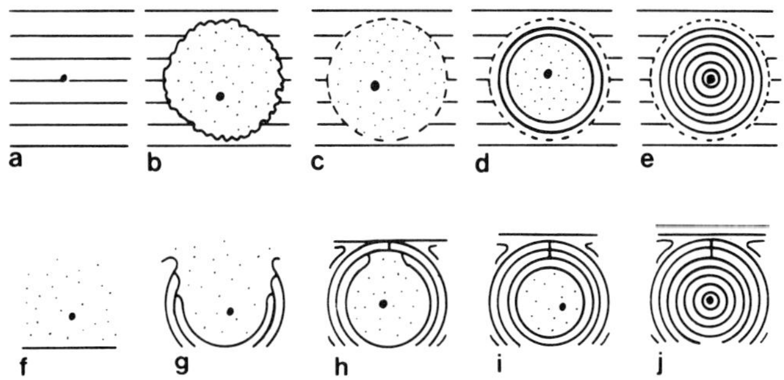

Figure 9.2. Two ways in which concentric cylinders are formed in bone, shown very schematically. (a–e) The formation of a secondary osteon, or Haversian system. (a) A blood vessel is surrounded by preexisting bone. (b) Osteoclasts resorb bone around the blood vessel (tissue fluid indicated by dots). (c) The edges of the cavity are neatened off, and a cement sheath (shown by interrupted line) is laid down. (d) The cavity begins to be filled in by lamellar bone. (e) The mature secondary osteon. Note that the course of the preexisting lamellae is interrupted by the osteon. (f–j) The formation of a primary osteon. This can take place only on a growing surface. The lamellae are not interrupted by the osteon, and there is no cement sheath. Illustration and legend reproduced from Currey (1984) and reprinted by permission of Princeton University Press.

this level. In *woven bone* the mineralized collagen fibrils show very little preferred orientation and are almost randomly distributed over distances greater than a few microns. *Lamellar bone* is much more ordered than woven bone. Its organization has been spectacularly revealed by an innovative preparation technique in which osteoclasts were allowed to "etch" polished bone surfaces (Reid 1986). SEM micrographs show that the lamellae comprise mineralized collagen fibrils that are aligned over large distances (tens of microns) to form sheets. The sheets have average thicknesses around 3 μm. The orientations of collagen fibrils in adjacent lamellae may be offset by anywhere between a few degrees to 90°. *Parallel-fibered bone* is structurally intermediate between woven and lamellar bone and is not as widely distributed (Currey 1984).

The next level of organization involves ordered structures that extend over millimeters. In mammals Currey (1984) recognizes four types, two of which are simply woven and lamellar bone textures that extend over much longer distances. The third type is the *Haversian system* or *secondary osteon.* The stages of formation of an Haversian system are schematically illustrated by Currey (1984) and are reproduced in Figure 9.2. It is important to note that Haversian systems are formed secondarily, that is, after the initial bone has been laid down. This process can continue over long periods of time, with the result that even a so-called "mature" bone can have recently mineralized collagen fibrils. The fourth type of organization at this level is *fibrolamellar bone.* Its formation process was also schematically illustrated by Currey (1984) and is reproduced here in Figure 9.3. The parallel fibered bone is laid down first and then the lamellar bone soon afterward.

The highest level of organization differentiates between compact and cancellous bone. These bone types can extend for centimeters or more. Compact bone is

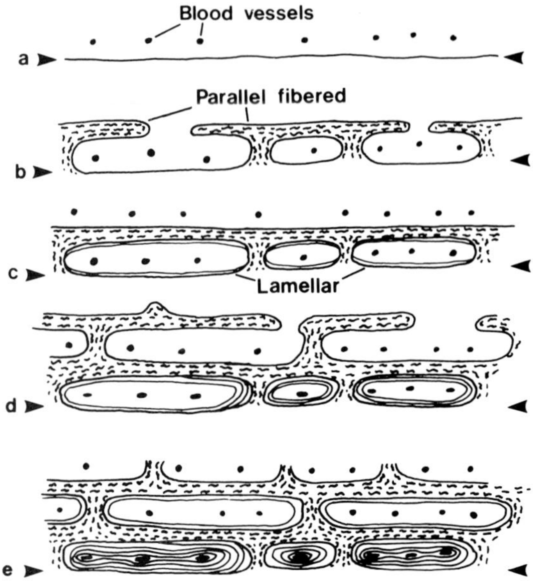

Figure 9.3. The formation of a fibrolamellar bone. These are cross sections of the outer surface of a rapidly growing bone. The arrowheads show the position of the original surface. Blood vessels are shown by black spots. (a) The original position. (b) Parallel-fibered bone, shown by squiggly lines, grows very quickly to form a scaffolding clear of the original surface. (c) Lamellar bone, shown by fine lines, starts to fill in the cavities left by the parallel-fibered bone. (d) As more lamellar bone is laid down, so is another scaffolding of parallel-fibered bone. (e) By the time the first row of cavities is filled in, the outer surface of the bone is far away. Illustration and legend reproduced from Currey (1984) and reprinted by permission of Princeton University Press.

solid whereas cancellous bone is spongy. Figure 9.4 is a low magnification SEM micrograph of a broken section through the tibia of a young rat. The distributions of compact and cancellous bone can be seen and it should be noted that there is no clear-cut differentiation between the two.

The manner in which bone grows at the macroscopic level has occupied scientists for hundreds of years, and still does. Figure 9.4 also shows the most active region of long bone growth, namely at the so-called epiphyseal plate, the border zone between the upper epiphysis and the lower metaphysis comprised mostly of spongy bone. Here the new bone forms on a scaffolding of preexisting cartilage. This process of bone formation is called endochondral ossification. The shaft of the long bone is thickened by the accretion of new bone directly onto the outer periosteal surfaces. This is known as membranous ossification (Sissons 1956). Detailed histological and immunocytochemical studies of bone formation in chick embryos show that the first bone is deposited outside the cartilage framework and, in fact, endochondral and membranous ossification may not be significantly different (Caplan and Pechak 1987). In general a representative sequence of temporal events can be recognized during the formation of bone. The first stage involves the synthesis and extracellular assembly of the organic matrix framework, which is followed by mineralization of the framework. Unlike many other mineralized tissues, bone continues to "form" as secondary mineralization processes are initiated. These involve the channeling out of existing bone and the remineralization of the channels to form the Haversian systems. The end result is that a mature bone is composed of a very complex mesh of bone "patches," each of which was formed at a different time. They also have slightly different densities, a property that has enabled Reid and Boyde (1987) to vividly illustrate this phenomenon using backscattered electrons in the SEM. In addition, bone is frequently extensively remodeled by being resorbed at certain surfaces by osteoclasts and formed on other surfaces by osteoblasts. To complicate matters even further some cells are imprisoned within the mineralized matrix (osteocytes) where they fulfill as yet unknown functions. Thus, the dynamic world of continuously forming and reforming bone represents an almost impossible situation for the student of biomineralization to unravel. Indeed our state of knowledge of bone formation reflects the difficulties involved.

Ideally we should be studying bones in which, for whatever reason, the internal and external remodeling processes do not take place and the different stages of mineralization are separated in space, e.g., the chiton radula of the vertebrates (Chapter 6). Although the ideal system has not been found, a variety of bone or bonelike mineralization processes are known that for some purposes may be more suitable for studying mineralization than "normal" bone itself.

Dentin appears to be similar to bone at the molecular level and does not remodel as such. Dentin is, however, complicated by the presence of extensive cel-

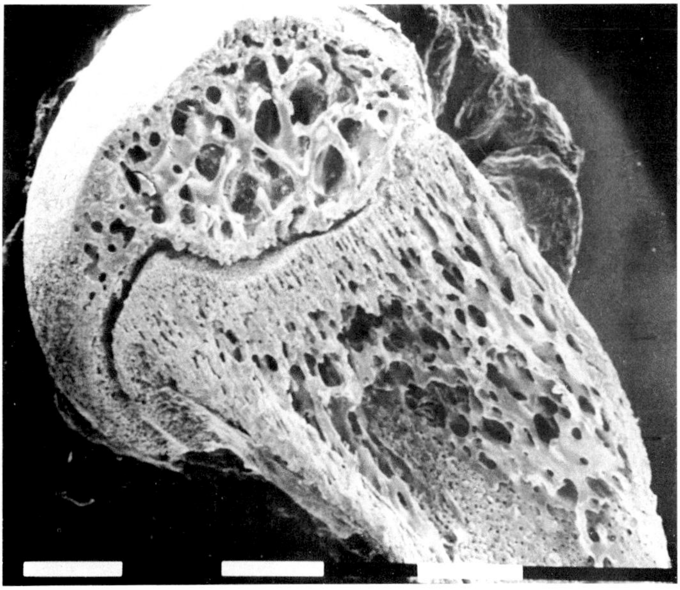

Figure 9.4. Scanning electron micrograph of a broken section through the end of the tibia of a young rat. The upper bulbous portion is the epiphysis and below that is the metaphysis. The zone bordering the two is the epiphyseal growth plate. Scale bar: 1 mm.

lular processes that are located in tubules that extend from the forming surface to the base of the dentin. The tubule walls mineralize (so-called peritubular dentin) with the result that the mature dentin also contains portions of tissue with somewhat different structures that may have mineralized at different times (Boyde 1971). For short-term studies, however, dentin has proved to be a very attractive system. The cells that form dentin, the odontoblasts, line the pulp cavity and as the tooth is growing they continuously lay down a collagenous organic framework that later mineralizes. Investigations of the events occurring during these early stages of mineralization have yielded some very important information. Two classic examples are the first identification of acidic proteins in mineralized tissues by Veis and his colleagues (Veis and Perry 1967) and the different biosynthetic pathways followed by collagen and the acidic proteins (Weinstock and Leblond 1973).

The bones of animals that have secondarily invaded the aqueous environment often have very unusual properties. Two possible reasons are that their bones are not subject to the same gravitational stress fields as those of terrestrial animals, and they often form particularly heavy bones to achieve a state of more or less neutral bouyancy in the water (Thompson 1942). Moss (1964) pointed out that in some cases the bones are not resorbed on their surfaces. Penguin long bones, for example, have a very narrow marrow cavity. Although no resorption of bone occurs inside the marrow cavity (Meister 1962), the bone contains extensive Haversian systems indicating that "internal" remodeling takes place. The long bones of the Florida manatee contain no marrow cavity at all. Haversian systems are, however, present in the earlier formed parts of the bone but are almost entirely absent in the outer last-formed regions. Thus, internal reorganization in these areas is reduced to a minimum so that parts of the adult manatee bone microscopically resemble fetal bone in other mammals (Fawcett 1942).

Fish scales are composed of layers of highly organized collagen fibrils arranged in a plywoodlike manner. Mineralization occurs only after the entire organic structure is in place. The hydroxyapatite crystals form between fibrils in one layer and in close association with collagen fibrils in other layers (Waterman 1970; Onozato and Watabe 1979; Yamada and Watabe 1979). Fish scales apparently do not undergo remodeling and more detailed studies of their molecular organization could well lead to important insights into basic processes of collagen-associated mineralization. Note that noncollagenous phosphoproteins have been isolated from fish scales (Sauk et al. 1984) and that matrix vesicles are also involved in the mineralization process (Schönbörner et al. 1979).

The calcified turkey tendon is another bonelike mineralization process that has proved to be very useful in understanding the molecular organization of mineralized collagen fibrils (Johnson 1960; Nylen et al. 1960). Turkey tendon comprises close-packed well-aligned collagen fibrils with intimately related hydroxyapaptite crystals. The tendon progressively mineralizes at one end and is not extensively remodeled. Studies of mineralized turkey tendon have provided some of our most important insights into bone molecular organization (see following). It is not known how good an analog turkey tendon is of bone.

We have emphasized these "alternate" systems for studying bone or bonelike mineralization. Each has one or more properties that makes the understanding of their basic mineralization processes just that much easier, and, because of this, we believe more attention should be devoted to them. In the following section we limit our discussion of bone almost entirely to the mineralization processes themselves and the very important and still largely unresolved issues of the molecular organization of bone. We do this at the expense of other important subjects in bone formation that are, for the most part, more peripheral to the major theme of this book.

9.3.1 Molecular Organization of Bone

The material bone is characteristically composed of type I collagen fibrils intimately associated in an orderly manner with calcium phosphate crystals. Minor constituents include an array of macromolecules as well as a series of small molecules associated mainly with the mineral phase. In this sense the material bone, as noted above, is also found in, for example, tooth dentin, cementum, fish scales, and mineralized tendons. Although the ultrastructural organizational patterns of these tissues differ from one another and from most bones, they all appear to have many properties in common at the molecular level of organization. Furthermore, during their formation they all follow the same basic developmental course, namely the major organic constituent collagen is first synthesized, extruded from the cell, and then self-assembles in the extracellular space before mineralization begins. For this reason, bone is a good example of an "organic matrix-mediated" mineralization process (Lowenstam 1981). In fact the term *matrix* was first coined by Le Gros Clark in 1945 for bone.

9.3.2 The Mineral

Perhaps the feature of bone that sets it apart from almost all other mineralized tissues is the exceedingly small size of the crystals. Bone crystals are certainly among the smallest biologically formed crystals known and, in fact, most crystallographers would intuitively not expect crystals just a few unit cells thick to be stable at all! An appreciation of the size of the crystals is important for understanding their mineralogy, their organization in relation to the framework constituents, as well as the many technical problems involved in studying a tissue of this type at the molecular level. It is also, in many respects, the key to understanding why so many gaps in our knowledge of the molecular organization of bone still remain. For good, thought-provoking reviews of bone mineral, see Brown and Chow (1976) and Posner and Betts (1981).

9.3.2.1 CRYSTAL SHAPES AND SIZES

The most direct means of studying these crystals is by imaging using the TEM. Two different preparation procedures have been used: (1) liberating the crystals from the organic framework and observing them directly on the microscope grid; or (2) dehydrating, embedding, and sectioning the bone and then viewing the ultrastructure through the "thin" sections. So-called "thin sections" are usually around 1000 Å thick! Both techniques are necessary to obtain a fairly complete picture, but both are fraught with real and potential artifactual difficulties. The dispersion approach was first used by R. A. Robinson in a landmark paper in 1952 and his

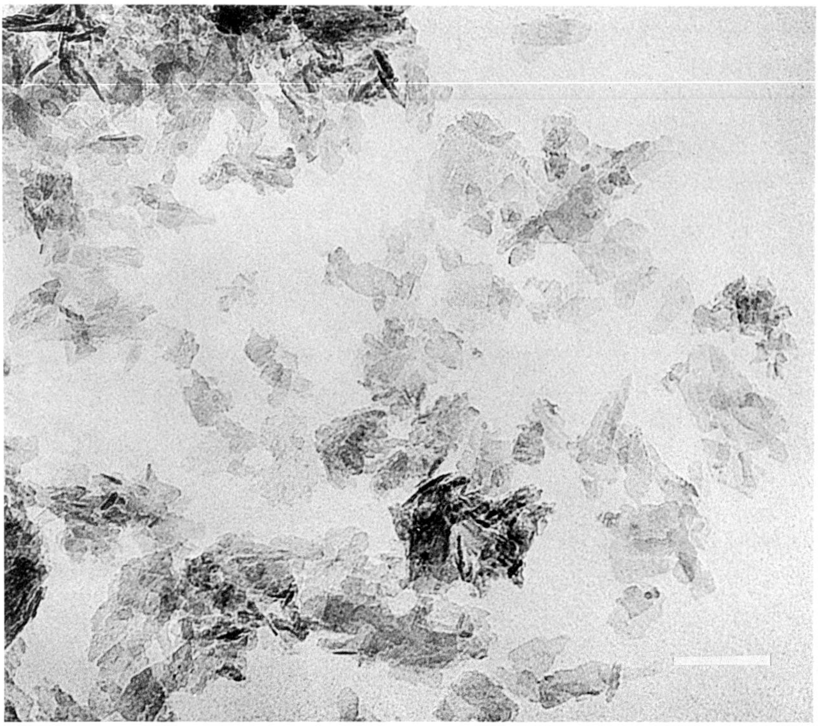

Figure 9.5. Transmission electron micrograph of whale bulla bone crystals after sodium hypochlorite treatment following the procedure by Weiner and Price (1986). Scale bar: 100 nm.

observation that human bone crystals are tablet or book shaped has been repeatedly confirmed for other bones using both preparation techniques (Johansen and Parks 1960; Steve-Bocciarelli 1970; Weiner and Price 1986). In fact the authors are of the opinion that in the absence of contradictory evidence it can be assumed that all bone crystals are of this shape. It is of interest to note that some ectopically formed crystals are not tablet or plate shaped, but needle shaped and, therefore, similar to synthetically formed hydroxyapatite (Robinson 1952). The outlines of the bone crystals when viewed in projection are very irregular, although they do tend to be somewhat elongated (Fig. 9.5). The crystals, when viewed on edge in embedded and sectioned preparations, are very thin and appear to be uniform in thickness.

Measuring the size ranges of bone crystals is not straightforward. Lengths and widths are most accurately measured on EM micrographs of dispersed crystals. Measurements of lengths and widths in embedded and sectioned preparations are smaller than on dispersed crystals (Jackson et al. 1978; Takuma et al. 1986), presumably because they are not oriented correctly. The methods used for removing the organic material may, however, affect some or all of the crystals. Table 9.2 lists the results of three studies of maximum crystal lengths and widths in which quite different techniques were used for removing the organic material with reasonably

Table 9.2 The Dimensions of Crystals from Mature Bones.[a]

Bone source	Bone type	Length (Å)	Width (Å)	Preparation procedure	References
Human	Cortical bone	500 (200–1500)	250 (100–800)	Autoclaving	Robinson (1952)
Human	Cortical femur	535 (225–1100)	285 (150–600)	NaOCl	Weiner and Price (1986)
Whale	Tympanic bulla	485 (150–1000)	260 (75–450)	NaOCl	Weiner and Price (1986)
Calf	Cortical tibia	490 (150–900)	210 (75–450)	NaOCl	Weiner and Price (1986)
Rat	Cortical tibia	405 (150–900)	215 (75–450)	NaOCl	Weiner and Price (1986)
Tuna	Vertebra	350 (150–675)	205 (75–375)	NaOCl	Weiner and Price (1986)
Turkey	Calcified tendon	355 (150–750)	205 (75–450)	NaOCl	Weiner and Price (1986)[b]
Turkey	Calcified tendon	290 (130–975)	165 (50–500)	Ethanol	Weiner and Traub (1986)

[a]The average and approximate ranges are reported.
[b]The results of Weiner and Price (1986) are of bone crystals extracted by their Method 3. This method minimizes exposure to NaOCl, but tends to extract a subpopulation of crystals about 25% larger than the average (Method 2). This explains the difference in turkey tendon measurements reported by Weiner and Traub (1986) with those reported by Weiner and Price (1986).

consistent results. Long bones from human, calf, and rat, as well as the tympanic bulla of the whale have crystals with average maximum lengths and widths of about 500 and 250 Å, respectively, just as Robinson reported in 1952. The crystals in tuna vertebra and calcified turkey tendon are considerably smaller (about 350 × 205 Å) (Table 9.2). Measurements of crystals that at least appear to be on edge in embedded and sectioned preparations show that the crystals are between 20 and 50 Å thick (Robinson 1952; Johansen and Parks 1960). Errors that arise in thickness estimates due to the crystals not being exactly aligned parallel to the beam will exaggerate the thickness! Thus 20 Å is probably a representative value. It should be noted that the order of magnitude difference between the thickness dimension and the other dimensions means that TEM micrographs of embedded and sectioned material will be dominated by crystals aligned with their flat surfaces more or less parallel to the beam. This, in fact, gave rise to the notion that bone crystals are needle shaped. Calculated surface areas for crystals of the dimensions given in Table 9.2 and having an average thickness of 30 Å are about 260–275 m^2/g, which is much larger than most measured surface areas of deproteinated bone material. These range from about 85 to 170 m^2/g (Wood 1947; Holmes et al. 1970; Misra et al. 1978; Weiner and Price 1986). A possible solution to this dichotomy is proposed at the end of this section on bone.

9.3.2.2 MINERALOGY OF BONE CRYSTALS

The small size and extreme thinness of the crystal plates have a profound effect on both the gathering of information on atomic structure and its interpretation. X-ray diffraction patterns of bone mineral have very broad reflections (see Fig. 1,

Glimcher 1984). They do, however, unequivocally show that the mineral belongs to the apatite series (de Jong 1926) and is more specifically hydroxyapatite [$CA_{10}(PO_4)_6(OH)_2$] (Klement and Trömel 1932). The presence of 4–6% by weight carbonate makes it a carbonate apatite and the accepted name for this mineral is dahllite (McConnell 1952, 1962). The broad nature of the reflections is wholly or in part due to the small size of the crystals. It may also be the result of dislocations and disorders due to lattice vacancies or the presence of other ions and/or molecules in the lattice. Significantly, direct observation of the atomic lattice by TEM lattice imaging does not reveal an unusually disordered structure, although some defects are certainly apparent (Selvig 1970; Jackson et al. 1978; Cuisinier et al. 1987). The crystal faces on the large surface of the plates have been identified as (100) faces (Selvig 1970; Cuisinier et al. 1987) based on studies of embedded and sectioned material. As the unit cell dimension in the *a* axis direction is 9.37 Å and the crystal thicknesses are between 20 and 50 Å (probably closer to 20 Å), the crystals are only two to five unit cells thick. This means a crystal has between 7 and 15 atomic layers and a very significant proportion of them is on or close to the surface. Clearly, a full understanding of the atomic structure of bone crystals requires a thorough examination of the structure of the surface itself (Neuman and Neuman 1953). Chemical analyses of the constituent ions show quite clearly that the mineral is not pure dahllite (Posner and Betts 1981). A careful analysis of age-related changes in the mineral of rat and bovine cortical bone shows that it is always calcium deficient, and contains CO_3^{2-} and HPO_4^{2-} ions in the crystal lattice. These parameters vary with age and the following formula is, therefore, needed to generally represent bone mineral: $Ca_{8.3}(PO_4)_{4.3}(CO_3)_X(HPO_4)_Y(OH)_{0.3}$; Y decreases and X increases with increasing age, whereas $(X + Y)$ remains constant and equal to 1.7! (Legeros et al. 1987). The crystal surfaces may well be able to account for much of the nonstoichiometry in ionic composition and accommodate a substantial amount of the carbonate and citrate molecules. Only some of the carbonate is known to substitute for phosphate inside the crystal (Baud and Very 1975).

Highly resolving methods for studying surface structure such as the scanning tunneling electron microscope have only recently become available. They offer, for the first time, the opportunity to begin to resolve problems that were already well defined more than 30 years ago (Neuman and Neuman 1953). Before they can be applied, however, we will have to learn much more about extracting crystals from the organic matrix without severely altering their surface structures. In vivo crystal surfaces are intimately associated with (and stabilized by?) the protein structures on and in which they grow. In vivo crystal surfaces might not be stable in an aqueous solution. Thus, examining crystals extracted from their in vivo milieu will always raise doubts about the true nature of their surfaces. We undoubtedly will need surface analytical techniques to study crystals in their natural environment with the protein present!

9.3.3 The Organic Matrix

The major component of bone, type I collagen, usually comprises between 85 and 90% of the total protein. The remainder is inauspiciously referred to as "noncollagenous proteins." This fraction comprises literally more than 200 proteins, the

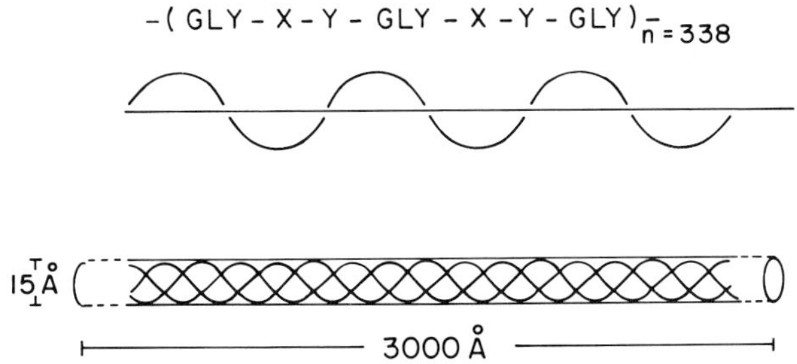

Figure 9.6. Schematic illustration of a collagen polypeptide chain with the (Gly-X-Y)$_n$ repeats (top) and the three polypeptide chains folded into the triple helical structure (bottom).

vast majority of which are serum proteins (Delmas et al. 1984). Of the quantitatively abundant proteins only about 12 are known to be in bone and not concentrated in bone from serum (Tracy et al. 1987).

9.3.3.1 COLLAGEN: THE MAJOR ORGANIC MATRIX CONSTITUENT

Since Robinson and Watson (1952) published TEM micrographs showing that mineralized bone collagen gives the same characteristic periodic banding pattern as unmineralized but stained collagen, it has been quite obvious that a very intimate association exists between the crystals and the collagen and, thus, in many respects *the* key to understanding the molecular organization of bone is the structure of collagen itself. In spite of the fact that since 1952 enormous strides have been made in the field of macromolecular structural determinations, our knowledge of the collagen structure is still incomplete. The major reason for this is that crystals of collagen are not available for X-ray diffraction analysis. Our current knowledge is based mainly on the known amino acid sequences of a series of collagen polypeptide chains, on X-ray and neutron diffraction patterns of collagen fibers, and on their electron microscopic images. In addition, some of the fine but important details have been obtained from careful and painstaking studies of the molecular cross-links between adjacent molecules. For a review of collagen structure, see Bornstein and Traub (1978).

Eleven different types of collagen are currently known (Miller 1985). Bone collagen is type I and curiously is the product of the same genes as skin, tendon, ligament, and aorta (Prockop et al. 1979). The polypeptide chains have a central region about 1000 amino acids long that is dominated by 338 contiguous repeating triplet sequence in which every third amino acid is glycine (Hofmann and Kuhn 1981) (Fig. 9.6). About one-third of the amino acids on the C-terminal side of the glycine residues are prolines and almost as many on the N-terminal side are hydroxyproline. The triplet repeating regions of three chains fold together to form the triple helical structure (Fig. 9.6) (Ramachandran and Kartha 1955). In bone collagen, two of the chains are identical [called $\alpha 1(\mathrm{I})$] whereas the third [$\alpha 2(\mathrm{I})$] is somewhat different, although it too has 338 continuous triplets (Hofmann et al. 1978). Both ends of the molecules contain short segments of polypeptide chains (telopeptides) that are not in the triple helical conformation. The N-terminal region

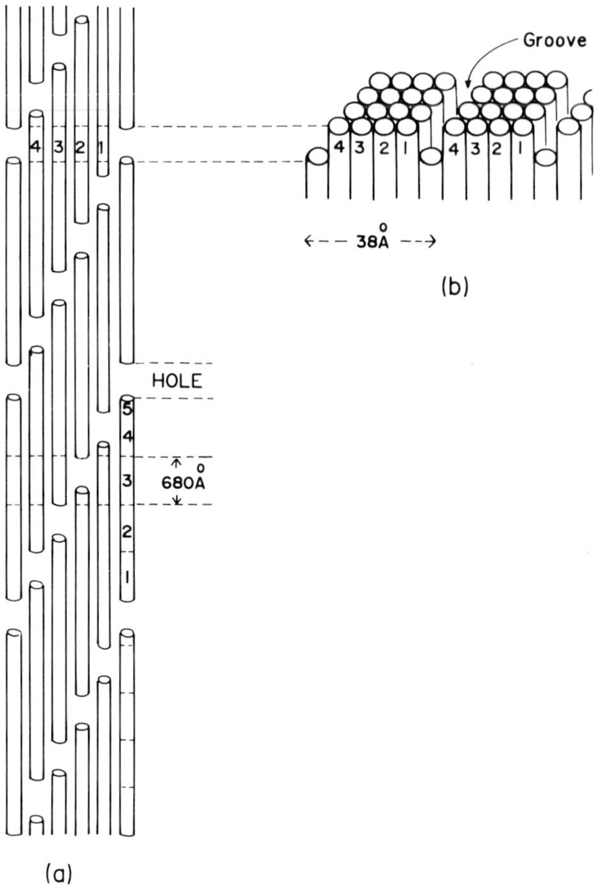

Figure 9.7. Schematic illustration of the collagen triple helical molecules (cylinders) according to the (a) Hodge–Petruska (1963) two-dimensional model and (b) extrapolated into three dimensions according to the Katz–Li model (1973) in which the hole regions are aligned to form grooves.

of the α1(I) chain, for example, may fold into a β-sheet structure (Helseth et al. 1979). The triple helical molecule is approximately 3000 Å long (Boedtker and Doty 1955), has a diameter of about 15 Å (Miller 1984) (Fig. 9.6), and is the basic building block of the large collagen fibril. Thus, understanding the structure of the fibril requires knowing the organizational pattern of the triple helical molecules. For an excellent and detailed discussion of collagen fibril structure and its relation to bone mineralization, see Miller (1984).

The most important discovery in collagen fibril structure was that of Hodge and Petruska (1963) who determined that the triple helical molecules are organized into linear arrays with the N-termini of one molecule not adjacent to the C-termini of the next, but separated by a hole or gap (Fig. 9.7a). They arrived at this conclusion by measuring the lengths of the triple helical molecules in an unusual form of collagen that precipitates with the molecules parallel and in register with each other (called SLS or segment long spacing). The lengths were 4.4 times longer than the repeat of a characteristic banding pattern (termed D) observed in TEM micrographs or measured more accurately by X-ray diffraction (D = 680 Å). From this they surmised that a space of 0.6D must exist between the N-termini of one triple helical molecule and the C-termini of the next. Furthermore, an adjacent array must be offset by 680 Å or 1D from its neighbor to account for the repeat pattern (Fig. 9.7a). This two-dimensional model forms the basis of our current understanding of collagen molecular organization. More accurate measurements of the gap length show that it is some 360 Å long (Miller 1984). What we do not know is whether or not the triple helical molecules in adjacent arrays always interact with their neighbors in the same specific manner, which would produce a three-dimensional crystalline structure rather than a two-dimensional "fiber" structure. We also do not know the packing arrangement of the triple helical molecules. A large and difficult to understand literature has developed on the three-dimensional structure of collagen, much of which is directly relevant to our particular problem—the mineralization of collagen. Indeed, Hodge and Petruska (1963) pointed out that the holes are the most likely locations for the crystals. However, individual holes are about 15–20 Å in diameter and some 360 Å long. Bone crystals are on the average 500 Å long, 280 Å wide, and around 20 Å thick (Table 9.2). The notion that a single crystal can fit into an individual hole (see, for example, the review by Glimcher 1984) cannot be correct (Weiner and Price 1986). There must be some special structural arrangement of the triple helical molecules in type I bone collagen that can accommodate the crystals. The most direct evidence for this could well be the organizational pattern of the crystals themselves (see next section).

9.3.4 Collagen–Crystal Relations

Polarized light (Schmidt 1936) and X-ray diffraction (Stuhler 1937) studies of bone show that the hydroxyapatite crystal c axes are well aligned with the collagen fiber axis. TEM micrographs of embedded and sectioned bone show that the mineral associated with the collagen fibril reveals the characteristic banded pattern of non-mineralized but stained collagen (Robinson and Watson 1952). In terms of the Hodge–Petruska model of collagen structure this implies that much of the mineral is localized at the level of the gap regions but does not necessarily prove that the crystals are located within the fibril itself. They could be arranged on its surface in a periodic pattern. X-Ray and neutron diffraction studies (White et al. 1977; Berthet-Colominas et al. 1979) refined the EM observations and favored the site for crystals within the fibril, but have provided very little new substantive information on the organizational arrangements of the crystals in relation to the fibril. Furthermore, none of the above-mentioned methods provides any information on whether or not all the crystals in bone are associated with collagen—a key question in trying to understand the organization of bone at the molecular level.

Katz and Li (1973a,b) published significant and thought-provoking papers on this aspect of bone organization. The major objective of their work was to quantitate the space available for crystal occupancy within fibrils and between fibrils. The basic strategy was to measure the space between fibrils with a radioactive polymer of polyethylene glycol. This polymer does not interact with the collagen nor does

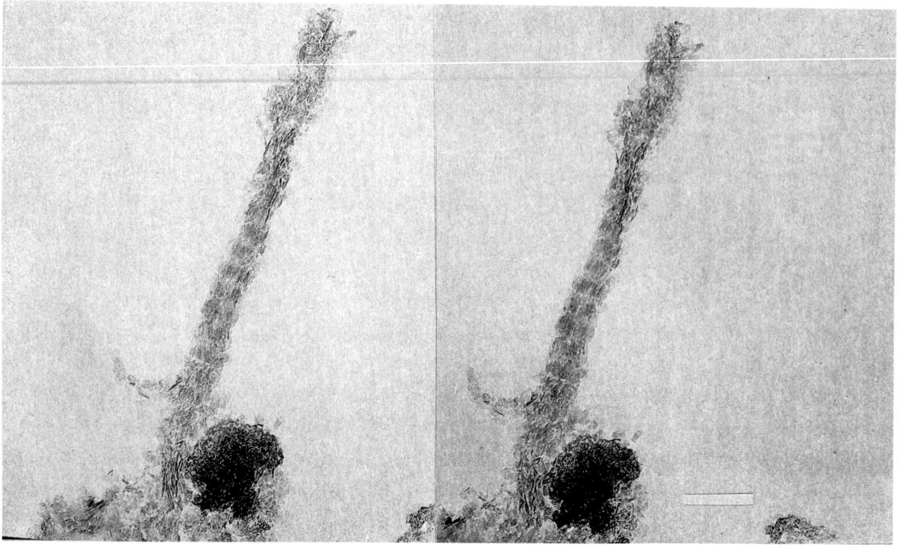

Figure 9.8. Stereoview of an unstained mineralized collagen fibril from turkey tendon showing the stacked arrangement of the plate-shaped crystals in the central ordered portion of the fibril. Reproduced from Weiner and Traub (1986). Scale bar: 200 nm.

it penetrate into the fibrils. The space inside the fibril can be calculated from the known dimensions of the triple helical molecules if the structure of the fibril is known. Even small errors in measuring key parameters such as the average distance between triple helical molecules (compare Katz and Li 1973b and Lees et al. 1984) lead to very different conclusions as to whether most of the mineral is located within or between fibrils. This important question has still not been answered. It is, however, worth noting that we do have good evidence (EM and diffraction) showing that at least a major portion of the crystals in bone is associated at the gap level within the collagen fibrils. In contrast, we have only circumstantial evidence to support the idea that some of the crystals are not associated with collagen, but are located somewhere between the fibrils. The possibility that all crystals form inside collagen fibrils should certainly not be discounted.

Questions relating to the very existence of crystals within the fibril as opposed to being located only on the fibril surface can be readily dispelled, at least in the case of the mineralized turkey tendon. TEM stereomicrographs of individual isolated collagen fibrils unequivocally show that plate-shaped crystals are within the fibril, that they are preferentially located at the gap level, and, most surprisingly, are stacked more or less like a deck of cards across the collagen fibril diameter (Fig. 9.8) (Weiner and Traub 1986). The fibrils were, of course, examined in a high vacuum in the EM and had as a result undergone considerable contraction. It is unlikely, however, that this "deck-of-cards" organizational motif could have been artificially induced. Figure 9.9 is a schematic illustration of how the crystals may be arranged in this way inside a collagen fibril. The molecular structure of type I collagen in turkey tendon must be such that it can accommodate crystal plates of average dimensions 290 × 165 × 20 Å stacked in a regular manner. Weiner and Traub (1986) point out that the models of collagen fibril organization first proposed by Katz and Li (1973a) and more recently by Fraser (1983) predict that adjacent gaps are in contact with each other at the same "height" in the collagen fibril and, as a result, extended "grooves" are formed whose lengths and thicknesses (about 360 and around 15 Å, respectively) are those of individual gaps, but whose widths are much longer (Fig. 9.7b). Hulmes et al. (1985) directly observed such grooves in nonmineralized rat tail tendon collagen fibrils. The models also predict that the grooves should be arranged in parallel rows. For this to occur the triple helical molecules in the fibril must be oriented in very specific ways with respect to each other to form local lateral crystalline regions. X-ray diffraction shows that crystalline three-dimensional packing is a characteristic of type I collagen fibrils including that of turkey tendon (Jesior et al. 1980). By far the most detailed information available on the three-dimensional structure of collagen fibrils is from the careful study of cross-links that bind different polypeptide chains together. By studying the exact locations of these cross-links, Mechanic, Katz, Yamauchi, and their colleagues proposed detailed structures of type I fibrils. Their results show that in periodontal ligament the fibrils have a tendency toward a crystalline packing arrangement (Yamauchi et al. 1986) and, furthermore, that skin collagen fibrils have a small but

Figure 9.9. Schematic illustration of the arrangement of crystals in a section through a collagen fibril, following the observations of Weiner and Traub (1986).

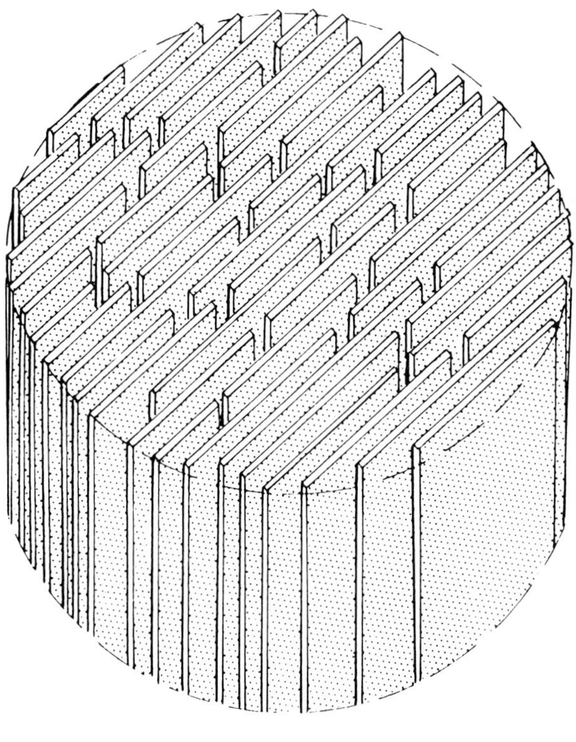

significantly different structure as compared to skeletal collagen fibrils (Mechanic et al. 1987). Even more detailed studies of the molecular locations of various cross-links and their precise stoichiometries could well result in a highly refined structural model of the fibril from which a first glimpse of the all-important wall structure of the grooves will be obtained. Will this provide a clue to the crystal growth properties of type I collagen or do we still have to evoke yet another component in the groove—a noncollagenous bone specific crystal nucleating protein?

9.3.4.1 THE NONCOLLAGENOUS PROTEINS (NCPS)

The realization that collagen associated with mineral in bone is very similar to collagen in nonmineralized skin, coupled with the discovery by Veis and his colleagues in the early 1960s of unusually acidic and phosphorylated proteins in dentin, provides the basis for the still widely held opinion that noncollagenous proteins (NCPs) play a very important role in mineralization of bone and dentin. Perhaps the more compelling argument in support of this is that acidic macromolecules, like those found in bone, are also found in every other mineralized tissue examined to date in which control is exerted over crystal growth (Table 2.3). The macromolecules that vary from tissue to tissue are the specific "framework" macromolecules, whose primary function is probably to provide the most appropriate biomechanical properties to the tissue (Weiner et al. 1983a). Collagen appears to fulfill such a "framework" function. For whichever reason, the NCPs are important. Much of the pioneering work on these proteins was done by Herring (1972) and they are still the subject of many investigations. Distressingly, however, we still know very little about the functions that they perform in vivo. One consequence of this is that the literature on this subject is confusing and difficult for the nonspecialist to understand. For good thoughtful "guided tours," see Fisher and Termine (1985), Termine (1986), Linde (1984), and Veis (1984). To address the issue of function there is an urgent need to develop conceptual approaches, as well as new techniques. With this in mind, we focus on the difficult questions of noncollagenous protein functions in mineralization.

Table 9.3 lists the known quantitatively abundant NCPs in bone and dentin and a few of their basic biochemical properties. The list does not include the macromolecules ("factors") that are known to be present in low abundance from their ability to induce various changes in cell behavior. It also does not include the enormous number of serum proteins that can be extracted from bone (Delmas et al. 1984), some of which [for example, alpha-2HS-glycoprotein and albumin (Ashton et al. 1976)] are present as quantitatively major constituents. The list is in a continuous state of flux and Table 9.3 should only be used to obtain an impression of the types of noncollagenous macromolecules present in bone and dentin. The table comprises mainly familiar "characters" well known from other mineralized tissues in which control is exerted over crystal growth. (Compare this table with Table 2.3.) They can be separated into three major classes: acidic glycoproteins, proteoglycans, and Gla proteins. Acidic glycoproteins (phosphorylated or not) and proteoglycans are the two classes of macromolecules found in various mineralized tissues of vertebrates, invertebrates and some protoctists. Their wide distribution emphasizes their basic importance in controlled mineralization systems. The third class of pro-

Table 9.3 Some of the Major Noncollagenous Proteins in Bone and Dentin

Class	Molecular weight[a] ($\times 10^3$)	Comments	Key references
Acidic glycoproteins			
Sialoprotein I (osteopontin, 44 KBPP)	44 (bovine)	Asp- and Glu-rich glycoprotein	Fisher et al. (1983); Franzen and Heinegard (1985a)
Sialoprotein II	~200	Glu-rich glycoprotein; Keratan sulfate absent in some species	Kinne and Fisher (1987)
Osteonectin	40	Major Asp- and Glu-rich phosphorylated glycoprotein	Termine et al. (1981)
Unnamed	80, 67, 60, 30 (bovine)	Asp- and Glu-rich glycoproteins	Sato et al. (1985a)
Unnamed	64	Glu-rich glycoprotein	Franzen and Heinegard (1985b)
Phosphophoryns	155 (bovine); 100 (rat and human)	Asp and phosphoserine-rich protein in dentin	Dimuzio and Veis (1978a)
Phosphoproteins	33, 28, 15, 14, 13 (chicken)	Asp- and Glu-rich glycoproteins	Uchiyama et al. (1985)
Phosphoprotein	24	Actually the N-propeptide of $\alpha 1$ type I collagen	Termine et al. (1981); Fisher et al. (1987b)
Phosphoprotein	44	Asp- and Glu-rich glycoprotein	Prince et al. (1987)
Proteoglycans			
Bone proteoglycan I	~350	Two chondroitin sulfate chains	Fisher (1985); Sato et al. (1985b)
Bone proteoglycan II	~200 (human)	One chondroitin sulfate chain	Fisher et al. (1983); Fisher et al. (1987a)
Cartilagelike proteoglycan	~1,000	Not unique to bone	Fisher (1985); Sato et al. (1985b)
Gla proteins			
Bone Gla protein (BGP or osteocalcin)	6	Contains 2 Gla residues	Hauschka et al. (1975); Price et al. (1976)
Matrix Gla protein (MGP)	15	Contains 5–6 Gla residues	Price et al. (1983)

[a]Estimates often vary significantly from species to species, between different methods, and even from one laboratory to another.

teins *may* be unique to vertebrates and within the vertebrates to bone and dentin. The γ-carboxyglutamic acid (Gla)-containing proteins have not been found in invertebrate phyla (King 1978) nor in the calcified cartilage of elasmobranchs (Price et al. 1976). They have, however, been reported to be present in hermatypic corals (Hamilton et al. 1982), although it is not proven that they are actually part of the mineralized tissue. Assuming that they are unique to bone and dentin, it seems reasonable to expect that the function(s) they perform is also specific to these tissues. Finally, there are some acidic glycoproteins, even quantitatively abundant ones, that are found in dentin but not in bone [the highly phosphorylated phosphophoryns (Linde 1984)]. These can most likely be expected to fulfill tissue-specific functions.

One direct indication that a protein fulfills an important function is that its amino acid sequence is conserved between different taxa. Bone Gla proteins from human, calf, chicken, and swordfish show remarkable conservation of sequence (Poser et al. 1980; Carr et al. 1981). Such sequence conservation implies a highly specific function for this protein. Unfortunately, no other information on comparative sequences is available for NCPs, although with the advent of new molecular biological techniques, this will no doubt change in the near future.

An array of different experiments aimed at elucidating NCP functions has been reported. Potentially some of the mose powerful in vivo experiments for this purpose are naturally occurring ones in which for some reason the gene product is altered or not expressed. The complement of NCPs in such diseased tissues show marked changes and deficiencies (Termine et al. 1984; Takagi and Veis 1981). However, no explanation that identifies a particular macromolecular function is available. An "unnatural" biologic experiment, but an ingenious one, was to grow rats on the rat poison warfarin! Warfarin is an antagonist of vitamin K which is necessary for the carboxylation of glutamic acid residues to form Gla. Rats can be grown on a warfarin-containing diet and the result is that their bones have only about 2% of their normal Gla-containing protein (BGP) levels (Price and Williamson 1981). Most surprisingly, the bone structure and mineralization, as far as could be discerned, were normal. The only major observed effect on the warfarin-treated rats was premature closure of the growth plate due to excessive mineralization (Price et al. 1982), a phenomenon that is difficult to interpret in terms of a unique protein function. It certainly does exclude functions for BGP (osteocalcin) that relate to de novo formation of mineral.

Experiments that provide information on the pathway and/or timing by which an NCP macromolecule enters the extracellular milieu in relation to mineralization can also potentially provide information about function. In one type of experiment a pulse of a radioactively labeled compound that tags a certain NCP is injected into experimental animals and then after various intervals the label is traced. Weinstock and Leblond (1973) performed an elegant experiment that graphically showed how the phosphophoryn in the continuously forming rat incisor bypassed the predentin collagen framework and was introduced directly into the mineralization front through odontoblast cell processes. These type of experiments were quantitated to actually give rates of macromolecular synthesis and assembly (Dimuzio and Veis 1978b). They were also used to show that some of the phosphophoryn forms a covalent conjugate with the collagen at the mineralization front, which occurs 2 hours after it is secreted (Maier et al. 1983). They provide good circumstantial evidence that phosphophoryn is somehow involved in mineralization, but little insight into the specific role it performs. A different "timing" type experiment makes use of the fact that subcutaneous implantation of demineralized bone matrix initiates a temporal sequence of events that includes the formation of cartilage, bone, and finally bone marrow. With a good assay for a particular NCP, it can be determined when the macromolecule is introduced into the mineralizing system. In this way Reddi and his colleagues (Price et al. 1981), for example, determined that BGP (osteocalcin) accumulates in bone only after the mineral is formed. Not much, however, can be concluded about the specific rule of BGP in mineralization, except, of course, that it is not likely to be involved in mineral nucleation.

A different approach to probing function is to determine in vitro whether a particular NCP is able to bind to collagen and to hydroxyapatite. Osteonectin was shown to bind to both (Termine et al. 1981), phosphophoryn to collagen (Stetler-Stevenson and Veis 1986), and BGP (osteocalcin) to synthetic hydroxyapatite (Poser and Price 1979). This information may be helpful in providing some guidelines about the nature of the molecules and their behavior in vitro, but caution should be exercised when extrapolating observations to the in vivo environment. Note too that the use of synthetic hydroxyapatite instead of bone crystals for these experiments introduces an additional unknown element. Another in vitro, function-type experiment is to add an NCP macromolecule to a saturated solution of calcium phosphate to determine how it affects the course of crystal nucleation and growth (e.g., Menanteau et al. 1982; Termine et al. 1980). These experiments generally show that they, in almost all cases, inhibit crystal nucleation, behaving in this way as do most nonspecific polyelectrolytes (Birchall and Davey 1981).

Good reasons exist as to why the identification of specific functions for these macromolecules has eluded investigators in the bone field. The primary difficulty is the very small size of crystals in bone as compared to other mineralized tissues. This, in turn, makes what we regard as probably the single most important clue to function in biomineralization, the identification of the precise location of a particular macromolecule with respect to the framework and to the crystals, very difficult to determine. In bone this has to be at the molecular level and it obviously requires a thorough understanding of the organization of this tissue at that level. It is clear from the earlier discussion on the organization of bone that wide gaps still exist in our knowledge of bone structure. It is our contention that until significant progress is made on this subject, the identification of NCP functions will be very difficult.

To illustrate the potential of detailed in situ mapping of NCP components, consider the following hypothetical situation in which the hypotheses that mineral formation inside the collagen fibril starts at the N-terminal side of the gap (Berthet-Colominas et al. 1979) and that mineralized type I fibrils contain extensive grooves (Weiner and Traub 1986) both prove to be true. Then the discovery of an acidic protein at this precise location would strongly suggest that it functions as a crystal nucleator. Its biochemical properties, genetic structure, biosynthetic pathways, behavior in vitro etc. could all provide additional evidence to support the conclusion that this is indeed a nucleating protein. Immunochemical mapping techniques using the TEM are capable of identifying the locations of these macromolecules at the required resolution, but we still do not have the conceptual framework to interpret the results.

Another potentially powerful technique for understanding function is to deliberately alter the gene of a particular matrix protein and examine its effect on the mineralized product. An elegant example of such an experiment is the work of Green and Warren (1985) in which the function of a bacterial protein as a nucleator of ice crystals was unequivocally established (Chapter 10). As a number of laboratories throughout the world are now using molecular biological techniques to characterize the NCPs and their genes, experiments of this type may well provide the first important breakthroughs on this subject.

In vitro experiments to assess whether or not a protein can interact specifically with certain crystal faces and not others (Addadi and Weiner 1985) or finding out if it is capable of nucleating in vitro when attached to a substrate (Addadi et al. 1987) can also be most informative. They cannot provide proof of a particular function in vivo, but they do provide important insights into the types of conditions and the necessary properties needed for a function to be fulfilled. For example, one can speculate based on the demonstrated cooperativity between sulfate and carboxylate ligands in in vitro calcite nucleation (Addadi et al. 1987) that the small bone proteoglycans, which contain both these ligands, may fulfill a nucleating function.

9.3.5 Stages of Bone Mineralization

The stages of bone mineralization are basically the same as all other organic matrix-mediated processes. The organic framework is initially formed and all the voids are filled with water. Mineralization starts and the crystals form in the voids, occupying the space of the water. This has been demonstrated in bone by careful analysis of the proportions of the major components of bone at different stages of mineralization. Figure 9.10 shows the relative proportions of mineral, collagen, and water as a function of degree of mineralization, as depicted by Doty et al. (1976). When the major components of bone are measured per unit volume, the mineral component increases at the expense of the water, whereas the collagen content remains constant. The weight percent relationships are misleading because of the different densities of the three major components.

An understanding of bone formation also requires detailed information on the stages of mineralization. In other mineralized tissues this mostly revolves around a description of the progressive growth of the crystals themselves. Bone crystals are so small that except for some isolated observations (Hohling et al. 1974) they have not actually been observed in the act of formation. In spite of the lack of available information some of the most fundamental and controversial issues in bone mineralization are concerned with related topics, such as the first mineral precipitates (where they form and their mineralogical identities), continued growth of crystals inside collagen fibrils (is this restricted to the gaps and/or grooves?), the possibility of direct crystal–crystal contacts, and the nature, or for that matter the existence of, crystals not associated with the collagen fibril. In this section we will very briefly discuss aspects of these and related issues and present some of our own speculative

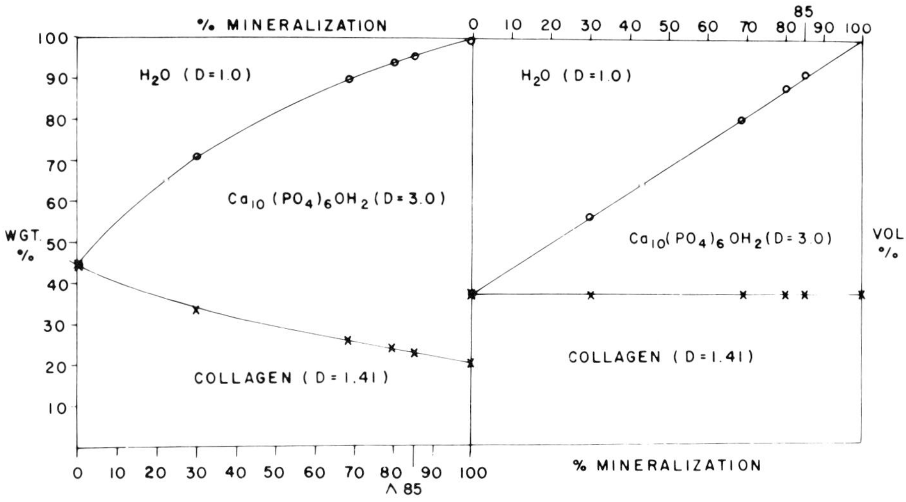

Figure 9.10. Left: The curvilinear *weight* relationships between the three major components of bone as it progressively mineralizes. Right: The straight line *volume* relationships that exist between the three major components of bone as it progressively mineralizes. Reproduced from Doty et al. (1976) by courtesy of the American Physiological Society.

ideas. From the discussion it will be apparent that new fresh approaches for studying these (very) old problems are urgently needed.

9.3.5.1 MATRIX VESICLES

In many bone and bonelike tissues the first minerals are actually not associated with collagen at all, but are present within membrane-bound vesicles (Bernard and Pease 1969). Some examples of boney tissues in which so-called matrix vesicles are found are predentin, woven bone, and calcified turkey tendon (Gay et al. 1978; Landis 1986). If these tissues have anything in common it is that they tend to form rather rapidly. The mineral bodies most often seen in mature vesicles have apparently needlelike shapes and are composed of crystalline hydroxyapatite (e.g., Dopping-Hepenstal et al. 1981). Amorphous calcium phosphate has also been reported (Gay et al. 1978). A quantitative study of some 40,000 vesicles showed that those furthest away from the mineralization front were devoid of any mineral, those somewhat closer contained amorphous looking mineral, and those close to the front contained crystalline deposits. Some of the latter had ruptured the vesicle membrane (Sela et al. 1987). It would be of great interest to know the precise morphology and size of these crystals and how they compare with collagen-associated crystals, as it is often asserted that these crystals are translocated from the vesicles onto and/or into the collagen fibril. For a good discussion of this controversial subject, read Butler (1985, pp. 368–372). It is interesting to note that protoctists from no less than seven different phyla form parts of their mineralized hard parts inside the cell and then translocate them to the cell exterior where they are assembled (Chapter 4). These mineralized bodies are, however, orders of magnitude larger than bone crystals and it is not difficult to visualize the cell's cytoskeleton fulfilling this translocation function. We have no evidence for any biological process in bone that transports the crystals and slots them into the collagen fibril, assuming that they are the right size. It seems even less likely that a physicochemical process could be responsible. The most likely function of these matrix vesicles in bone, in our opinion, is as a temporary storage site for calcium and phosphate ions needed for collagen mineralization, in which case the vesicle crystals would dissolve and be transported to the fibril in a soluble form. For comparison, see the section on the chiton radula and how iron is stored and then transported to the teeth (Chapter 6) and/or Chapter 7 on Arthropoda that have also evolved sophisticated temporary storage mechanisms for minerals in vesicles. On the other hand, it is quite conceivable that the role of matrix vesicles in cartilage may well be as an initiator of mineralization (see section 9.4).

9.3.5.2 FIRST PRECIPITATES

The mineralogical identity of the first bulk (presumably collagen-associated) mineral deposits is another widely discussed topic and many different minerals, aside from the first detectable mineral, which is hydroxyapatite (Bonar et al. 1985), have been proposed [see the summary by Boskey (1985)]. The transient nature of the phase, if it exists at all, makes it very difficult to detect, let alone verify its mineralogical identity. To our knowledge, there is no compelling reason why in bone a first-formed precipitate other than hydroxyapatite has to be invoked. In enamel, in which the earliest formed crystals are more easily sampled than bone, only hydroxyapatite has been found (Robinson et al. 1981). The one documented biological case of a first precipitate being different from the mature hydroxyapatite phase is in the radula of the chitons (Lowenstam and Weiner 1985). The first precipitate is amorphous calcium phosphate (ACP) (Chapter 6). In this case, however, ACP is stabilized for a number of days, a situation that clearly does not exist in bone. An ACP precursor phase in bone has been proposed (Termine and Posner 1967). Intuitively it seems a logical choice as X-ray diffraction patterns (Menczel et al. 1965) as well as infrared spectra (Termine and Posner 1966) of maturing bone tend to show sharper features with age. A simple extrapolation to the youngest first-formed phase could imply complete disorder, in essence ACP. Regardless of whether ACP is the first-formed precipitate, the extrapolation is not justified. Growth of small ordered crystals to larger ordered crystals could also account for the changes in X-ray and infrared patterns.

Brown (1966) proposed that the initial precipitate is in the form of octacalcium phosphate (OCP) and that the OCP then acts as a template upon which hydroxyapatite nucleates. Brown and his colleagues have amassed a large body of evidence to support this proposal [reviewed in Brown and Chow (1976) and Brown et al. (1981)], much of it being circumstantial, but taken together most convincing. Direct evidence supporting this idea is now available. Using high-resolution electron microscopy Nelson et al. (1986) have shown that synthetic carbonated apatite crystallites have an unusual core of some 7–15 Å thick with many structural defects. Computer-simulated lattice images of the defect structure are most consistent with it being a two-dimensional octacalcium phosphate inclusion, implying that indeed OCP is the first-formed precipitate. These defect structures (also called "central dark lines") are present in bone, dentin, and enamel crystals (Travis and Glimcher 1963; Marshall and Lawless 1981; Nakahara and Kakei 1984; Cuisinier et al. 1987) and, thus, support the notion that the first precipitate in vivo is also octacalcium phosphate.

9.3.5.3 CRYSTAL GROWTH

Only indirect evidence of crystal growth is available. X-ray diffraction patterns of bones from different aged animals show that the reflections become sharper with increasing age. Interestingly this effect is more enhanced in the *a* crystallographic axis (310) reflections as compared to the *c* axis (002) reflections (Burnell et al. 1980). A likely interpretation is that crystal growth in the *a* axis direction continues for a longer period of time than for the *c* axis direction. The (002) face is fast growing in hydroxyapatite and presumably the forming crystals inside collagen rapidly achieve their maximum length along the *c* axis direction. Crystal growth in the *a* axis direction is slower but should not be restricted, if indeed bone crystals are located in "grooves" similar to those postulated for mineralized turkey tendon (Weiner and Traub 1986). An alternative, but less plausible explanation, is that the layers of ions in the lattice that give rise to the reflections in the *a* axis direction become more ordered over a longer period of time than those required for ordering the *c* axis parameters.

Many other changes (e.g., increase of calcium content, decrease of HPO_4^- content) occur in bone mineral with age (Pellegrino and Biltz 1972; Legeros et al. 1987). Age studies are inherently very insensitive monitors of crystal growth, particularly in developing animals, as the turnover due to remodeling of bone is so rapid that at any given time an individual bone sample contains an assemblage of differently aged crystals. Alternatives are to work with bones in which turnover does not occur or is minimized, or to try and separate the differently aged crystals from a given bone. Density fractionation has, to some extent, provided a means of achieving the latter option (Richelle and Onkelinx 1969; Bonar et al. 1983). Bone formation processes in which the stages of mineralization are separated in space as noted are few. Probably the best studied bonelike process in this regard is that of mineralizing turkey tendon. Mineralization of the tendon starts at about 15 weeks of age and progresses steadily along the tendon (Johnson 1960). Excellent TEM (Nylen et al. 1960), X-ray, and neutron diffraction studies (Berthet-Colominas et al. 1979) of the mineralizing zone indicate that not only are the first-formed crystals in the collagen gap regions, but show that mineralization must start at a sharply defined locus within the gap. Furthermore, the mineral is not evenly distributed throughout the gap and there is even a hint that the locus is toward the N-terminus side of the gap (Berthet-Colominas et al. 1979; Miller 1984).

9.3.5.4 THE FINAL ORGANIZATIONAL STATE OF THE CRYSTALS

Ironically we seem to know almost as little about the final organizational state of bone crystals as we do about the very first-formed precipitates. We have discussed the evidence supporting the idea that a portion of bone crystals is located inside fibrils and have pointed out that even though it is commonly believed that a fraction of the crystals is formed between fibrils, evidence for their existence is only circumstantial. There are so many other confusing observations, that all options should remain open. For example, the average lengths of "normal" bone crystals are about 500 Å (Table 9.2) and they range in size from about 200 up to 1000 Å or more. Thus, most of the crystals in mature bone are significantly longer than the gap region. There is no indication that the longer crystals are wider and thicker, which might suggest that they do not form in the fibrils at all. It should be emphasized, however, that very good size measurements to prove or refute this statement are not available. Another even more puzzling observation is that the calculated surface areas of crystals with average sizes reported in Table 9.2 are two or three times larger than the surface area values reported in the literature for deproteinated bone. They are not, however, very different from surface area measurements of "purified" single crystal preparations (Weiner and Price 1986). Yet another unusual observation, and one that may help explain this surface area dichotomy, is that deproteination of bone using sodium hypochlorite does not completely disaggregrate whale bulla, rat, human, and calf long bones into their constituent crystals, even though turkey tendon and fish bone are completely disaggregated (Weiner and Price 1986). Significantly, the fraction that does not disaggregate has a low surface area and still contains some organic matter (mostly collagen) that is protected from the oxidizing agent. It is interesting to note that whale bulla, which is composed almost entirely of these aggregates, turns completely black when ashed, apparently because the organic material is so well protected by the crystals that even molecular oxygen cannot penetrate. Upon sonication in the presence of sodium hypochlorite, the aggregates fall apart into the "normal" plate-shaped crystals (Weiner and Price 1986). These aggregates may form as a result of some bone crystals undergoing an additional growth phase and ultimately contacting each other to form tight fused boundaries. This is similar, in some respects, to an earlier proposal by Ascenzi and Chiazzoto (1955).

To complicate matters even further, it is quite apparent that the collagen structure itself is not static. It can swell or contract depending upon the nature of the aqueous environment. It has been demonstrated using cartilage collagen that increases in osmotic pressure brought about by higher concentrations of proteoglycans cause the collagen fibril to contract. This is shown by a decrease in an equatorial collagen X-ray diffraction spacing (Katz et al. 1986). The same effect occurs in bone as a function of increased mineralization (Lees et al. 1984). The changes are most significant and can markedly affect the amount of space available for accommodating crystals.

Much still remains to be understood about bone mineralization. Some possibilities worth considering are that the crystal growth phase in which the newly formed crystals fill the gap/groove volumes inside fibrils corresponds to the very rapid mineralization phase, which accounts for about two-thirds of the final mineral content (Amprino and Engström 1952). Additional growth of these crystals, or, at least theoretically, growth of crystals between fibrils, corresponds to the slow secondary mineralization phase, in which aggregates are formed. The observed differences in bone density between species, between bones of one species, and even within a single bone (Baud and Very 1982) may, in part, be due to variations in the proportions of "aggregates" and monomer crystals. Furthermore, the well-documented break down of NCPs during late phase mineralization and maturation (Fisher and Termine 1985) may also be related somehow to aggregate formation. Confusion still reigns and it seems that some of the most important aspects of bone mineralization still remain to be discovered!

9.4 Cartilage

9.4.1 Introduction

Cartilage is widely distributed in the animal kingdom, where it is usually but not exclusively part of the endoskeleton (Person 1983). Mineralized cartilage, however, is confined to the vertebrates (Schaffer 1930) and even then it is by no means the common form. Cartilage can be identified histologically as being "composed of polymorphic cells suspended in highly hydrated, metachromatic gel matrices of varying rigidity, composition and abundance. Chemically, cartilage is characterized by its high content of collagen, glycosaminoglycan complexes, and water" (Person 1983, p. 34). The mineralized form is at least histologically similar to the nonmineralized form, except, of course, for the presence of crystalline apatite. Furthermore, in some pathological situations, such as rickets, cartilage that should normally be mineralized is virtually nonmineralized (Follis 1960). This is in sharp contrast to other vertebrate mineralized tissues such as bone and enamel that, almost by definition, are mineralized. Cartilage is in this respect similar to vertebrate keratin, or in the invertebrates to arthropod cuticle, both of which are mineralized only under certain circumstances and commonly exist in the nonmineralized form. Bearing this in mind, some key questions in understanding cartilage mineralization are why some forms of cartilage do mineralize and whether or not the mineralization process is controlled by an array of specialized macromolecules commonly found in other biologically controlled mineralized tissues?

Cartilage mineralizes in two different circumstances: as a transient form during a sequence of events leading to bone formation, and as a final mineralized product. The latter occurs only in elasmobranchs (sharks) and certain other fishes (Moss and Moss-Salentijn 1983), whereas the former situation, known as endochondral ossification, occurs in almost all vertebrates. The mineralized cartilage formed during endochondral ossification is covered with bone and, at a later stage, is partially replaced by marrow and vascular elements (Caplan and Pechak 1987). In either case the mineralization of cartilage and the mineralization of bone are two quite different events (Caplan 1984). It should be noted that in boney fishes, anura (frogs and toads), and small lizards secondary centers of mineralized cartilage in the endochondral zone do persist (Moss and Moss-Salentijn 1983). The histology and ultrastructure of mineralized shark cartilage have been described (Kemp 1977; Kemp and Westrin 1979; M. Takagi et al. 1984), but unfortunately little is known about the organization of this tissue at the molecular level. The mineralized cartilage is found in the central portions of the vertebra and close to the surface of the jaws, where it seems to serve as points of attachment of uncalcified perichondral collagen fiber bundles (Moss 1977). For a good overview of cartilage and cartilage mineralization, see Caplan (1984) and Poole and Rosenberg (1987).

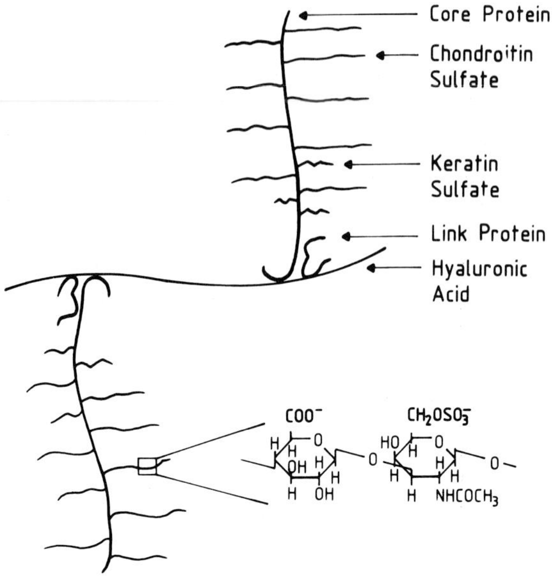

Figure 9.11. Schematic illustration of two proteoglycan monomers bound to a central polymer of hyaluronic acid core by means of a link protein. The enlarged area shows the structural formula for a repeating disaccharide unit of chondroitin 6-sulfate.

9.4.2 Cartilage in the Unmineralized Form

Cartilage is primarily composed of two types of extracellular macromolecular complexes—proteoglycans and type II collagen fibrils, associated with large amounts (60–80% by weight) of water. The cells that synthesize and deposit these complexes around themselves are called chondrocytes. The extracellular constituents act in concert to produce a tissue that has both tensile strength and resilience. The primary biological function that cartilage performs during development is to provide a shape or mould around which the major bones of the body, except for those of the skull, form. Throughout life cartilage covers portions of bones allowing them to move smoothly over each other as well as acting as a cushion for compressive forces.

Proteoglycans are large macromolecular complexes that contain "a core protein to which at least one glycosaminoglycan (GAG) chain is covalently bound" (Hascall and Hascall 1981, p. 39). Proteoglycans adopt a wide variety of structures differing with respect to their core proteins and GAG chains. They are found, in one form or another, in almost every tissue. They are also present in all mineralized tissues of vertebrates and probably invertebrates as well in which control is exerted ove the mineralization process (Table 2.3).

Proteoglycans are the major macromolecular constituents of cartilage. Of all the proteoglycans known, the structure of cartilage proteoglycans has been studied most thoroughly. This subject is reviewed by Hascall and Hascall (1981) and Rosenberg and Buckwalter (1986) and is schematically illustrated in Figure 9.11. Briefly, a proteoglycan monomer comprises a protein core some 300 nm long to which are covalently attached, through serine or threonine, linear polymers of the repeating disaccharides, chondroitin sulfate or keratin sulfate. Shorter branching polysaccharides are also linked to the protein core either through serine or threonine, or through asparagine. The chemical nature of the proteoglycan is dominated by the abundant chondroitin sulfate chains. On the average a proteoglycan monomer contains 80 such chains each with an approximate molecular weight of 20,000. The repeating disaccharide unit of chondroitin sulfate contains a carboxylate as well as a sulfate group (Fig. 9.11). The chains are, therefore, highly charged and electrostatically attract cations (Donnan effect). The affinity of cations varies with the square of their valence and, as a result, calcium is concentrated around the chains to a much greater extent than sodium. Sodium is much more abundant than calcium in the extracellular medium and is still the dominant ion in the immediate vicinity of the chondroitin sulfate chains. It is important to note that the ions are not bound to the polysaccharide, but are free to move around (Weinberg et al. 1986).

The proteoglycan monomers are themselves bound noncovalently with the aid of a special "link" protein to another polysaccharide polymer composed of hyaluronic acid. Some 100 or so monomers can be arranged in this way to form a giant complex. Aggregation into this giant complex occurs outside the chondrocytes, whereas monomer formation occurs within the cells.

Type II collagen is the other major macromolecular constituent of cartilage. As with type I collagen in bone and skin, type II collagen also forms fibrils with a characteristic triple helical structure. The fibrils are, however, much thinner than those in bone (Robinson and Cameron 1956) and often do not show the characteristic banding pattern (Bonucci 1967). Those that do are the thicker ones and they have the characteristic 670 Å periodicity (Bordas et al. 1978). Type II collagen fibrils are composed of three identical polypeptide chains that differ in their amino acid sequences when compared to the two chains in bone (reviewed by Miller 1976). Furthermore they are more highly glycosylated and contain three to four times as much hydroxylysine as is found in the type I bone collagen polypeptide chains (reviewed by Miller 1976).

The structure of type II collagen fibrils (reviewed by Brodsky and Eikenberry 1982) is not easily determined as they usually do not have a preferred orientation in cartilage. In one exceptional case, the human intervertebral disk, oriented type II fibrils do exist. These were studied by X-ray diffraction and were found to have basically the same structure as type I fibrils with some important differences (Grynpas et al. 1980). The equatorial reflection, which provides information on the side-by-side packing arrangement of the triple helical molecules, was 16–17 Å whereas in bone it is always 15 Å or less. This makes a very large difference in packing density and Grynpas et al. (1980) speculate that the abundant sugar residues may be responsible. Although type II collagen is the major type in cartilage, small amounts of types IX and X are also present. Their functions are, however, not known (reviewed by Poole and Rosenberg 1987).

9.4.3 Mineralized Cartilage

Mature mineralized endochondral cartilage comprises basically the same assemblage of major macromolecular constituents as nonmineralized cartilage, together with hydroxyapatite crystals. Significantly, however, mineralized cartilage also contains considerable amounts of the amino acids, phosphoserine, phosphothreonine, and γ-carboxyglutamic acid, whereas nonmineralized cartilage does not. In bone these amino acids are present in various acidic glycoproteins thought to be involved in regulating crystal formation. Their presence in cartilage may be indicative of the existence of analogous cartilage proteins (Glimcher et al. 1979).

The molecular organization of cartilage macromolecules and crystals has not been well studied. TEM micrographs of embedded thin sections generally give the impression that the crystals are needle shaped (Robinson and Cameron 1956). Tilting of the microscope stage demonstrates that the crystals in clusters examined from the endochondrial plate of a rat tibia are actually plate shaped and not needle shaped (Landis and Glimcher 1982). Note that in bone all the crystals that have been examined are plate shaped (Robinson 1952; Weiner and Price 1986). The average crystal thicknesses reported in one study of cartilage are 50 to 75 Å (Robinson and Cameron 1956) and in another 18 Å (Bonucci 1967). The range of crystal lengths reported are 250–750 Å (Robinson and Cameron 1956) and 500–1600 Å (Bonucci 1967). It should also be noted that nothing is known about the widths of the crystals, as this dimension cannot be measured in embedded thin sections. Clearly a more detailed study of the crystal dimensions in cartilage from different vertebrates is needed.

The crystals show no obvious preferred orientation and are certainly not organized in any regular manner with respect to the collagen fibrils (Robinson and Cameron 1956). Significantly, the crystals do not form in association with the collagen fibrils, which is probably *the* characteristic feature of bone. The basic structure of the type II fibrils is the same as in bone, although, as noted above, significant differences do exist. It would be interesting to determine whether some of the differences are responsible for the prevention of intrafibrillar mineralization. Interestingly, TEM micrographs of fixed, decalcified, and stained sections of mineralized cartilage show the outlines of the crystals as if mineral was present (Bonucci 1967). The phenomenon has been graphically referred to as "ghosts" by Bonucci (1967) and may indicate that the crystals are coated by a thin layer of organic material. The significance of this observation with respect to mineralization is not known.

Much more remains to be understood about the molecular organization of cartilage. The above mentioned properties, in particular the haphazard arrangement of the crystals, do, however, imply that cartilage crystal formation is a poorly controlled process, as compared, for example, to bone. The possible presence of acidic macromolecules, however, may indicate that there is some modulation of the crystal formation process.

Cartilage is often compared to bone as they coexist in close proximity in the endochondral plate, have some similar macromolecular constituents, and both contain hydroxyapatite crystals. The striking differences in molecular organization (Robinson and Cameron 1956) make it most unlikely that the mineralization processes in the two tissues are similar. Indeed a more detailed examination of the stages in cartilage mineralization reveals some of these differences.

The continuously growing endochondral growth plate is relatively convenient for studying the progression of mineralization, as various stages (articular, resting, proliferating, hypertrophic, and hypertrophic–mineralizing) can be separated in space (and time). The first histological descriptions of the growth plate were made by Fell (1925). Studies using the electron microscope revealed many more details (Martin 1954; Scott and Pease 1956) and, in a recent scanning electron microscopic study, zones of the growth cartilage were divided into 20 working levels (Shapiro and Boyde 1984). In mammals the chondrocytes are arranged more or less into columns with the forming cells at one end and the degenerating cells at the other. Mineralization in the extracellular matrix begins only after the cells have gone through a proliferative stage and have entered into a hypertrophic stage. Figure 9.12 shows a composite diagram of the zone in which mineralization occurs in a rabbit epiphyseal growth plate. We will describe some of the metabolic changes that occur in the cells prior to and during mineralization and then compare these to the associated extracellular matrix.

9.4.3.1 MITOCHONDRIAL MINERALIZATION

Prior to the onset of mineralization, the chondrocytes load up with calcium and phosphorus. An important site for the concentration and accumulation of these ions is within the mitochondria (Shapiro and Greenspan 1969; Mathews et al. 1970; Lehninger 1970) possibly in the form of amorphous calcium phosphate granules (Betts et al. 1975). Mitochondria actively control the influx and efflux of calcium ions and in this way participate in regulating the concentration of calcium in the cytosol. This subject is reviewed in relation to mineralization by Lehninger (1983). The flux is dependent upon the redox state of the cell. Kakuta et al. (1986) used a sophisticated microfluorimetric system to investigate the chondrocyte redox state by mapping the distribution of reduced pyridine nucleotides. In growth plate cartilage they found that coincident with the onset of mineralization, the concentration of NADH (nicotinamide adenine dinucleotide) increased markedly. This promotes efflux of calcium and phosphate from the mitochondria and would favor extracellular mineral formation (Kakuta et al. 1986). Analyses of the ion contents of individual isolated cells, using an SEM equipped with a microdissection apparatus and an electron probe X-ray emission microanalysis (EDX) attachment, show that as mineralization begins in the extracellular matrix the cells unload calcium, but still remain viable (Shapiro and Boyde 1984).

9.4.3.2 THE EXTRACELLULAR MATRIX

This is the site at which mineralization occurs. Prior to any crystal formation, the matrix itself loads up first with calcium and then just prior to mineralization with phosphorus (Shapiro and Boyde 1984). The fact that the transport of the calcium and the phosphate is decoupled may indicate that fine tuning of the timing of mineralization may be the controlled increase in phosphate concentration (Shapiro and Boyde 1984). Electron spectroscopic imaging also shows that the distributions of calcium and phosphorus in the about-to-be-mineralized matrix are spatially completely separate, with the calcium being intimately associated with sulfur-containing moieties (proteoglycans?) and the phosphorus in some unidentified granules (Arsenault and Ottensmeyer 1983). The matrix at this stage is poised for mineral nucleation.

Two quite different sites of initial mineralization have been observed: within membrane-bound vesicles (matrix vesicles) that are present in abundance in the extracellular matrix at some distance from the chondrocytes (Bonucci 1967; Ander-

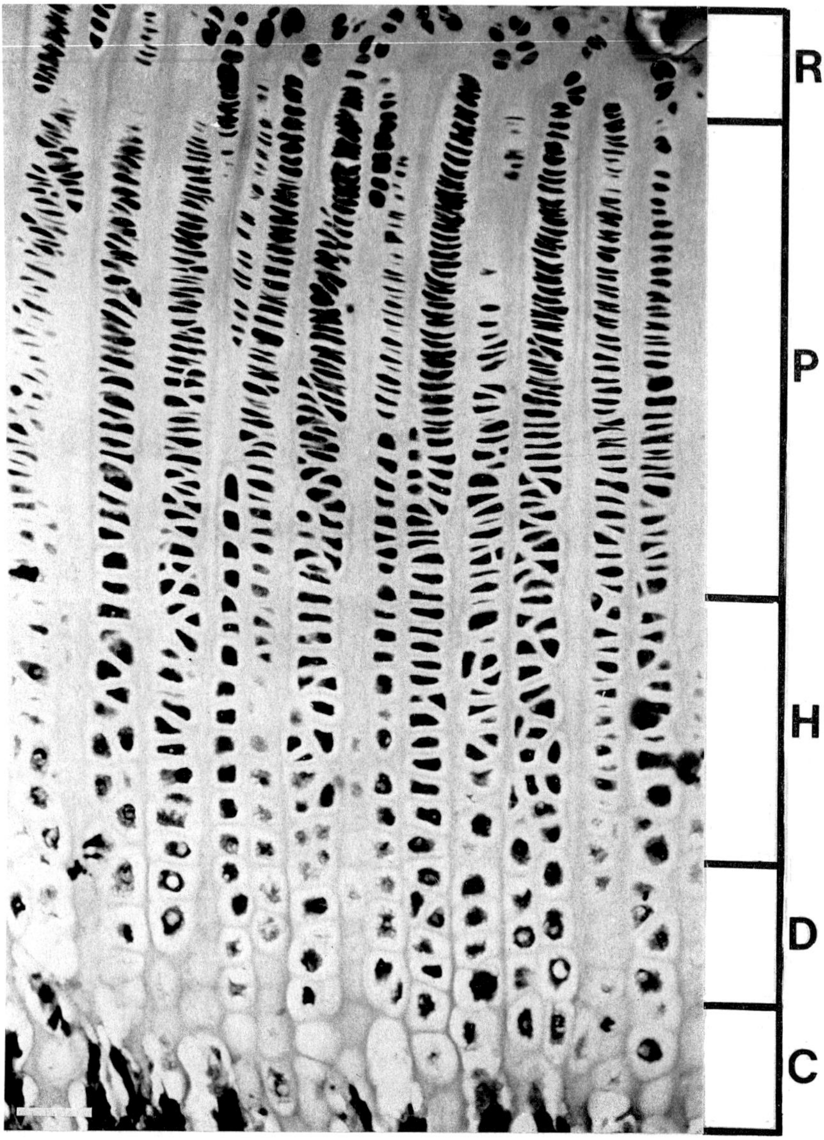

Figure 9.12. Light micrograph of rabbit femoral epiphyseal cartilage showing columns of chondrocytes separated by longitudinal septa. The growth plate is conveniently divided into the reserve zone of resting cells (R), the proliferative zone (P), the hypertrophic zone of mature cells (H), and the degenerative cell zone (D), which coincides with the calcified septum region (C). The invading capillaries from the metaphyseal side are seen at the bottom. Scale bar: 0.5 mm. Reproduced from Ali (1983) by courtesy of Academic Press.

son 1976), and in the matrix itself at loci unassociated with the vesicles. The latter possibility, proposed by Poole et al. (1984), associates the first mineral precipitates with a fairly acidic protein that they initially called chondrocalcin, but later was found to be the C-propeptide of type II collagen (van der Rest et al. 1986). Following the initial mineralization events, the mineral clusters rapidly enlarge and eventually coalesce to form a solid mineral phase.

Matrix vesicles appear in the extracellular matrix long before mineralization starts, but do not contain mineral. They first begin to load up with calcium and then only with phosphate (Ali et al. 1977). Many studies [reviewed and well illustrated by Ali (1983)] have shown that the first crystals of hydroxyapatite that form in cartilage are located inside these vesicles. The crystals seen in TEM micrographs appear to be long needles. Unfortunately, no good morphological studies or size measurements of all three dimensions of the crystals are available. Crystals continue to form inside the vesicles and then rapidly proliferate into the matrix outside the vesicle to form clusters of randomly oriented crystals. These clusters fuse to form the mature mineralized cartilage structure, which, as noted, is characterized by a haphazard arrangement of crystals (Robinson and Cameron 1956). We emphasize that the fabric of randomly oriented crystals within a vesicle is similar to that of the mature product, also essentially a network of randomly oriented crystals. It is, therefore, not difficult to envisage that the crystals in the vesicles act as centers of mineralization. Mineral-containing matrix vesicles are also present during the formation of some, but not all, bones (see section 9.3.5.1). As the majority of crystals in mature bone are associated with collagen and as a result are highly ordered, it is difficult to envisage a similar sequence of events for bone.

The contents of cartilage matrix vesicles have been relatively well characterized ever since a procedure for isolating them from the tissue was found (Ali et al. 1970). Alkaline phosphatase is an enzyme that is known to be widely associated with vertebrate mineralization. Some 80% of all the alkaline phosphatase activity in cartilage is localized in the vesicles, providing support for the hypothesis first proposed by Robinson (1923) that this enzyme might be involved in concentrating phosphate. It is now known to be a polyfunctional enzyme not confined to mineralizing tissues and the role it fulfills in the vesicles is still enigmatic (Register et al. 1986). One other outstanding feature of matrix vesicles with regard to mineralization is the presence of large amounts of acidic phospholipids as compared to the chondrocyte membranes (Peress et al. 1974). Phosphatidylserine is the major acidic lipid present in the vesicles where it forms stable complexes with calcium and phosphate (Cotmore et al. 1971; Boskey and Posner 1977). The complex acts in vitro as an ionophore that can transport calcium into the vesicle. The process is modulated by phosphate (Yaari et al. 1984). An additional or complementary role postulated for these calcium–phospholipid–phosphorus complexes is as a substrate for crystal nucleation (Vogel and Boyan-Salyers 1976). These complexes are found in many other vertebrate mineralized tissues (Boskey 1981), and a similar complex associated with a proteolipid is found in the membrane of a bacterium present in the oral cavity that induces hydroxyapatite crystal formation. The bacterial complex has also been shown to induce hydroxyapatite formation in vitro (Boyan and Boskey 1984) and complexes extracted from bone and reimplanted in rabbit muscle pouches can induce apatite mineral formation under physiologic conditions (Raggio 1986). A detailed understanding of the mechanisms of crystal nucleation and growth in these vesicles is not available, although a most impressive body of information has accumulated about them. Much of this is well reviewed by Anderson (1976), Ali (1983), and Wuthier (1984). Crystal nucleation on lipid membranes is also discussed in Chapter 3.

The C-propeptide of type II collagen (chondrocalcin). In 1983 Choi and his colleagues isolated a 35,000 molecular weight protein from fetal epiphyseal cartilage that binds strongly to hydroxyapatite and increases in concentration at the time when secondary centers of mineralization appear in the epiphysis. Animo acid sequencing of the protein revealed that it is very similar to or identical with the cleaved off C-terminal end of type II collagen (van der Rest et al. 1986). Both immunofluorescence and immunoelectron microscopy showed that the protein was first detectable within the chondrocytes and later appeared in the extracellular matrix at precisely those locations at which mineral was being deposited. These sites were reported to be unrelated to the mineral-containing matrix vesicles. A similar relationship between the protein's distribution and mineralizing loci was observed in mandibular condylar cartilage (Livne et al. 1987). Perhaps an even more impressive observation is that this protein is present only in areas of chick embryonic sternum that undergo calcification, but is absent in adjoining areas (Caplan and Pechak 1987). These observations demonstrate that the C-propeptide is somehow involved in cartilage mineralization but operates apparently independently of matrix vesicles.

Proteoglycans. The large majority of crystals that are present in the mature mineralized tissue form in the extracellular matrix outside matrix vesicles. They bear no obvious ultrastructural relationship to the collagen fibrils and these presumably have no direct function in cartilage mineralization. The environment in which these crystals form is dominated by proteoglycans, and, as a consequence, much of the literature on cartilage mineralization mechanisms focuses on proteoglycans. More specifically it addresses the paradox that in vitro proteoglycans are efficient inhibitors of hydroxyapatite nucleation (Cuervo et al. 1973; Blumenthal et al. 1979), but in vivo crystals grow in association with proteoglycans. One proposal to resolve the paradox is that proteoglycans are removed or modified prior to mineralization (Glimcher 1959; Hirschman and Djiewiatkowski 1966). A number of papers compare proteoglycans in mineralized and nonmineralized cartilage (reviewed by Poole and Rosenberg 1987) and conclude either that no changes occur (e.g., Poole et al. 1982; Buckwalter 1983) or that some degradation does occur (e.g., Campo and Dziewiatkowski 1963; Lohmander and Hjerpe 1975; Baylink et al. 1972; Campo and Romano 1986). Either way the fact remains that in vivo crystals form in the proximity of proteoglycans.

9.4.3.3 MINERALIZATION IN CARTILAGE: A PROPOSAL

The postulated roles of matrix vesicles, the collagen type II C-propeptide, and proteoglycans in mineralization appear to be mutually exclusive and in some respects contradictory. This is unlikely to be the consequence of misinformation and, in our opinion, with more detailed understanding of the processes involved, a synthesis is likely to emerge. The purported role of the proteoglycans as inhibitors of mineralization is most questionable. This is inferred from in vitro experiments, but with no verification in vivo. It is known that most flexible polyelectrolytes inhibit crystal formation (Birchall and Davey 1981) in vitro, but that in vivo they might well function differently. A good example is the sulfated acidic glycoproteins of mollusk shells that inhibit calcite nucleation when they are in solution, but adsorbed on a rigid substrate act as efficient nucleators (Addadi and Weiner 1985). Proteoglycans in vivo might function as concentrators of calcium to be used for crystallization, based on the observation that phosphate can effect the binding of calcium to proteoglycan (Hunter 1987). However Addadi et al. (1987) have shown that in vitro the calcium-concentrating ability of admittedly rather crude analogs of proteoglycans is insufficient to induce oriented calcite nucleation. An additional ordered substrate such as an acidic polymer (polyaspartate) in the β-sheet conformation is necessary. When combined the sulfonate ligands and the ordered carboxylates cooperate to induce oriented calcite nucleation. A similar phenomenon occurs in vitro with adsorbed natural sulfated polysaccharides covalently bound to aspartic acid-rich proteins. It is *conceivable* that in cartilage the sulfate moieties of the proteoglycans concentrate calcium and that the nucleation sites could be the surfaces of the crystals released from the matrix vesicles, or, for that matter, some other ordered acidic macromolecule (L. Addadi and S. Weiner, personal communication). The C-propeptide is certainly one possibility. In fact, calcification sites rich in proteoglycan occur exactly when and where the C-propeptide appears in the matrix in relatively high concentrations (Poole and Rosenberg 1987).

9.5 Enamel and Enameloid

One feature that distinguishes enamel and enameloid from all other vertebrate mineralized tissues and possibly even from all other known mineralized tissues is the most unusual shape of the carbonate–apatite (dahllite) crystals. They are tens of microns long, but sometimes only about one-twentieth of a micron wide—in other words something like spaghetti! (Nylen et al. 1963; Daculsi et al. 1984). They are quite different from the much smaller dahllite crystals of bone, dentin, and cartilage, which are very thin (about 20 Å) plates with lengths and widths averaging only about 500 and 250 Å, respectively. Synthetic dahllite crystals grown at atmospheric temperatures and pressures never achieve the shape and size of enamel crystals. However, dahllite crystals grown at 450°C and 2 kbar pressure are longer than enamel crystals (up to 3.7 mm) but are also very much thicker (Arends and Jongebloed 1979). There is no doubt, therefore, that the biological controls exerted over crystal growth in this tissue must be quite unique. In this section we will try to reconstruct what is known about the environment in which these crystals form. However, as with many other fascinating mineralizing processes in the biological world, this tissue has also not yet revealed the key secrets to its crystal growth processes.

The enamel crystals are generally organized into parallel arrays. These arrays may be local and not well defined, or may extend over very large distances. This occurs, for example, in the most superficial layer of mammalian enamels, which is composed entirely of a homogeneous sheet of crystals oriented perpendicular to the surface (Boyde 1971). Only in the bulk inner portion of mammalian enamel are the crystal arrays grouped into various higher order structures such as "prisms" (also known as "rods") and sheets. Mammalian enamel, which is composed mostly of hexagonally packed prisms (called pattern 1 by Boyde 1965), is characteristically found in the Chiroptera, Insectivora, Cetacea, Sirenia, Tipiridae, and some, if not all, Lemuroidea. In pattern 2 enamel the prisms are organized in long rows that are separated by sheets. Pattern 2 is characteristically found in Marsupialia, Artiodactyla, and the Equidae. The pattern 3 prisms have a keyhole-shaped cross section and these pack together with the narrower segments all more or less aligned in a preferred direction. This type of enamel is characteristically found in Primates, Proboscidea, Pinnipedia, and Carnivora (Boyde 1965, 1971). It is, however, important to note that the different mammalian enamel types are not restricted to particular taxa. One individual tooth may have all three patterns present and even some intermediate forms (Boyde 1971)! Mammalian enamel ultrastructure is complex and can be confusing. To complicate matters even further, the inner enamel of rodents (a favorite among investigators in this field) does not fit any one of these patterns. The Wistar rat incisor enamel illustrated in Figure 9.13a has two sets of prisms or rods set at an angle of some 64–70° from each other, with a third array of crystals (called "interrod") oriented more or less perpendicular to the major rods (Jodaikin et al. 1984). Together these combine to form one of the most impressive ultrastructural motifs known in biomineralization. At higher magnification, parallel arrays of the long spaghetti-shaped crystals that make up the rods can clearly be seen (Fig. 9.13b).

Enamel structures aside from the mammals are relatively poorly understood. The organization of the crystal arrays into prisms and/or sheets appears to be a characteristic of mammals (Moss 1968). Reptiles have a relatively homogeneous structure (Poole 1971) still composed essentially only of enamel crystals and of course organic matrix. The outer tooth layers of some amphibians, almost all boney fishes (Osteichtyes), and all sharks (Chondrichthyes) have a rather different structure in which the characteristic enamel-shaped dahllite crystals are present, still in parallel arrays, but are associated with collagen fibrils (Poole 1967). The latter are never present in reptilian and mammalian enamel. This organizational motif is called *enameloid* (Ørvig 1967; Poole 1967). By comparing enamel and enameloid, it seems that these long thin dahllite crystals form in two quite different environments: between collagen fibrils in enameloid and in the absence of collagen in reptilian and mammalian enamel.

In this section we will compare the crystal growth processes in enameloid and enamel to gain better insight into the basic factors that control this mineralization process. Much of the literature on enameloid focuses on the ontogenetic and evolutionary relations and differences between enamel and enameloid. Intriguing questions are addressed, for example, about the different roles of ectodermal and mesodermal cells in the formation of these tissues (Slavkin et al. 1984), or when in the geologic past enamel and enameloid evolved (Ørvig 1967; Poole 1967). We will, however, concentrate primarily on the crystal formation processes in the belief that good comparative information can provide a better understanding of basic mechanisms.

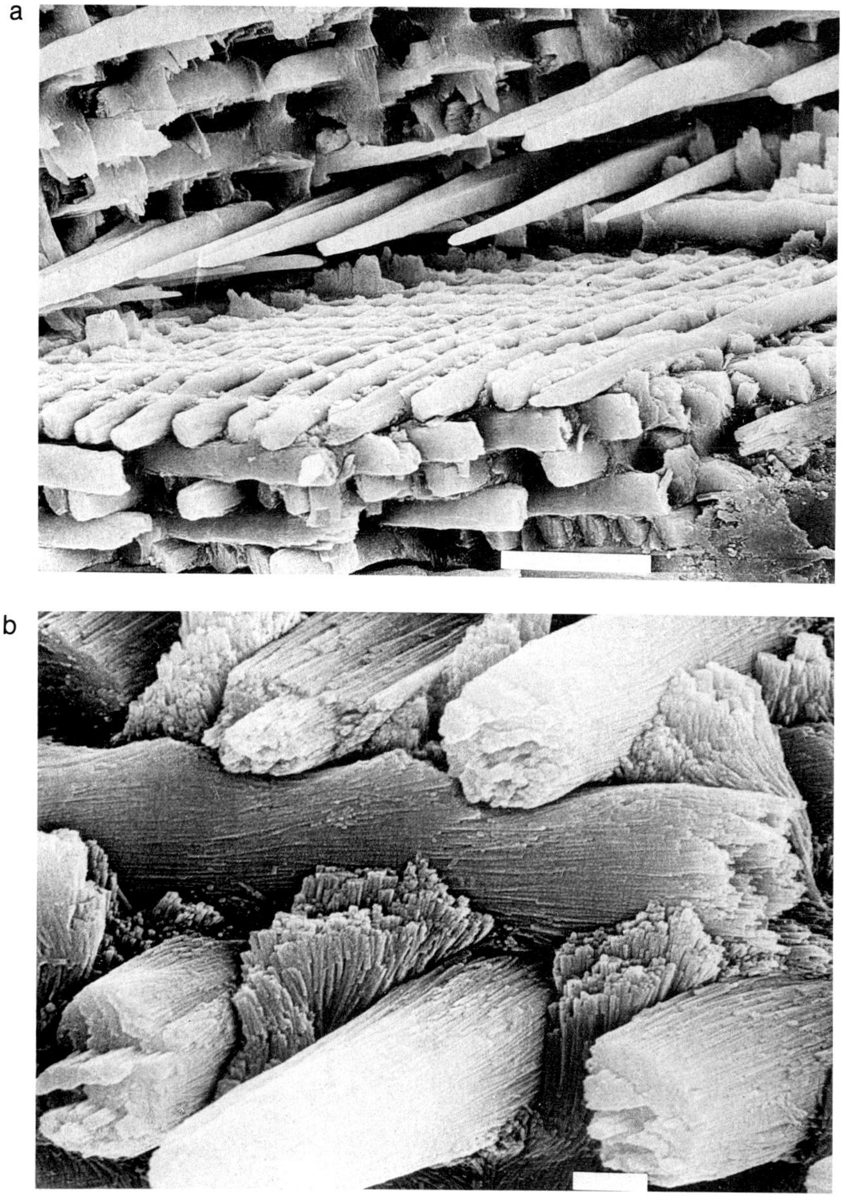

Figure 9.13. Scanning electron micrographs of (a) the forming enamel of continuously growing rat incisor showing the rod (horizontal layers) and interrod relations. Scale bar: 10 μm. (b) Rat incisor enamel rods and interrods at higher magnification showing the individual crystals. Scale bar 1.0 μm. Prepared as described by Jodaikin et al. (1984). Courtesy of Dr. A. Jodaikin.

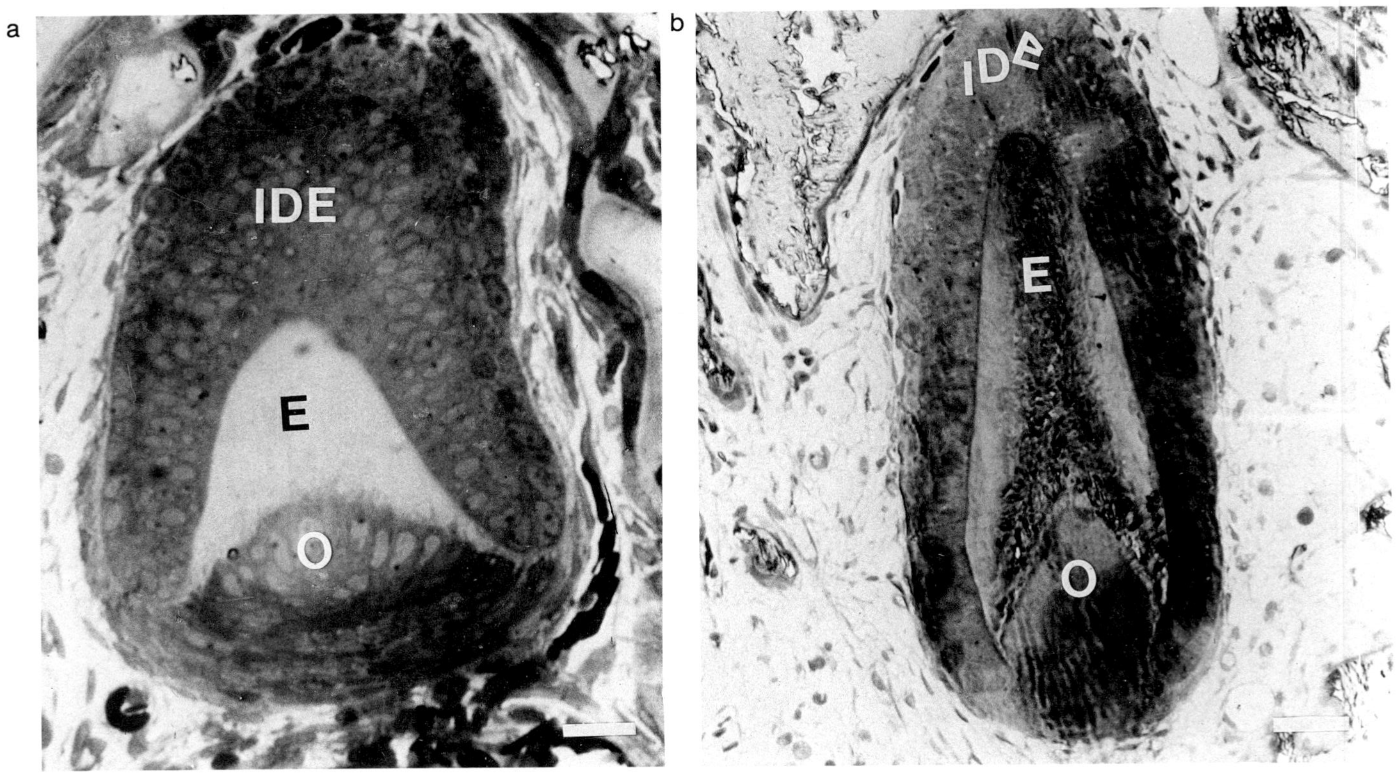

Figure 9.14. (a) Light micrograph of a secretory stage tooth bud from the teleost fish, *Cichlasoma cyanoguttatum.* The enameloid matrix (E) is primarily an ectodermally derived collagen synthesized by the inner dental epithelium (IDE). The newly secreted matrix is not mineralized at this stage. The odontoblasts (O) will later form the predentin of the tooth shaft. Scale bar: 20 μm. (b) Light micrograph of an early mineralization stage tooth bud from *C. cyanoguttatum.* Mineralization of the enameloid matrix (E) is initiated near the odontoblasts, then proceeds centrifically towards the IDE cells. Toluidine blue deeply stains portions of the enameloid matrix that are mineralizing. Scale bar: 25 μm.

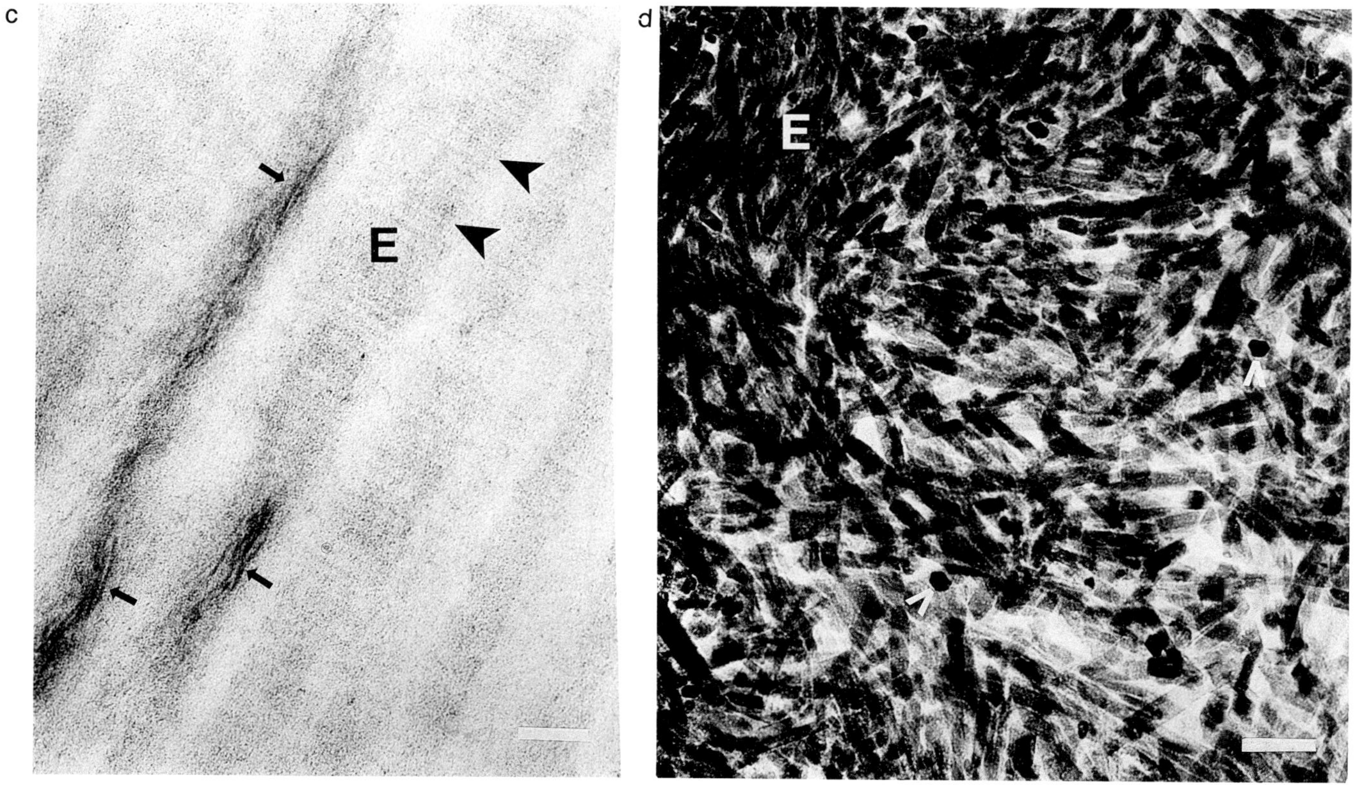

(c) Micrograph of initial mineralization in the enameloid matrix of *C. cyanoguttatum*. The 660-Å periodic banding of the ectodermally derived enameloid collagen is indistinct at this mineralization stage (arrowpoints). Nascent enameloid crystallites (arrows) are aligned with the long axis of the collagen fibers. Scale bar: 45 nm. (d) Micrograph of nearly mature enameloid from *C. cyanoguttatum*. Large enameloid crystallites (up to 200 nm wide) sometimes appear hexagonal in tangential section (arrows). Collagen fibers are no longer present within this stage of mineralization. Scale bar: 47 nm. Courtesy of Dr. K. Prostak.

9.6 Enameloid

Enameloid mineralization has not been thoroughly investigated in different fish and most generalizations cannot be well substantiated. One of the most detailed recent studies is that of Prostak and Skobe (1986) on the mineralization of enameloid in unerupted tooth buds of juvenile teleost cichlids *(Cichlasoma cyanoguttatum).* The basic sequence of observed events described below appears to be representative of enameloid formation in other teleosts and to some extent in sharks as well.

The full thickness of cichlid enameloid matrix is formed prior to the initiation of mineralization, an excellent example of the widely observed preformed organic matrix. The enameloid matrix is surrounded on its upper surface by a layer of epithelial cells (called inner dentintal epithelial cells or IDE cells) and on its lower surface by predentin followed by a layer of odontoblasts (Fig. 9.14a). Mineralization of the enameloid matrix is initiated near the odontoblasts then proceeds centrificially toward the IDE cells (Fig. 9.14b). The major matrix component is collagen. The fibrils show indistinct 660-Å periodicities and are some 300 Å in diameter (Fig. 9.14c). They are formed by the IDE cells that are of ectodermal origin (Prostak and Skobe 1985). Mineralization starts in the enameloid and not in the predentin. The first crystals are seen close to the junction between dentin and enameloid. With the onset of mineralization the IDE cells almost double their height, indirectly suggesting that they are indeed orchestrating the process. A similar change in cell morphology occurs in enamel, where it is well established that the "tall" cells are actively involved in the elaboration of matrix (Reith and Butcher 1967). The first crystals form between collagen fibrils and are well aligned with the fibril axes (Fig. 9.14c). The crystals are some 1500 Å long and 30 Å wide. There is no evidence that mineral forms inside the fibrils, as is observed to occur when the dentin begins to mineralize. Additional enameloid mineralization occurs at increasingly greater distances from the dentin–enamel junction and progresses toward the layer of IDE cells. Crystals are always closely associated with collagen fibrils. The mature crystals are very long, about 2000 Å wide, and are usually not symmetrically hexagonal in cross section (Fig. 9.14d). Electon diffraction patterns show that they are hydroxyapatite, although it is not known whether the crystals in this particular species are fluorinated or not. Other cichlid enameloid crystals from mature teeth do contain 3.2% fluoride (Suga et al. 1980). Prostak and Skobe (1986) also show that at some stage during mineralization the collagen fibrils alter their structure and are removed.

Many similar observations have been made in other teleosts and to some extent in sharks as well (Poole 1967). In sharks enameloid also mineralizes before dentin, an observation already noted by Tomes in 1898. Note that the opposite is true for mammalian tooth mineralization (Frank and Nalbandian 1967). Collagen is a major matrix constituent of all forming enameloids (Poole 1967). The collagen in cichlid teeth does not mineralize, although Fearnhead (1979) states that in general collagen in enameloid does mineralize. In some teleosts the enamel crystals are fluorapatite and in others hydroxyapatite [actually dahllite as they contain carbonate (Legeros and Suga 1980)]. The distribution of these two minerals is taxonomically determined (Suga et al. 1980; Suga et al. 1981) and does not reflect environ-

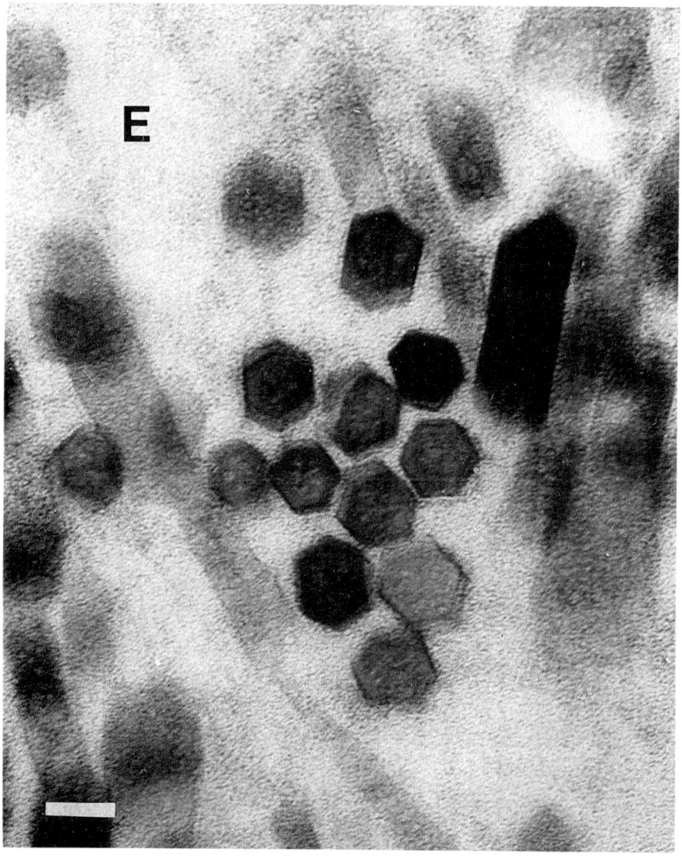

Figure 9.15. Micrograph of nearly mature enameloid crystallites from an elasmobranch fish, *Raja erinacae.* Fluorapatite crystallites of the enameloid are symmetrically hexagonal. Scale bar: 33 nm. Courtesy of Dr. K. Prostak.

mental fluoride concentrations (Legeros and Suga 1980). All sharks have fluorapatite crystals (Trautz et al. 1952; Glas 1962). All enameloid crystals are very long and thin, but in sharks the cross-sectional shape tends to be symmetrically hexagonal (Garant 1970; Kemp and Park 1974) (Fig. 9.15). The range in diameters of the shark crystals is very large (Daculsi et al. 1979; Kemp 1985) when compared to teleosts or mammals. In fishes little is known about the cross-sectional crystal shape, except in the cichlid species analyzed by Prostak and Skobe (1986) and the eel (Shellis and Miles 1976), where they are not symmetrically hexagonal. Although it is commonly assumed that the symmetrical hexagonal shape is a direct result of the crystals being fluorapatite (Glas 1962; Kemp 1985), this is by no means obvious and certainly not proven. It would be most interesting to know if, indeed, this is the case. The alignment of the crystal long axes with the collagen fibril axes is well established in the enameloid of the ray, *Raia clavata* (Poole 1967), and, in fact, appears to be a general property of all enameloids (Schmidt 1958). The removal of collagen from the matrix following mineralization has also been documented in eel enameloid (Shellis and Miles 1976).

Acidic glycoproteins appear to be found almost universally in mineralized tissues in which control is exerted over crystal growth (Weiner et al. 1983a). It is no surprise that they are also present in enameloid. Amino acid composition analyses of unerupted teeth of a hammerhead shark and of partially erupted teeth of a Bermuda black tip shark show that these proteins are rich in the acidic amino acids, glycine and serine (Levine et al. 1966). They have amino acid compositions very similar to proteins extracted from mature mammalian enamel (see following). Protein biosynthesis studies show that these acidic proteins, now known as *enamelins* (Eastoe 1979), are secreted by the ectodermal cells (Graham 1984). They have been isolated and partially purified from the enameloid layer of the blue shark, *Prionace glauca* (Graham 1985). The enamelins are extractable only when the mineral phase is dissolved suggesting an intimate relationship with the crystals and implying a functional role for these proteins in regulating crystal growth. Their locations in this tissue are not known, but in mammalian enamel they are indeed found in close proximity to the crystal surfaces (see following). To our knowledge, comparable information is not available for teleost enameloid.

What then can be surmised about the environment in which crystals grow in enameloid? One outstanding feature that needs to be understood is the unparalleled length of these crystals. The observations that the crystals are always well aligned with the collagen fibril axes, form in their close proximity, but are located between fibrils suggest that the packing of the fibrils forms a framework for crystal growth. As many crystals appear to form in parallel arrays in the space between a group of fibrils without fusing and penetrating each other, this space is probably further subdivided. Candidates for this latter role are the enamelins. Garant (1970) and Kemp (1985) have, in fact, observed organic sheaths surrounding individual crystals in developing shark enameloid. Their observations are consistent with the idea that the crystals grow into the preformed space created by the organic sheath material.

9.7 Enamel

The enamel organ is composed of four layers, the innermost of which is called the inner dental epithelium. The cells of this layer are responsible for matrix elaboration and mineralization. When they begin this process they become highly elongated and the resulting "tall" cells are referred to as *ameloblasts* (Reith and Butcher 1967). The distal ends of the cells adjacent to the extracellular matrix are composed of highly invaginated cell membranes, called *Tomes processes.* In contrast to enameloid, in which the entire extracellular matrix is preformed prior to the onset of mineralization, enamel mineralization starts almost immediately after matrix elaboration (Reith and Butcher 1967). So much so that good proof of this postulated sequence of events is still not available. With ongoing enamel formation, the cells recede, but always maintain very close contact with the zone of active mineralization. For detailed descriptions of the development of the enamel organ and the cellular transitions during enamel formation, see Reith and Butcher (1967), Frank and Nalbandian (1967), and Warshawsky (1985).

Enamel crystals are similar in overall shape to those of enameloid. They are composed of carbonate hydroxyapatite or dahllite (Table 9.1). Fluorapatite is not found in enamel. The crystals are arranged in parallel arrays (Fig. 9.13). Enamel, as with enameloid, also contains a class of acidic glycoproteins, the so-called *enamelins* (Eastoe 1979). [The term *enamelin* was actually coined by Mechanic (1965) to describe other proteins.] The enamelins constitute only 5–10% of the matrix protein in forming teeth (Belcourt et al. 1982). The major matrix constituents of forming enamel are relatively hydrophobic proteins called *amelogenins* (Eastoe 1965) that are apparently absent in enameloid, or, if present, have not as yet been unequivocally detected. Collagen is not present in enamel (Fearnhead 1979). Collagen fibrils are easily identified in the TEM, whereas amelogenins appear as featureless globular particles (Bai and Warshawsky 1985), making even a basic understanding of the molecular organizational relationships of the amelogenins and the mineral in enamel very difficult. In the section below we will focus on the molecular organization of enamel, and in particular the possible functional roles of the amelogenins in an effort to gain more insight into crystal formation in this tissue.

Note that enamel formation is extremely dynamic in the sense that not only is the whole tissue assembled and mineralized, but it also "matures." This refers to the process in which the matrix is partially broken down and removed, and the crystals continue to form. It is, therefore, important to identify the developmental zone from which the information is derived. Robinson et al. (1977) describe the zones in continuously growing rat incisor, which are also more generally applicable to other teeth. A tabulation of many of the basic properties of enamel matrix and crystals can be found in Weiner (1986). The two volumes entitled *Structural and Chemical Organization of Teeth* edited by Miles (1967) are an excellent source of information.

9.7.1 The Crystals

Daculsi et al. (1984) have shown that in forming mammalian enamel the crystals are at least 100 μm long! This astonishing observation supports an earlier proposal (Boyde 1965; Warshawsky and Nanci 1982) that each crystal extends from the dentinoenamel junction through to the enamel surface. Thus, crystal growth, which progresses from the dentinoenamel junction to the surface, involves the continuous elongation of these crystals. This can occur either by multiple renucleation events with the incipient crystals growing together to form a macro-single crystal or by additional crystal growth at one end (Weiner 1986). Whatever the mechanism, the process is under the strictest control, because at every stage the crystals are arranged in well-ordered parallel arrays and are all remarkably uniform in cross-sectional size (Rönnholm 1962; Nylen et al. 1963). The very earliest formed crystals are flat thin ribbons (Rönnholm 1962; Nylen et al. 1963), reminiscent of the shape of octacalcium phosphate crystals precipitated in vitro (Brown and Chow 1976). The crystals widen and thicken with additional growth, but do not lose this elongated hexagonal symmetry (Nylen et al. 1963). A similar crystal growth sequence has been observed in vitro using enamel crystals as seeds (Doi and Eanes 1984). The crystal faces expressed in enamel are always the (100) and at the ends presumably (001) (Selvig 1972), which are the ones usually found in synthetic hydroxyapatite (Terpstra et al. 1986). The centers of enamel crystals characteristically contain a most unusual linear structure known as the "central dark line" (Rönnholm 1962; Nylen et al. 1963; Nakahara and Kakei 1983) based on its appearance in the TEM. This same phenomenon has been observed in dentin (Nakahara 1982) and bone crystals (Cuisinier et al. 1987) and has been the subject of much speculation and debate. Nelson et al. (1986) have convincingly resolved the puzzle by showing that the structure is also present in synthetic apatites, provided they contain carbonate. Their lattice images show that the structure is due to defects arising from the presence of a thin plate, probably a single unit cell thick, of octacalcium phosphate (OCP) embedded inside the apatite crystal lattice. This is consistent with a proposal made by Brown (1966) that OCP is the first formed deposit in in vivo apatite formation and that hydroxyapatite grows epitaxially upon the surface of the OCP. This view of crystal elongation and growth does not directly involve matrix constituents at the site of octacalcium phosphate nucleation, except possibly the initial site of de novo crystal nucleation close to the dentinoenamel junction. The major role of the matrix would appear to be controlling the cross-sectional dimensions and the continual elongation of the crystals. One curious observation that must represent some sort of fine biological control either at the matrix or at the cellular level is that at a particular stage of growth, single crystals split into two crystals and these continue to grow independently (Daculsi et al. 1984). We conclude that the observations of crystal organization, shape, and growth all imply that each crystal grows in a well-defined space, a hypothesis that should be borne in mind when examining matrix structure and function.

9.7.2 The Organic Matrix

An important clue about the different functions of amelogenins and enamelins is that amelogenins can be extracted from forming enamel with a denaturing reagent (guanidine hydrochloride), whereas most of the enamelins require the dissolution of the mineral before they are released (Termine et al. 1980a; Fukae and Tanabe 1987) This strongly suggests that the enamelins are intimately associated with the mineral phase, whereas the amelogenins are not (Termine et al. 1980a). This was confirmed directly by differential extraction of these components followed by examination in the TEM, which showed that the enamelins, or at least some of them, surround individual crystals, whereas the amelogenins fill in the spaces between crystal arrays (Bai and Warshawsky 1985).

Enamelins are characterized by their acidic nature, being rich in glutamic acid, aspartic acid, and glycine (Belcourt et al. 1982; Limeback 1987). They are also associated with covalently bound polysaccharide (Termine et al. 1980). The molecular weights of the most abundant enamelins in bovine teeth are estimated to be 56,000 and 72,000 and in pig 63,000 and 67,000 (Belcourt et al. 1982; Limeback 1987). Relatively little is known about their biochemistry and none has yet been sequenced. X-ray diffraction and infrared spectroscopy show that the enamelins tend to adopt the β-sheet conformation (Jodaikin et al. 1987). Furthermore, X-ray (Jodaikin et al. 1985) and electron diffraction (Jodaikin et al. 1988) show that the bulk of the enamelin polypeptides are more or less perpendicular to the long axes of the crystals, which are aligned with the crystallographic *c* axes. This demonstrates that a well-defined spatial relationship exists between some of the enamelins and the crystals, strongly suggesting that the enamelins fulfill a specific function in regulating crystal formation (Jodaikin et al. 1988).

The most obvious role for the enamelins, possibly together with other matrix constituents, is to form at least part of a tubule or sheath around the crystal that *somehow* controls crystal elongation and growth. Tubule structures have been observed in forming enamel (Jessen 1968; reviewed by Nylen 1979). The problem is that these tubules have not been observed as preformed structures into which the crystals grow, leaving open the possibility (not likely, in our opinion) that the organic macromolecules are adsorbed onto the crystal surfaces at a later stage (Nylen 1979). One reason for this ambiguity is that crystal growth elongation takes place in very close proximity to the ameloblast cell membrane (Nanci and Warshawsky 1984), a zone of very complex and highly invaginated structures (Sasaki and Higashi 1983). In fact, the membrane structures themselves might well be fulfilling structural roles, together with the enamelins, in ordering crystal growth.

Amelogenins are relatively hydrophobic when compared to the enamelins, being rich in proline, glutamine, leucine, and histidine (reviewed by Fincham and Belcourt 1985). They are also much smaller molecules with the most abundant ones in forming enamel having molecular weights around 25,000. There are indications that higher molecular weight precursor forms exist (Christner et al. 1985). Many of the different sized amelogenins present even in the earliest formed enamel might all be fragments of a few gene products (Fincham and Belcourt 1985). The amino acid sequence of the major amelogenin gene product has been determined both at the protein level (Takagi et al. 1984) and at the gene level (Snead et al. 1985) for a number of different mammals. The sequences from different species are highly conserved. They do not show any striking regularly repeating segments, as was observed in the sea urchin larval spicule matrix protein (Sucov et al. 1987). The sequence itself provides no obvious clue about the functions(s) of this protein. Efforts by Jodaikin et al. (1987) to determine its secondary conformation using X-ray diffraction and infrared spectroscopy have not revealed the presence of any regular conformation. In contrast Renugopalakrishnan et al. (1986a) and Zheng et al. (1987) have tentatively proposed an unusual secondary structure including β-turns, β-sheets, and novel β-spiral based on circular dichroism, infrared spectroscopy, and Raman spectroscopy. The protein has an unusual amino acid sequence and, therefore, an unusual structure could be expected.

At this stage we can only speculate about possible general functions for these important matrix constituents. Their hydrophobic nature, their conserved amino acid sequences, and the fact that they are not in direct contact with the mineral phase are all consistent with the postulate that they function primarily as a structural framework (Weiner 1986).

9.7.3 Maturation

Maturation refers to the changes that take place in the mineral and the matrix once it is laid down. This occurs in enamel to an unparalleled extent when compared to all other mineralized tissue we are aware of. In fact, the process appears to start almost concurrently with the first deposits that can be dissected for analysis. The

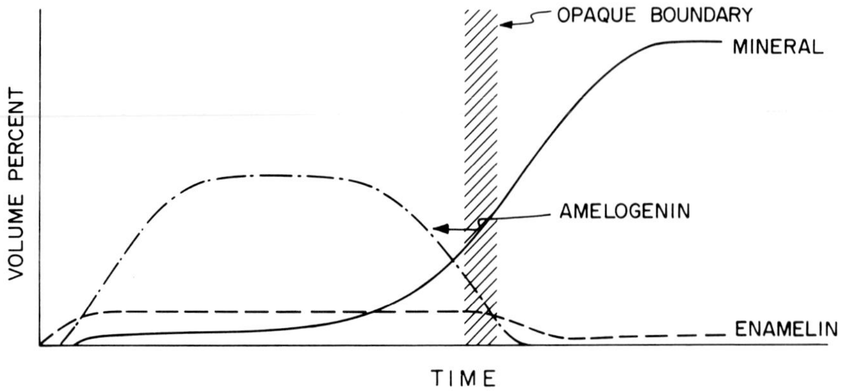

Figure 9.16. Schematic illustration of the trends, in volume percent, of the mineral, amelogenin, and enamelin constituents with time during enamel formation. The visable so-called opaque boundary is the zone in which rapid changes in the proportions of these constituents take place.

basic pattern that has been well documented is that the matrix breaks down and is removed from the tissue (Deakins 1942). It is replaced by water and then by growing crystals (Hiller et al. 1975). Many of the details of these changes have been determined by most impressive microdissection and microanalytical techniques, exemplified by the studies of Hiller et al. (1975), Robinson and Kirkham (1984), and Deutsch et al. (1985). The basic trends are schematically illustrated in Figure 9.16. Protein amino acid compositional changes show that whereas the quantitatively abundant matrix components at the very earliest stage are enamelins (Robinson et al. 1977), amelogenins rapidly accumulate to the point that they comprise 15–25 weight percent of forming enamel (Glimcher et al. 1977; Deutsch et al. 1985). Even as this is happening, amelogenins are breaking down into lower molecular components and are completely absent in mature enamel. The enamelin content of the tissue remains more or less constant (Termine et al. 1980a) except during the final maturation phase when some break down and removal occurs. The end result is that enamelins are the major matrix constituents in mature enamel, and in bovine teeth constitute about 0.1% by weight (Glimcher et al. 1977).

The crystals are continuously forming during the entire process. X-Ray diffraction shows that the more recently formed mineral is less well ordered than the more mature mineral (Nylen et al. 1963). The crystals rapidly elongate along their *c* axes and become wider and thicker. During the forming stages crystals are always separated from each other by organic material and have very well-defined and uniform shapes (Nylen et al. 1963). However, concurrent with the massive efflux of protein that occurs at a boundary, which is even visible to the naked eye (opaque boundary), the crystals undergo a growth spurt (Deutsch and Pe'er 1982), lose their regular shapes, and eventually form a massive structure with a fused appearance (Fig. 9.17). It seems likely that this occurs as a result of the organic tubules losing their coherent structure (Weiner 1986), although this too has not been proved.

9.8 A Perspective

The crystals in enamel and enameloid are similar, sharing unique properties not noted in other biogenic dahllite crystals. As the crystals are the products of the combined activities of cells and extracellular matrix components, the presence of similar crystals in enamel and enameloid is a good indication that similar crystal formation processes are involved in both tissues. The acidic enamelins are also present in both tissues and there is now a well-substantiated body of information that directly implicates them in crystal growth regulation. Therefore, it seems likely that they are responsible in some, as yet, undetermined ways for the production of the characteristically shaped crystals and for their organization into parallel arrays. The only major difference in terms of extracellular tissue organization is the presence of amelogenins in forming enamel and collagen in forming enameloid. From the observed alignment in enameloid of the long axes of the crystals with the collagen fibril axes, we can infer that collagen is somehow actively involved in directing crystal formation. Viewed from the broad perspective of matrix-mediated mineralization processes in general (Chapter 3) it seems reasonable to assign the collagen a "framework" role in which the acidic macromolecules are specifically positioned on the framework and in turn direct crystal formation. Comparable information about amelogenin structure and its relationship to the crystals is not available. We can only suggest that in the absence of solid evidence to the contrary,

Figure 9.17. Scanning electron micrograph of the more mature coronal zone of the continuously growing rat incisor where the enamel cannot easily be separated from the dentin. Note that the rods and interrods have partially fused. Scale bar 10.0 μm. Courtesy of Dr. A. Jodaikin.

a good working hypothesis can be that amelogenins fulfill a function in enamel similar to that of collagen in enameloid.

It has been suggested by Fearnhead (1979) that the collagen-associated mineralization of enameloid actually occurs in all enamel in the zone close to the boundary between enamel and dentin. This is known as the "dentinoenamel junctional region." This attractive suggestion, we feel, emphasizes the importance of exploiting enameloid to probe the basic processes involved in controlled crystal growth of these tissues. The fact that the collagen structure is, at this stage, easier to recognize than amelogenin provides a considerable advantage to the investigator interested in understanding the molecular organization of this tissue.

10

Some Nonskeletal Functions in Biomineralization

The functions of mineralized hard parts are often self-evident. In many of the tables throughout the book we note the assigned or very often assumed functions of many different mineralized bodies. Often, however, assumed functions do not stand up to closer examination. A good example is the study of the cells of the hepatopancreas of gastropods (Howard et al. 1981). These glands have numerous cells containing intracellular mineralized granules. It was generally assumed that they all functioned as transient storage sites for calcium ions, until it was found that a subpopulation forms granules of a different type, which are used for heavy metal detoxification. Granules can be used in other ways as well. Certain polychaete worms, for example, strengthen their muscles by packing them with granules (Gibbs and Bryan 1984). Spicules are also commonly formed by many organisms and their functions are often not understood. They tend to have elaborate morphologies and mineralogies that are species specific, implying that they do perform specialized functions. These are just a few of many examples in which the functions of mineralized bodies still need to be determined.

In this chapter we describe four different cases in which the functions are fairly well established. They have been investigated in some detail and, thus, provide good guidelines as to the various approaches by which function can be investigated. Some gravity receptors have been closely examined with respect to neuroanatomy and function, but not with respect to the specific adaptations of structure and mineralogy of the ubiquitous "heavy bodies." Studies of biologic magnetic field receptors, in contrast, have focused on the mineral, and virtually nothing is known about the neuroanatomy. The molecular structure of the iron storage molecule ferritin is known with a resolution of a few Ångstroms. Ferritin provides us with a glimpse of the insights that can be gained into function from such detailed structural information. Finally, some studies on the control of proteins on ice crystal formation represent the first application of the powerful techniques of molecular biology to determining function in biomineralization. These are undoubtedly the forerunners of many function-oriented studies using molecular biological techniques.

10.1 Gravity Perception

Members of three eukaryotic kingdoms, the plants, proctoctists, and animals, are able to perceive gravity. Table 10.1 is a compilation from the literature of most of the documented cases of organisms that have gravity receptors and shows how widespread this phenomenon is. It is of particular interest that the "modes of life" of organisms possessing gravity sensors range from sessile to highly mobile (Table 10.1). Plants are the major group of sessile organisms. Their gravity sensors are in their root systems and function to orient the growing root tips downward into the soil (Rawitscher 1932; Nemec 1964). This phenomenon is known as geotropism. The one sessile animal known to have a gravity sensor is a marine hydrozoan (Cnidaria) that anchors itself into the mud with rootlike "holdfasts." The two proctoctists listed in Table 10.1 are both sessile, being anchored by their rhizomes to the substrate. All the remaining organisms known to have gravity sensors are mobile animals. They presumably use the field of gravity as a frame of reference for orientation. Among the highly mobile animals the basic gravity receptor system has been further developed to also detect linear as well as angular accelerations. Their receptor systems are generally referred to as the vestibulary apparatus. These are found in the highly mobile group of cephalopod mollusks, the coleoids (cuttlefish, squids, octopii) (Barber 1968), and the vertebrates (Carlström 1963; Marmo et al. 1981, 1983).

A survey of the literature demonstrates a simple, but somewhat surprising fact: all the known biological gravity sensor systems employ the same basic principle. Downward gravity forces exerted by specially formed "heavy" bodies are detected by the surrounding tissues. Movement of the whole organism realigns these tissues relative to the forces of gravity exerted by the heavy bodies and it is this movement that is detected and interpreted in terms of a change in orientation (Ross 1983, 1984). To improve their effectiveness, the heavy bodies are generally mineralized. The major exception is the plants, which use starch granules (Rawitscher 1932; Nemec 1964). The specific gravity of starch is 1.5 (Schröter et al. 1975) as compared to about 1.0 for the fluid in which they are located.

The inventory of minerals used by organisms for gravity perception (Table 10.1) is still far from complete. Their specific gravities range from 2.2 in gypsum to 4.5 in barite. We know even less about their structures and how they are formed. It is noteworthy that almost all the minerals formed by animals for gravity perception are calcium minerals although the reasons for this are not known. The presence of calcium is not a necessity, as certain polychaete worms, crustaceans, and sharks actually make use of sedimentary particles obtained from the environment for their gravity sensors. In the polychaete worm, *Arenicola,* the detrital grains are coated with a chitinoid material (Wells 1950), apparently to blunt the sharp edges of the grains. In some sharks the detrital material is already present in the embryos that are still in the amnion (Fitsch, personal communication to H. A. Lowenstam).

The manner in which these particles enter the embryo is most puzzling, as the amnion is completely free of detrital material.

Three different types of gravity sensors can be recognized.

Type 1. Detection of the "heavy body" movements is by cells without cilia and without any connections to surrounding nerve cells. The response is confined to the local tissue surrounding the sensor. Plants with their intracellular starch granules are the best documented examples of this type. To emphasize that this type of sensor is not confined to plants, we describe in more detail the gravity-sensing organ of the sessile hydrozoan, *Corymorpha palma* (or more graphically known as the "fairy palm") as reported by Campbell (1972). It is a solitary hydroid polyp that lives on marine mud flats. Its stalk is anchored in the mud by "holdfasts." The tip of each holdfast contains an extracellular cavity (statocyst) formed by the plasma membranes of two endodermal cells. A heavy body, the statolith, is located within the cavity. The endodermal cells contain no cilia and there are no known connections to nerve cells. In fact even severed tips respond to gravity, and presumably the endodermal cells themselves transmit the signal to other cells in the local area.

Type 2. The statocyst is a fluid-filled cavity internally lined with ciliated receptor cells that all appear to be identical or undifferentiated. The cilia detect the movements of the heavy body. There are a large number of small heavy bodies in the statocyst of many adult animals and these are called statoconia. The signal is transmitted by mechanosensory transduction to the nervous system (Wiederhold et al. 1986). The gastropod, *Aplysia,* is a good example of this type. The ciliated cells of its statocyst all seem to be identical with no evidence of differentiation. The statoconia have round shapes and are mineralized by an as yet unidentified calcium mineral. They have a concentrically layered internal structure made up of alternating layers of mineral and organic material. The exact location of statocyst formation in *Aplysia* is not known. In other mollusks the statoconia are thought to form either extracellularly within invaginations of the supporting cells of the statocyst (Tsirulis 1974), or within vacuoles inside these cells (Cragg and Nott 1977; Kuzirian et al. 1981) or in vacuoles on their surfaces (Geuze 1968).

Type 3. The third type of sensor is a highly differentiated and specialized system, functioning both as a gravity sensor and as a sensor of linear and angular accelerations. Hair cells act as sensors of movement and the signals are transmitted to the central nervous system. A well-documented example of a vestibulary apparatus is that of rats (Salamat et al. 1980). The heavy bodies are termed otoconia rather than statoconia as they were originally thought to be involved in sound perception. As with the statoconia of *Aplysia,* the rat otoconia have a large range of sizes. In vivo, however, they are loosely arranged according to size into a "solid" structure. The different sized otoconia apparently allow the system to be finely tuned for detecting a large range of linear accelerations. The otoconia therefore fulfill very specific functions and, not surprisingly, have unusual structures. In spite of the fact that their external morphologies superficially resemble those of single crystals, high-resolution TEM photographs show that they are polycrystalline bodies (Mann et al. 1983a) in which the crystals are sheathed by organic material containing acidic glycoproteins (Salamat et al. 1980). Note too that the otoconia of the mollusk, *Nautilus* (Lowenstam et al. 1984), and those of the mouse (Nakahara and

Table 10.1 The Distribution, Localization, and Mineralogy of the Mineralized Hard Parts of Gravity Receptors

Phylum	Lower taxa	Location	Hard parts	Mineralogy	Mode of Life	References
Protoctista						
Gamophyta	*Spirogyra*	Rhizoid	Statoconia	Barite	Sessile benthos	Kreger and Boere (1969)
Charophyta	*Chara*	Rhizoid	Statoconia	Barite	Sessile benthos	Schröter et al. (1975)
Plantae	General	Root caps	Amyloplasts	Starch*	Sessile	Rawitscher (1932); Nemec (1964)
Animalia						
Cnidaria (Coelenterate)	Hydrozoa					
	Corymorpha	Holdfast	Statoconia	Indeterminate	Sessile benthos	Campbell (1972)
	Obelia	Statocysts	Statoconia	Amorphous Ca-Mg-phosphate	Plankton	Chapman (1985)
	Lovenella	Statocysts	Statoconia	Amorphous Ca-Mg-phosphate	Plankton	Chapman (1985)
	Phyalidium	Statocysts	Statoconia	Amorphous Ca-Mg-phosphate	Plankton	Chapman (1985)
	Aglantha	Statocysts	Statoconia	Amorphous Ca-Mg-phosphate	Plankton	Chapman (1985)
	Chiropsalmus	Statocysts	Statoliths	Amorphous Ca-Mg-phosphate	Plankton	Chapman (1985)
	Scyphozoa					
	Auralia	Statocysts	Statoconia	Gypsum	Plankton	Spangenberg and Beck (1968)
	Chrysaora	Statocysts	Statoconia	"Gypsum"(?)	Plankton	Chapman (1985)
	Cubozoa					
	Chiropsalmus	Statocysts	Statoconia	"Gypsum"(?)	Plankton	Chapman (1985)
Ctenophora	*Pleurobrachia*	Apical sense organ	Statoconia	Amorphous Ca-Mg-phosphate	Plankton	Chapman (1985)
Platyhelminthes	*Turbellaria*	Statocyst	Statoliths	Indeterminate	Vagrant benthos	Summarized by Hyman (1951)
Rhynchocoela (Nemertinea)	*Ototyphlonemertes*	Statocyst	Statoliths	Indeterminate	Vagrant benthos	Bürger (1895)
Brachiopoda	*Lingula*	Statocyst	Statoconia	Indeterminate	Benthos, sedentary burrowers	Yatsu (1902)

Annelida	Polychaeta					
	Arenicola	Statocyst	"Statoconia"	Detrital particles with chitinoid cover	Benthos, sedentary burrowers	Wells (1950)
	Orbiniidae	Statocyst	Statoconia	Indeterminate	Benthos, sedentary burrowers	Barnes (1980)
	Terebellidae	Statocyst	Statoconia	Indeterminate	Benthos, sedentary burrowers	Barnes (1980)
	Sabellidae	Statocyst	Statoconia	Indeterminate	Benthos, sedentary burrowers	Barnes (1980)
Mollusca	Monoplacophora	Statocyst	Statoconia	Indeterminate	Vagrant benthos	Lemche and Wingstrand (1959)
	Gastropoda					
	Aplysia	Statocyst	Statoconia	Indeterminate carbonate	Vagrant benthos	Morton (1958) Wiederhold et al. (1986)
	Heteropoda	Statocyst	Statolith	Calcite?	Plankton	Barber (1968); Tschachotin (1908)
	Bivalvia	Statocyst	Statoconia	Indeterminate	Vagrant benthos	Buddenbroek (1915)
	Scaphopoda	Statocyst	Statoconia	Indeterminate	Vagrant benthos	Morton (1958)
	Cephalopoda					
	Nautilus	Statocyst	Statoconia	Aragonite and ACP	Nekton	Lowenstam et al. (1984)
	Coleoidea	Statocyst	Statoliths	Aragonite	Nekton (highly mobile)	Lowenstam et al. (1984)

Table 10.1 (*Continued*)

Phylum	Lower taxa	Location	Hard parts	Mineralogy	Mode of Life	References
Arthropoda	Crustacea	Statocyst	Statoconia	Detrital particles	Vagrant benthos	Barnes (1980)
	Mysidacea	Statocyst	Statolith	Fluorite, vaterite	Nekton	Lowenstam and McConnell, (1968): Ariani et al. (1983)
	Copepoda	Statocyst	Statoliths	Indeterminate Ca mineral	Nekton	Lowenstam (unpublished)
	Isopoda	Statocyst	Statoliths	Indeterminate	Vagrant benthos	Rose and Stokes (1981)
Echinodermata	Holothuroidea					
	Elpidiidae	Statocyst	Statoconia	Indeterminate	Vagrant benthos	Théel (1876); Danielsson and Koren (1882)
	Molpadia	Statocyst	Statoconia	Indeterminate	Vagrant benthos	Théel (1876; Yamanouchi (1929)
Chordata	Cyclostomata	Otocyst	Otoconia and otoliths	ACP	Nekton	Carlström (1963); Lowenstam (unpublished)
	Elasmobranchii	Otocyst	Otoconia	Aragonite, calcite, monohydrocalcite, ACP	Nekton	Carlström (1963) Lowenstam and Fitch(1978)

Teleostei	Otocyst	Otoliths and Otoconia	Aragonite, calcite, vaterite	Nekton	Carlström (1963); Lowenstam and Fitch (1988)
Amphibia	Otocyst	Otoconia	Aragonite, calcite	Nekton	Carlström (1963); Marmo et al. (1983)
Reptilia	Otocyst	Otoconia	Aragonite, calcite	Nekton	Carlström (1963); Marmo et al. (1981)
Aves	Otocyst	Otoconia	Calcite	Highly mobile	Carlström (1963)
Mammalia	Otocyst	Otoconia	Calcite	Highly mobile	Carlström (1963)

*Starch fulfills the function of a mineral in this gravity receptor.

Bevelander 1979) are also polycrystalline. The statoconia of the scyphozoan medusae, *Auralia,* are each composed of single calcium sulfate crystals (gypsum). With growth these also "fuse" into polycrystalline aggregates (Spangenberg 1976).

The limited information available on the formation of otoconia and statoconia suggests that some may be formed in vacuoles within cells. They are then exocytosed and transported to the cyst where further growth takes place (Spangenberg 1976; Salamat et al. 1980; Dunkelberger et al. 1980).

In teleost fishes the otocyst is really involved in sound reception, in addition to gravity sensing and detecting acceleration (Morris and Kittleman 1967; Sand 1974). The otocysts contain single mineralized otoliths. The teleost fishes, as with all nonplacental vertebrates, have three separate otic chambers: the sacculus, the utriculus, and the lagena. The sacculus is usually the largest and its otolith is composed of aragonite. These otoliths are widely used in taxonomic studies and for determining the age of the fish by counting their growth rings (Pannella 1974). The second largest otic chamber is the utriculus, which also has an aragonitic otolith. In contrast to the other two chambers, the otolith in the third and usually smallest chamber, the lagena, is always composed of vaterite (Lowenstam and Fitch 1981). Eighteen species were examined and only the taxonomically quite distinct sun fishes were found to be exceptions to this rule; all chambers contain vateritic otoliths. Figure 10.1 shows the ultrastructural fabrics of otoliths from the three chambers.

In amphibians, two chambers (the sacculus and lagena) and the endolymphatic sac have aragonitic otoliths, whereas the otolith in the third chamber (utriculus) is calcitic (Marmo et al. 1983). In the lizard, *Podarcis sicula,* the utriculus and lagena contain calcitic otoliths, and the sacculus has an aragonitic otolith, with traces of calcite. Crystals of aragonite also occur in the endolymphatic sac (Marmo et al. 1981). The observed mineralogical differences in the above examples all involve the three polymorphs of calcium carbonate and, in the large majority of cases, the observed mineralogical differences between otic chambers are highly conserved. We therefore suspect, but as yet have no proof, that the mineralogical differences are functional. An alternative, but less likely possibility is that the environments of crystal growth are different and, hence, one polymorph forms instead of another.

10.2 Functions of Biologically Formed Magnetite Crystals

The discovery of biogenic magnetite in the teeth of chitons by Lowenstam (1962a) was greeted with both skepticism and amazement, as it was then believed that magnetite could not be formed under temperatures and pressures found in the biosphere. The fact that these mollusks were able to form magnetite was subsequently firmly established (Lowenstam 1967), but was generally regarded as a fascinating curiosity. This changed, however, when Blakemore (1975) discovered magnetotactic bacteria. Much interest in the subject was aroused and it rapidly blossomed into a very active field. The book *Magnetite Biomineralization and Magnetoreception* edited by Kirschvink, Jones, and MacFadden contains most of what was known about this field up to 1985.

Unlike all other biogenic minerals magnetite can be detected remotely using a magnetometer. The development of the SQUID (superconducting quantum device) magnetometer (Fuller et al. 1985) made possible the detection of even picogram quantities of magnetite. Although very small particles of biologically formed magnetite can be detected in this way, the possibility of error due to contamination greatly increases. Magnetite has been identified in the following organisms using only the magnetometer: honeybee (Gould et al. 1978), various crustaceans (Buskirk and O'Brien 1985), monarch butterfly (MacFadden and Jones 1985), amphibians and reptiles (Perry et al. 1985), many birds (Presti 1985), bats (Buchler and Wasilewski 1985), cetaceans (Bauer et al. 1985), primates (Kirschvink 1980, 1981), and questionably humans as well (Kirschvink 1980; Baker et al. 1983). The presence of biogenic magnetite, assuming it is always an essential part of any magnetic field receptor, can also be inferred from behavior experiments in which the ability of animals to detect the presence of a magnetic field is determined. However, isolation and identification by X-ray diffraction, electron diffraction, or Mössbauer spectroscopy are still the most reliable approaches. Magnetite has been identified directly in one species of magnetotactic bacteria (Frankel et al. 1979), a species of algae (Torres de Araujo et al. 1986), chiton teeth (Lowenstam 1962a), fish (Walker et al. 1985; Mann et al. 1988b), the green turtle (Perry et al. 1985), and homing pigeons (Walcott et al. 1979).

Organisms are known to use magnetite for three primarily different functions:

1. It is formed as a metabolic end-product by some iron-reducing bacteria (Lovley et al. 1988).
2. It is used to harden the outer surface of chiton teeth to allow them to scrape rocks for endolithic food extraction purposes (Lowenstam 1962a).
3. Magnetite crystals are used to detect the earth's magnetic field.

Only the latter is discussed in this section.

Magnetotactic bacteria form linear chains of magnetite crystals each of which is sheathed and held together by a lipid bilayer admixed with proteins (Gorby et al. 1988). The entire structure is called a magnetosome (Fig. 10.2). The crystals have various, often species-specific shapes and, most significantly, are always in the size range around 0.1 μm, with the result that they form single-domain crystals. In this way the entire magnetosome acts as one "bar magnet" (Kirschvink 1988). The function of the magnetosome is to align the bacterium along magnetic field lines; then using the flagellae it propels itself in this preferred direction. Thus, magnetotactic bacteria in the northern hemisphere are north seeking and those in the southern hemisphere are south seeking. The bacteria, however, live in mud and the apparent usefulness of the magnetic guidance system is to allow them to follow magnetic lines of declination so that they can move toward their optimal habitat. This is at a depth in the mud at which the oxygen tension is low. This is also the environment most conductive to the formation of inorganic magnetite (reviewed in Frankel 1984).

The magnetite-containing euglenoid algae live in an environment similar to that of the magnetotactic bacteria and are also mobile. However, they are much larger than bacteria (20 $\times$ 12 μm) and have numerous (about 3000) chains of single-domain-sized magnetite crystals (Fig. 10.3). It has been suggested that these algae also use their magnetosomes as a magnetic guidance system (Torres de Araujo et al. 1976). The presence of so many magnetite crystals, however, raises the pos-

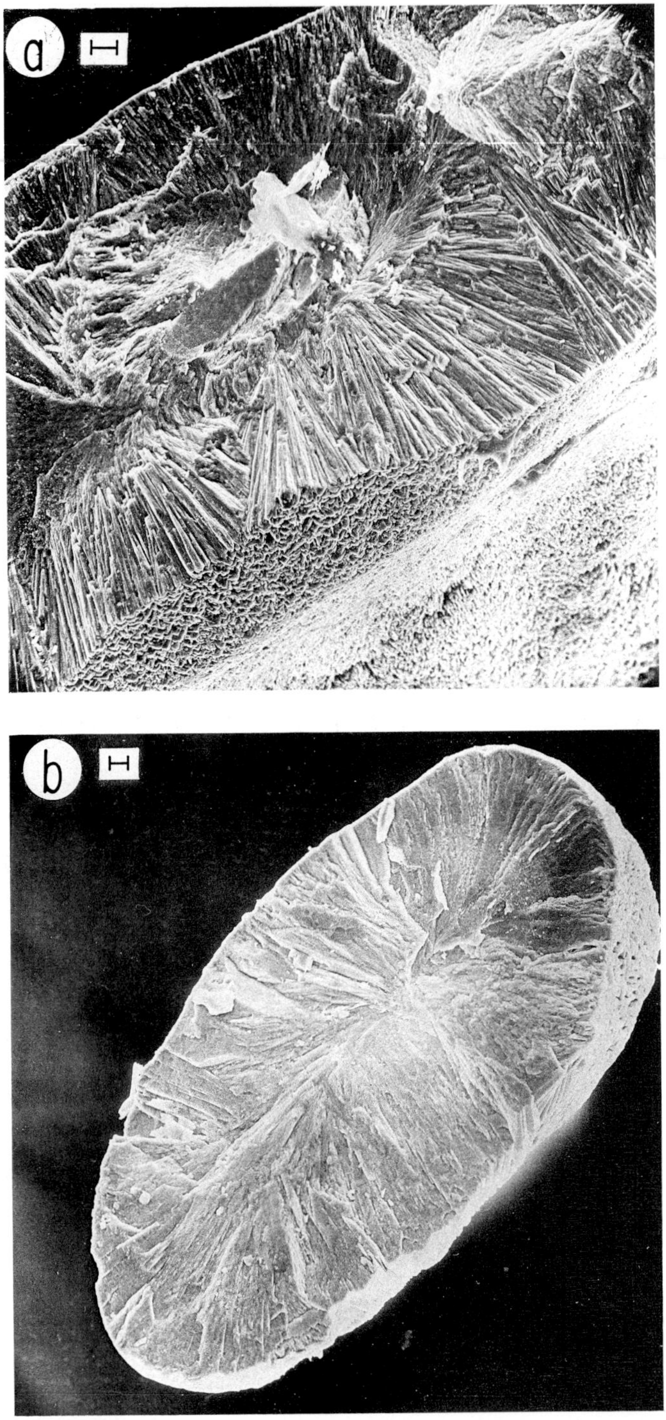

Figure 10.1. Scanning electron micrograph of the crystal fabrics of fractured otoliths from Tarleton beania crenularis (a) aragonitic sagitta, (b) aragonitic lapillus.

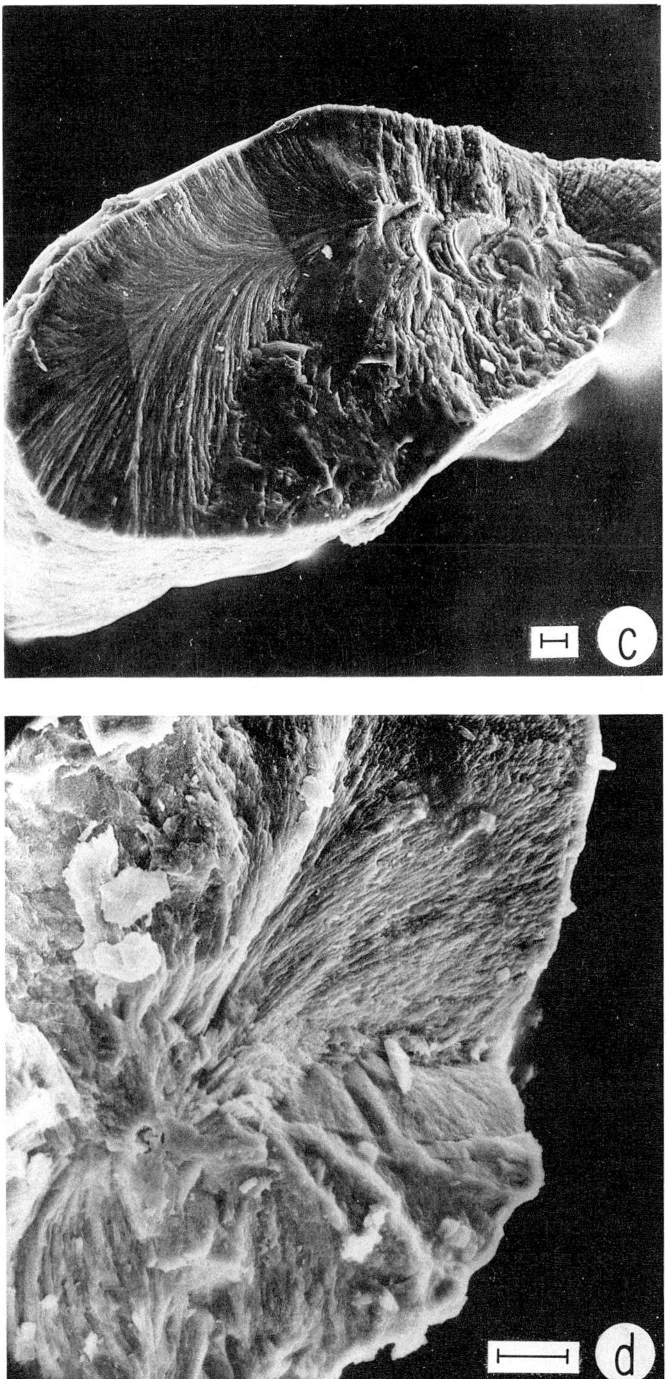

Figure 10.1. (*Continued*) (c) vateritic asteriscus and (d) vateritic otoconium of the sunfish *Mola mola*. Scale bars: 10 μm.

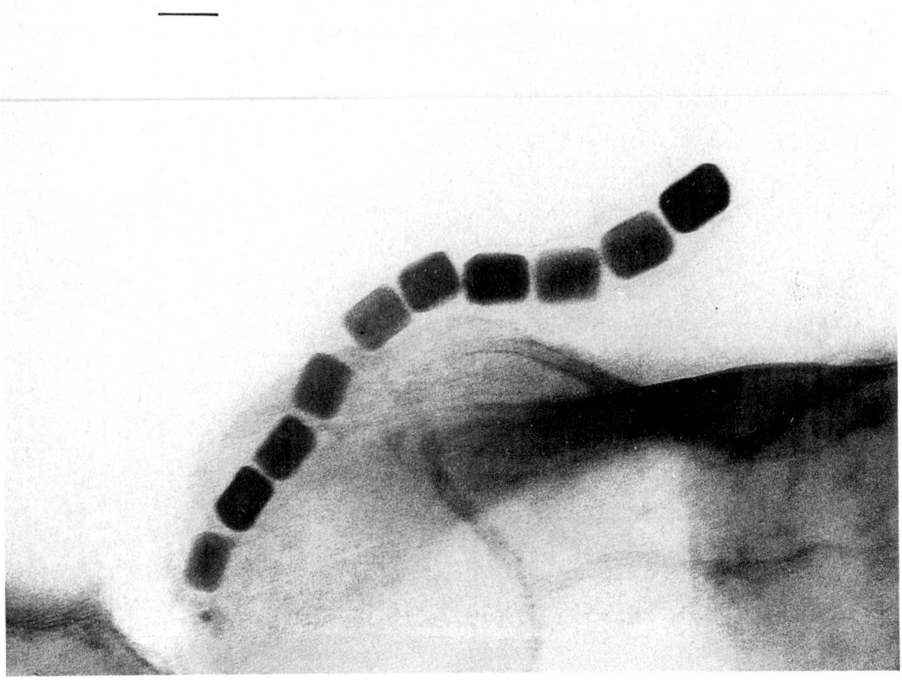

Figure 10.2. Transmission electron micrograph of the magnetosome of a magnetotactic bacterium. Scale bar: 100 nm. Courtesy of Dr. J. Stolz.

sibility that they may, in addition, function as a gravity receptor in the manner described in the previous section. Note that this is unlikely to be the case for the magnetotactic bacteria as their overall densities, including the magnetosome, are close to unity (Blakemore et al. 1981).

Animals with magnetic field receptors for the most part have one common trait: they have known abilities to navigate relatively long distances. This may, however, reflect the fact that these animals have been preferentially singled out for study. In only one case, the sockeye salmon *Oncorhyncus tshanytcha,* has an actual magnetosome been found (Kirschvink et al. 1985). In the green turtle the magnetite crystals or possibly crystal aggregates are much larger (10–50 μm) than those of the salmon and are spherical in shape (Perry et al. 1985).

The most direct and convincing evidence that these animals use the magnetic field for navigation is obtained from conditioning experiments in which, for example, a change in magnetic field intensity is reflected in a change in behavior. Using this approach it has now been established that many different fish can detect the earth's magnetic field (e.g., Tesch 1974; Kalmijn 1978; Quinn 1980). Of particular interest is the discovery that honeybees can detect and orient themselves in the direction of the earth's magnetic field lines, but cannot distinguish the polarity directions (Towne and Gould 1985).

Much, of course, still remains to be done in this field. One area of particular importance in animals is the determination of the anatomy and particularly the

Figure 10.3. Transmission electron micrographs of a magnetotactic alga showing (bottom) chains of magnetite crystals just inside the cell. The cell itself is approximately 12 × 20 μm. Scale bar: 1 μm. (Top) Higher magnification view of the magnetite crystal chains. Scale bar: 3.3 μm. Courtesy of Dr. R. Frankel.

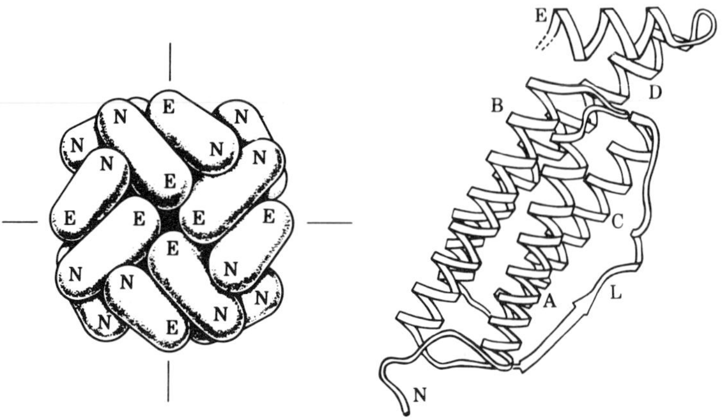

Figure 10.4. 1. Schematic representation of the ferritin molecule viewed down a molecular fourfold axis. The hydrophobic channel is located at the center of the structure. The N-terminal of the polypeptide chain lies close to the end labeled N, and the E-helix close to that labeled E. 2. Ribbon diagram of the α-carbon backbone of an apoferritin subunit. Reproduced from Ford et al. (1984) by courtesy of the Royal Society, London.

neuroanatomy of these systems, which in turn will provide the basis for a systematic investigation of this most interesting phenomenon.

10.3 Ferritin: An Iron Storage Macromolecule

Iron is an element essential for all living organisms. Free iron is, however, potentially toxic. Thus, cells have a need to store it safely, but in a form that is readily available for remobilization when the need arises. In the animal, fungi, and plant kingdoms this function is carried out by ferritin. Ferritin is composed of a multisubunit protein shell with an outside diameter of some 12.5 nm, which surrounds an inner core containing iron minerals (Ford et al. 1984). Bacteria also have an analogous molecule called bacterioferritin. The protein that makes up the shell of bacterioferritin is quite different from that of ferritin and comparative amino acid sequence studies suggest that it evolved independently (Tsugita and Yariv 1985). The iron core of bacterioferritin also differs in mineralogy from ferritin, primarily in its elevated phosphate content and poor crystallinity (St. Pierre et al. 1986). In this section we focus on the better studied mammalian ferritins, and particularly on the molecular organization of a ferritin derived from horse spleen. This is probably better understood than any other biological mineralization system primarily because Harrison and her colleagues have solved the crystal structure of the protein shell to 2.8 Å resolution. It provides a unique opportunity to examine the molecular structure of ferritin in terms of its functions: the sequestering, deposition, and remobilization of iron.

The structure of horse spleen apoferritin, the name given the protein shell, is described by Ford et al. (1984). It is made up of 24 identical protein subunits each with a molecular weight of around 20,000. Figure 10.4a is a schematic representation of how the protein subunits are arranged to form the protein shell and Figure 10.4b shows the α-carbon backbone structure of a single protein subunit. The space inside the protein shell can accommodate up to 4,500 Fe^{3+} atoms, although on the average only some 3,000 or so are usually present. Chemically the iron core consists of a ferric–oxyhydroxy–phosphate complex. X-ray diffraction patterns of the core show that the mineral is a microcrystalline ferrihydrite (Towe and Bradley 1967; Massover and Cowley 1973). High-resolution electron microscopy shows that the cores consist mainly of single-domain crystallites associated with apparently non-crystalline regions (Mann et al. 1987). The phosphate, it seems, is not part of the ordered domain. In fact, the phosphate contents of the iron core vary considerably and it has been noted that the crystallinity of the core is in inverse proportion to the phosphate content (St. Pierre et al. 1986). For detailed descriptions of the structures of both the apoferritin and the iron core see Ford et al. (1984), Harrison (1986), and, in particular, Harrison et al. (1987).

The protein shell can self-assemble and even in vitro can take up iron (Treffry et al. 1984). Thus, iron atoms find their own way into the protein shell and at the same time become oxidized to Fe^{3+} during ferrihydrite formation. An examination of the structure of the inner surface of the protein shell shows that it contains a large number of acidic amino acids (Ford et al. 1984). If these carboxy groups are modified, the iron uptake potential of apoferritin is eliminated (Wetz and Crichton 1976). Thus, as with many other biomineralization processes, carboxyl ligands appear to play a key role. Comparisons of the amino acid sequences of the protein subunits from different species show that three glutamic acid residues are conserved and may well be involved in iron core nucleation. There is, as yet, no direct evidence to confirm this. Once a small ferrihydrite crystal is established inside the core it will successfully compete for iron against the various other possible protein-binding sites, and continue to grow (Clegg et al. 1980).

The molecular structure of the protein shell reveals the presence of two different types of channels that connect the inside of the shell with the outside. There are six equivalent channels located around the fourfold symmetry axis (Figure 10.4a) and these are lined with hydrophobic leucine residues. There are eight channels located around the threefold symmetry axes, which are lined with hydrophilic amino acid residues. These channels are obvious candidates for entry and exit points of iron and other molecules. The channel favored by Harrison et al. (1987) for iron entry is the hydrophilic one. Spectrophotometric evidence suggests that at least under the in vitro conditions used, Fe(II) is oxidized inside the channel and then migrates into the core as Fe(III) (Treffry and Harrison 1984; Harrison et al. 1986). The hydrophobic channel may function in the iron release process although this too has not been confirmed. In vitro experiments do show that release of iron occurs in the presence of reductants and small chelators (Aisen and Listowsky 1980; Clegg et al. 1980). How they enter and leave the core still remains to be determined.

Temporary storage of ions in the solid state is common and widespread particularly in the animal kingdom. Almost all known cases involve lipid bilayer vesicles usually located inside the cell. Ferritin with its protein shell is a major exception in this regard. However, it is significant to note that detailed examination at the molecular level of the manner in which it functions emphasizes the roles of various specifically positioned carboxylate groups in both sequestering iron and in nucleation. Although much still remains to be learned about the manner in which it functions, one lesson is clear: information at the molecular level is essential for understanding the details of mineralization.

10.4 Biological Control over Ice Formation

Organisms usually cannot tolerate the presence of ice crystals in their tissues and certainly not within their cells. Thus, various organisms living in very cold environments have learned to control ice crystal formation and, in some cases, to manipulate it to their advantage. Two examples of this activity have been relatively well studied. Certain plant bacteria deliberately induce ice crystals to form in order to destroy the plant tissue and provide themselves with an environment conducive for growth. Fish living in the polar or subpolar regions of the oceans have evolved the ability to prevent ice crystals from forming within their tissues.

Biological control over ice formation is a subject that is not normally regarded as being an integral part of the field of biomineralization. This brief review, however, clearly shows that the processes involved have much in common with other biologically controlled mineralization processes and, furthermore, much can be learned from these studies that is directly relevant to biomineralization.

10.4.1 Induction of Ice Crystals by Certain Plant Bacteria

The discovery of ice nucleation activity by bacteria began in a rather circumspect manner when Schnell and Vali (1972) were searching for possible sources of atmospheric ice nuclei that initiate the phase transition to ice at temperatures between 0°C and −10°C. Small volumes of water free of heterogeneous ice nuclei will generally not freeze until the temperature approaches −40°C. The presence of dust particles can reduce this temperature to about −10°C and even the cloud seeding agent, sodium iodide, causes ice nucleation only at −8°C (Phelps et al. 1986 and references therein). Schnell and Vali (1972) showed that one source of nuclei active in the 0°C to −10°C range was from actively decomposing plant vegetation. Arny and co-workers (1976) first demonstrated that the ice nucleating particles in the vegetation were produced by the bacterium *Pseudomonas syringae.* Since then two other bacteria, *Pseudomonas fluorescens* and *Erwinia herbicola,* have been found to have ice nucleation ability (Lindow 1982). These bacteria have a worldwide distribution and are found in soils, on plant leaf surfaces, and in leaf mulch. There is much interest in these bacteria not only as a source of atmospheric ice nuclei, but because they are responsible, in part, for frost damage to agricultural crops.

Sprang and Linlow (1981) isolated the gene responsible for the ice nucleation activity of *P. syringae.* It codes for a protein that is located in the outer cell membrane. *E. herbicola* is also able to release the protein into the growth medium (Phelps et al. 1986). Green and Warren (1985) sequenced the gene and found that it has an interesting structure in terms of its predicted function. It contains 122 imperfect repeats of the consensus polypeptide Ala-Gly-Tyr-Gly-Ser-Thr-Leu-Thr. All the repeats are contiguous. Significantly four of the eight amino acids in the repeat have hydroxyl groups as part of their side-chain structures. It is also interesting to note that the only other amino acid sequence determined of a protein known to be intimately involved with mineralization is from the sea urchin larval skeleton (Sucov et al. 1987). It too has a repeating structure (Chapter 8).

To test the hypothesis that the repeating units act cooperatively to bind water and induce ice nucleation, Green and Warren (1985) performed an elegant experiment. They isolated genes for the ice nucleation protein in which portions of the repeating structure were missing. They found that the deletions resulted in considerably lower ice nucleation temperatures and, therefore, demonstrated that indeed the repeating units act cooperatively to produce an efficient "high temperature" ice nucleating substrate. To our knowledge this is the first use of this powerful technique in biomineralization. Green and Warren (1985) also indicate that the 48-residue periodicity corresponds to a sixfold repetition of the octapeptide building block. This could conceivably be folded into a structure that could mimic the hexagonal structure of ice. Information on the protein–ice crystal spatial relations is needed to test this idea.

10.4.2 Inhibition of Ice Crystal Formation by Glycoproteins from Polar Fish Blood

Polar fishes live in waters that are near their freezing points and are covered with ice for most of the year. They avoid freezing by introducing a unique assemblage of glycoproteins into their blood (DeVries and Wohlschlag 1969). These antifreeze glycoproteins together with the salts normally present in the blood of fishes reduce the serum freezing point sufficiently to prevent ice crystals from forming.

The antifreeze glycoproteins in the blood of the Antarctic cod *Trematomus borchgrevinki* range in molecular weight from 2,600 to 33,700 (DeVries 1982). They are composed in part of unusual repeating tripeptides, Ala-Ala-Thr, with the threonine residue being linked covalently through its hydroxyl group to a disaccharide. The latter contains six hydroxyl groups. This or similar repeating structures are always found in all Antarctic notothenoids and are present only during the winter in northern coldwater fishes (DeVries 1982). Conformation studies show that the antifreeze glycoproteins are neither entirely random coil polymers nor rigid rods, but have locally defined structures, including a threefold left-handed helix (reviewed in Feeney et al. 1986). Computer modeling studies are consistent with this interpretation (Atkins 1985).

A key study partially elucidating the manner in which these antifreeze glycoproteins function is that of Knight et al. (1984). They showed that ice crystals grown in the presence of microgram per milliliter quantities of antifreeze glycoproteins change their morphology. Control ice crystals grown in the absence of glycoprotein are elongated along their *c* axes. Those grown in the presence of the glycoproteins are elongated along their *a* axes. This can be attributed to the glycoproteins adsorbing specifically onto the crystal faces more or less perpendicular to the *c* axis. Direct evidence for adsorption has been obtained (Brown et al. 1985). The organic polymers, polyethylene glycol, dextran, and polyvinyl pyrrolidone, do not produce such growth habit modifications. The mechanism by which the antifreeze glycoproteins inhibit crystal growth is not known. The experiments of Knight et al. (1984) raise the possibility that they adsorb onto incipient crystal nuclei and in so doing prevent further growth and increase the chance that they will redissolve. Feeney et al. (1986) list the observations that support this type of growth inhibition mechanism, and Atkins (1985) discusses the type of match that might occur at the molecular level between the glycoproteins and ice crystals.

It is interesting to note that the hydroxyl groups of both the bacterial ice nucleating proteins and the fish antifreeze proteins appear to be the key functional ligands, an observation consistent, of course, with the structure of ice itself. In contrast, the proteins that control calcium carbonate and calcium phosphate mineralization are all dominated by carboxyl ligands.

11

Environmental Influences on Biomineralization

For the most part, the minerals that organisms form do not form abiologically in the environment in which the organism lives. In fact, some of them do not form abiologically anywhere in the biosphere. A striking example is the crystalline strontium sulfate test of the Acantharia, a group of planktonic proctoctists that is found in most of the world's oceans. Seawater is undersaturated with respect to strontium sulfate. Thus, the biological environment of mineral formation is isolated from the surroundings. This is, in fact, a generally observed phenomenon (Chapter 3) and it is, therefore, surprising to find that in a wide variety of organisms, the environment does still in some ways influence these biogenic minerals. This can manifest itself in the particular mineral deposited (for example, the deposition of the calcium carbonate polymorph calcite as opposed to aragonite or vice versa), the concentration of minor and trace elements, the stable isotopic composition of the mineral, and, at the ultrastructural level of the tissue, the distribution of growth lines.

The degree to which the environment can affect biomineralization is a function of how isolated the process is from the outside world. Poorly controlled mineralization processes are expected to be affected more than well-controlled processes, although as we show in this chapter each case must be examined individually. There are examples of a clear environmental effect occurring in one species of a particular genus, whereas another species or even subspecies of the same genus appears to be unaffected (Lowenstam 1954c). The environmental effect can and very often is filtered out by the physiological processes of the organism, which either completely or partially determine the solution characteristics of the microenvironment in which the crystals grow. A lack of appreciation of this fact has caused considerable confusion in the literature, with some investigators concluding that an environmental influence on biogenic minerals does not exist in a whole taxonomic group, because they found that it was absent in one or a number of species

(Bornhold and Milliman 1973; Taylor et al. 1969). The first published documentation of this phenomenon (Lowenstam 1954a) clearly showed that closely related organisms can differ in this respect and that extrapolations and sweeping conclusions are not justified.

The subject of the environmental influences on biogenic minerals provides some thought-provoking insights into biomineralization. Its major importance, however, lies in the fact that the organisms that are affected in this way become, in a sense, monitors of environmental change. As we are dealing with the relatively durable hard parts, this record can be retrieved from unusually well-preserved fossils and can provide invaluable information about past environments. Perhaps the most important application of this type was based on a proposal by Urey (1947). He predicted correctly that the oxygen-18 abundance in marine calcium carbonate shells is temperature dependent and that this could be used to determine the paleotemperatures of ancient oceans, provided the isotopic composition of the water of these oceans can be determined (Urey et al. 1951; Epstein et al. 1951, 1953). Environmental influences on biogenic minerals have resulted in many other applications in fields as diverse as planetary science, geophysics, archaeology, and population dynamics. In this chapter we will describe four related topics that together represent the major body of knowledge of the subject, but by no means the entire subject. They are (1) change in the amount of mineral formed by organisms in different environments, (2) changes in mineralogy from one environment to another, (3) the influence of the environment on trace element content and oxygen isotopic composition, and (4) growth lines as a monitor of environmental stimuli.

11.1 Increase in the Amount of Biogenic Mineral Formed in Marine Warm Waters as Compared to Cold Waters

In several groups of closely related eukaryotes cold and warm water species form the same mineral, but the rate of mineralization and the volume of mineral deposition in the tropical species are noticeably greater than in the cold water ones. The scleractinian corals, a single family of red algae, and the body surface scales of teleost fishes are three examples of this type of influence of the environment on biomineralization.

In the scleractinian corals, hermatypic reef-building species are limited to the circumequatorial surface waters of the subtropics and tropics (see Chapter 5). This results in an abrupt, spectacular increase in species diversity, coupled with a noticeable increase in the rate and volume of aragonite formation in ocean surface waters bounded approximately by the 15°C isotherm for the coldest months of the year. Growth rates in ahermatypic corals do not follow temperature gradients. Instead they appear to depend primarily on each taxon's unique growth rate, which transcends temperature gradients. Note also that the increase of reef-building hermatypic corals from subtropical to tropical waters is small when compared to the magnitude of the change at the 15°C isotherm.

Aragonite is also the only mineral present in the red algae family Peyssoneliaceae. These encrusting algae are cosmopolitan, ranging from Arctic to tropical waters and are particularly common on coral reefs (Denizot 1968). Mineralization occurs intracellularly within the cell wall or extracellularly on the underside of the thallus. In some species it occurs at both sites (Denizot 1968). The extent of mineralization within a species may differ widely between individuals (James et al. 1988). Crystal fabrics in weakly calcified cell walls show a randomly oriented meshwork of acicular aragonite crystals, and on the underside of the thalli discontinuously merged spherulites are present (James et al. 1988). This suggests that biologically induced mineralization occurs throughout. Some cold water species do not calcify, and those that do are less mineralized than temperate water and tropical species. Among the latter, reef-dwelling species are most heavily mineralized and in New Caledonia, for example, may constitute as much as 60% of the reef front carbonate mass (Denizot 1968). Although only limited data of this kind are, as yet, available, it appears that a noticeable increase in the extent of mineralization occurs in temperate as compared to cold waters. This is in contrast to the corals described previously, in which the abrupt increase in mineralization occurs at the temperate–tropical water boundary.

Numerous green algae, certain red algae, and species of one brown alga all form needle-shaped aragonitic crystals (Lowenstam 1955). They are characteristically found in circumequatorial shoal water habitats and, in particular, on reef slopes and in lagoons, where they are a major source of the skeletal aragonite that accumulates in the sediments (Stockman et al. 1967; Neumann and Land 1975). In terms of ocean water temperature, almost all the species are restricted to waters in which the temperature remains above 15–16°C for the coldest months of the year, the same temperature boundary as that of the hermatypic corals. Among green algae, the representative genera are *Halimeda, Udotea, Pennicilus, Neomeris, Cymopolia, Acetabularia, Acicularia, Bornetella,* and *Rhipocephalus. Galaxaura* and *Liagora* are the red algae and *Padina* is the one common mineralizing brown alga. The aragonite crystals are commonly formed extracellularly or occasionally between cells and are thus more or less exposed to the external environment. In some cases, such as *Acetabularia* and *Galaxaura,* some of the aragonite is found within the cell wall. Interestingly, an unidentified species of *Padina* found off the Pacific coast of Mexico extends into waters that are colder than 15–16°C (Dawson 1946), but the algae do not mineralize (H. A. Lowenstam, personal observation). Another interesting case is that of *Liagora californica,* which is endemic to an insulated embayment of Catalina Island, California. It remains dormant until water temperatures rise above approximately 15°C, at which time it grows rapidly and mineralizes (Dawson 1966). *Avrainvillea* is a tropical green alga that is remarkably similar in gross morphology to *Penicillus, Rhipocephalus,* and *Udotea,* but does not mineralize. It has a conspicuous covering of mucus and this may possibly act as an inhibitor of mineralization (H. A. Lowenstam, personal observation).

The last example of increased degree of mineralization in relation to the environment occurs in marine fish with crystalline carbonate hydroxyapatite mineralized scales, as noted by Moss (1963). The extent of scale mineralization progressively increases from 26.9% in Arctic species to 33.2% in temperate water species and 50.3% in tropical species. Moss (1963) also noted that the bones of Arctic fishes were considerably less calcified than those from temperate and tropical waters that are mineralized more or less to the same extent. He attributed both these effects to decreased calcium concentrations in the low-salinity waters of the Arctic basin. It seems that the almost linear increase in degree of scale mineralization from the Arctic to the tropics is more likely to be related to temperature and not salinity.

11.2 Different Minerals Formed in Response to Environmental Changes

There are many well-documented cases in which the minerals formed by species living in warm waters are different from those precipitated by closely related species in colder waters. Almost all known examples involve the two polymorphs of calcium carbonate, calcite and aragonite. One exception is one of the microarchitectural units of the mature teeth of certain chitons (mollusks) that in temperate waters is composed of amorphous hydrous ferric phosphate (Lowenstam 1972b). Species of a genus living in the tropics have two analogous layers, one of which is composed of the iron oxide mineral lepidocrocite and the other carbonate hydroxyapatite (dahllite) or fluorinated hydroxyapatite (francolite) (Lowenstam 1967; Lowenstam and Weiner 1985). (For more details see Chapter 6). Another example of a mineralogical change occurring in response to environmental variations that does not involve carbonate occurs in the Molpadiidae, a group of benthic eurybathial Holothuroidea. The adult animals have dermal granules composed of amorphous hydrous ferric phosphate and opal (Lowenstam and Rossman 1975) (see Fig. 3.5). In *Molpadia intermedia* living in shallow waters (less than 30 m depth) the opal content is minimal (about 0.5% as silicon) but increases linearly to 2.5% (as silicon) in individuals living at about 300 m depth. Specimens from 2000 to 2900 m in depth have silicon contents ranging from 5.4 to 8.7%. In *Molpadia musculus* from deep water the silicon content is even higher (6–10% from individuals living between 1500 and 3600 m) (H. A. Lowenstam, unpublished data) (Fig. 11.1). These changes in opal content grossly parallel the distribution of dissolved silicon with increasing depths in the oceans and this may well be the environmental factor responsible for the variations.

The two polymorphs of crystalline calcium carbonate, aragonite and calcite, are widely used by organisms for forming mineralized exoskeletons. In general, however, there is no consistent distribution pattern and, for the most part, the polymorph formed is determined completely by the organism, irrespective of the environment in which it lives. There are, however, many exceptions to this rule. These "exceptions," which are the subject of the remainder of this section, are listed in Table 11.1. In general, all the known cases show a tendency to form more aragonite in warm waters and more calcite in cold waters.

Cnidaria. Many Cnidaria contain both calcite and aragonite in their mineralized tissues including one of the scleractinian corals (Constantz and Meike 1988) (see Chapter 5). Aragonite is the predominant form in scleractinians, but it is not known if the proportions of calcite and aragonite vary.

The *Holoxonia* are a group of cnidarians that lives in a wide range of environments. They are colonial animals that are attached firmly to the substrate. They have elongated rodlike axial skeletons surrounded by a cortex or rind. The cortex contains calcitic spicules formed under well-controlled conditions (organic matrix mediated), whereas the base of the axis contains variable amounts of calcite and

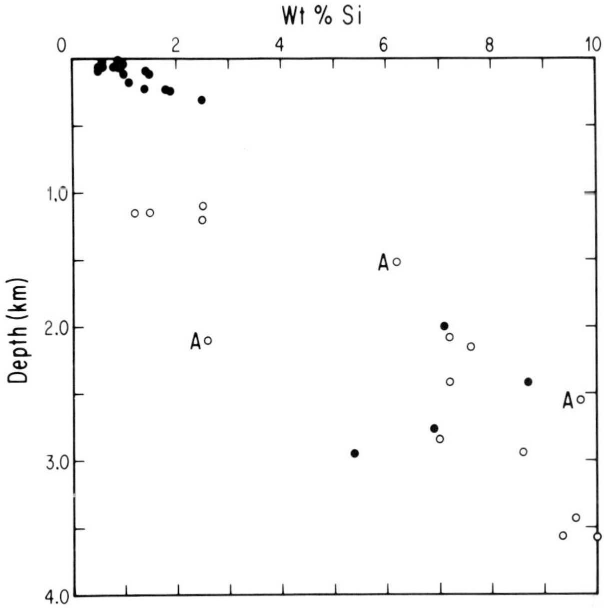

Figure 11.1. Graph showing the increase of silicon content in the amorphous hydrous ferric phosphate granules of *Molpadiidae* with increasing depth habitat in the oceans. Solid circles: *Molpadia intermedia.* Open circles: *Molpadia musculus.* A, Specimens from the Atlantic Ocean. Remainder from the Pacific Ocean.

aragonite, formed, it seems, under rather poorly controlled conditions (for more details see Chapter 5). Comparisons of the proportions of calcite and aragonite in the attachment base of some of these animals from different environments show a tendency to form more aragonite in warmer waters (Table 11.1).

Ectoprocta or bryozoa generally have skeletons that are composed entirely of calcite or less frequently entirely of aragonite (Rucker and Carver 1969; Poluzzi and Sartori 1975). There are two known cases in which both calcite and aragonite are present in the same individual, only one of which, the skeleton of *Schizoporella unicornis,* has been well studied (Lowenstam 1954a; Rucker and Carver 1969). The frontal wall is composed of aragonite, whereas the remainder of the skeleton is calcitic. The thickness of the frontal wall increases with the age of the individual. Comparisons of mature individuals in the tropics with those in colder waters show that they have up to three times as much aragonite as a function of total mineral formed. This is a result of their frontal walls being more massive in the tropics (Rucker and Carver 1969).

The Cirripedia or barnacles in the arthropod phylum almost always have shells composed entirely of calcite. Species belonging to the common genus *Tetraclita* (Table 11.1) living in the tropics and in subtropical waters are exceptions as they have a base plate that is composed of aragonite. The thickness of the base plate increases with increasing environmental temperature and, as a result, the overall proportions of aragonite as compared to calcite also increase. In temperate waters

Table 11.1 Mineralogy–Temperature Relations as Weight Percent Aragonite

Taxon	Taxa reported	Mineralization site	Mineralogy–temperature relations as percentage aragonite: Cold temperature environment	Subtropical–tropical environment	References
Cnidaria	*Allopora* div. sp.	Skeleton	0–83% at 1.5°C; 100% from 4 to 13.5°C	No data	Lowenstam (1964a)
	Holoxonia div. genera	Attachment base	0–5% from 5.6 to 18°C	18–100% from 18 to 28°C	Lowenstam (1964a)
	Plexaura flexuasa	Base + cortex spicules	No occurrence	70–90% from 23 to 28°C	Lowenstam (1964a)
Ectoprocta	*Schizoporella unicornis*	Colonial skeleton	25–58% from 11 to 17°C	58–78% from 17 to 30°C	Lowenstam (1954a); Rucker and Carver (1969)
	Stylopoma div. sp.	Colonial skeleton	No record	89–100% from 16 to 28°C	Lowenstam (1964a); Poluzzi and Sartori (1975)
Arthropoda	*Tetraclita* div. sp.	Skeleton	0–3% up to 16°C	10–20% up to 28°C	Lowenstam (1964a)
Annelida	*Spirorbis* div. sp.	Skeleton	0–82% from 5.2 to 17°C	100% from 18 to 28°C	Lowenstam (1954a)
	Eupomatus gracilis[a]	Skeleton	No data	49–86% from 16 to 30°C	Lowenstam (1954b)
Mollusca	*Mytilus* div. sp. and sub sp.	Shell	18–76% from 5 to 23°C	100% from 23 to 30°C	Lowenstam (1954a); Dodd (1963)
	Mytilus californianus[b]	Shell	33–53% from 14 to 21°C	No occurrence	Lowenstam (1954a); Dodd (1963)
	Brachidontes div. sp.	Shell	94–96% from 5 to 13°C	100% from 13 to 28°C	Lowenstam (1954a)
	Chama div. sp.	Shell	22–65% from 11.5 to 14°C	100% from 18 to 28°C	Lowenstam (1954a); Taylor and Kennedy (1969)
	Lima div. sp.	Shell	29–36% from ? to 18°C	62–88% from 23 to 28°	Lowenstam (1954a)
	Pedalion alatum[b]	Shell	No occurrence	42–76% from 15 to 30°C	Lowenstam (1954a)
	Haliotis div. sp.	Shell	50–90% from 16 to 18°C	100% at 28°C	Lowenstam (1954a)
	Littorina div. sp.	Shell	5–64% from 3 to 17°C	95%–100% from 15 to 28°C	Lowenstam (1954a)

[a]Skeletal growth increments.
[b]Growth series at one collecting site.

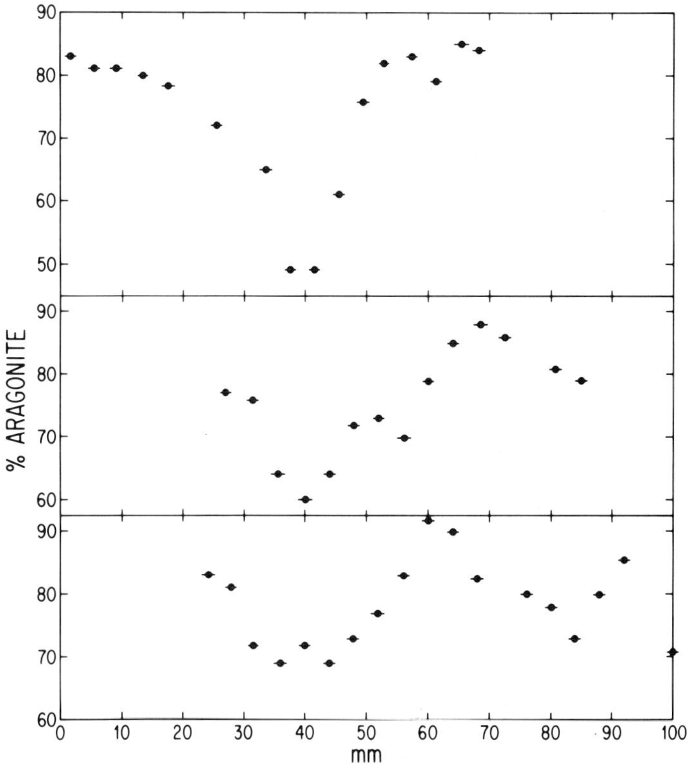

Figure 11.2. Graph showing the varying proportions of aragonite as compared to calcite (weight percent) in 3-mm increments of the tubes of three coexisting individuals of the serpulid worm, *Eupomatus gracilis.* The specimens were collected alive in Bermuda. Top figure was previously published in Lowenstam (1954b).

a very thin aragonitic plate is occasionally formed by one species. In cold water species, the base plate is not formed at all (Lowenstam 1964b).

Annelida. A few taxa within the Annelida have mineralized hard parts. The serpulid polychaete worms form mineralized tubes of either calcite, aragonite, or both. In some species in which both minerals are found it has not yet been shown that each polymorph occupies a separate layer as is known to occur in related fossil species.

One of the most detailed and informative studies on the variations in the proportions of calcite and aragonite in relation to environment was performed on the tube of the serpulid worm, *Eupomatus gracilis.* The tubes of three individuals living in the same microenvironment in Bermuda were analyzed (Fig. 11.2). The rhythmic changes in the proportions of polymorphs are surprisingly similar in the three individuals. The last formed growth increment (at the extreme left of each curve) was laid down at the beginning of July at a water temperature of 27°C, which is 3°C lower than the summer maximum of the previous year. Even this small difference is reflected in the analyses. The total seasonal temperature range is from 15°C in the winter to 30°C in the summer. The calcite–aragonite proportions of these tubes, therefore, faithfully record the seasonal temperature fluctuations. Note too that curves of this type can be used to determine the ages and growth rates of individuals. Members of the genus *Spirorbis* also show variations in calcite–aragonite proportions that correlate with changes in environmental temperature (Table 11.1). The corresponding ultrastructural variations have, however, not been studied.

Mollusca. Mollusk shells are usually composed of either calcite or aragonite. In some shells calcite is found in certain shell layers and aragonite in others. Among the latter are cases (Table 11.1) in which the overall proportions of calcite and aragonite in the whole shell vary with changes in environmental temperature. The best studied examples are among the *Mytilacea,* which in cold and temperate waters have shells in which the outer layer and sometimes an innermost layer as well are calcitic, and the intermediate nacreous layer is aragonitic. In tropical waters the Mytilidae and some related species have shells that are composed entirely of aragonite and, in fact, even the outer layer has an aragonitic nacreous microarchitecture (Lowenstam 1954a; Taylor et al. 1969). In the cold and temperate water species, the calcite and aragonite proportions have been shown to vary with seasonal water temperature changes (Lowenstam 1954a; Dodd 1963). The variations are due to changes in the relative amounts of different shell layers (Dodd 1964) (Fig. 11.3). Some mytilid species also live in brackish waters and their shell calcite–aragonite proportions can also vary in relation to salinity. A subspecies of *M. edulis* living in low-salinity brackish waters has shells with relatively more aragonite than those living at the same temperature in the normal marine environment (Lowenstam 1954a; Dodd 1963; 1966).

Species of the genus *Chama* are common in the tropics, where their shells are composed entirely of aragonite. The shell has three layers: an outer crossed-lamellar layer, a middle layer also with a crossed-lamellar ultrastructure, and an inner complex crossed-lamellar layer. Only a few species of *Chama* live in temperate waters and, of these, the best studied are those off California and Chile (Bernard 1976). Their inner layers are aragonitic, but their outer layer is composed of prismatic calcite. Figure 11.4 is a comparison of sections cut through tropical and temperate water *Chama* shells to illustrate these differences. Thus, *Chama* species living in colder water have a calcitic layer that is either homologous to the outer aragonitic layer in tropical species (Lowenstam 1954a, 1963) or is an additional layer not found in the tropical water species (Kennedy et al. 1970). An analysis of the proportions of aragonite relative to calcite in a temperate water species, *Chama arcana,* shows that with increasing mean annual temperature the proportions of aragonite increase most conspicuously in the 16–18°C zone (Fig. 11.5). This corresponds more or less to the transition between the temperate and tropical ocean areas.

The Connection between Environmental Temperature and Mineralogy. Table 11.1 lists the known examples of this phenomenon among carbonate-precipitating organisms. These constitute the large majority of known cases. Bearing in mind that calcite and aragonite are very common biogenic minerals, the environmental effect is not the rule, but rather the exception to it. For the most part the organisms determine the polymorph formed. In fact, close inspection of all the examples in Table 11.1, except possibly the *Holoxonia,* shows that their polymorph contents

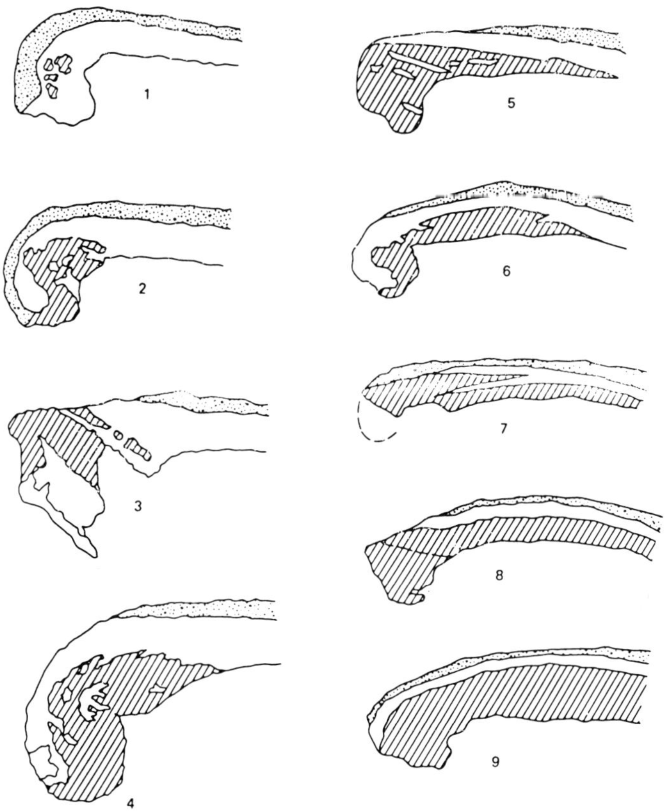

Figure 11.3. Tracings of photographs showing structural types in *Mytilus californianus.* The higher numbered structural types are characteristic of colder growth temperatures. Stippled areas are the calcitic outer prismatic layer, clear areas are the aragonite nacreous layer and myostracum layers, and inner lined areas are the calcite inner prismatic layer. ×2. Reproduced from Dodd (1964) by courtesy of Society of Economic Paleontologists and Mineralogists.

are also determined by the organism and not directly by the environment. A good illustration of this is that species with both polymorphs in their skeletons/shells living in the same environment almost invariably do *not* have the same proportions of calcite and aragonite. The overall proportions of calcite and aragonite in the skeletons are influenced only *indirectly* by temperature, presumably through differential growth rates of skeletal layers. It is not known how this occurs, although it must be under strict genetic control to account for the fact that even closely related genera sometimes do not exhibit this phenomenon at all. Interestingly, the *Holoxonia* is the only taxon in which the phenomenon is common, although here too there are exceptions. Significantly the mineralization process of *Holoxonia* appears to be poorly controlled and it is conceivable that temperature might directly affect the proportions of calcite and aragonite formed. The situation is in

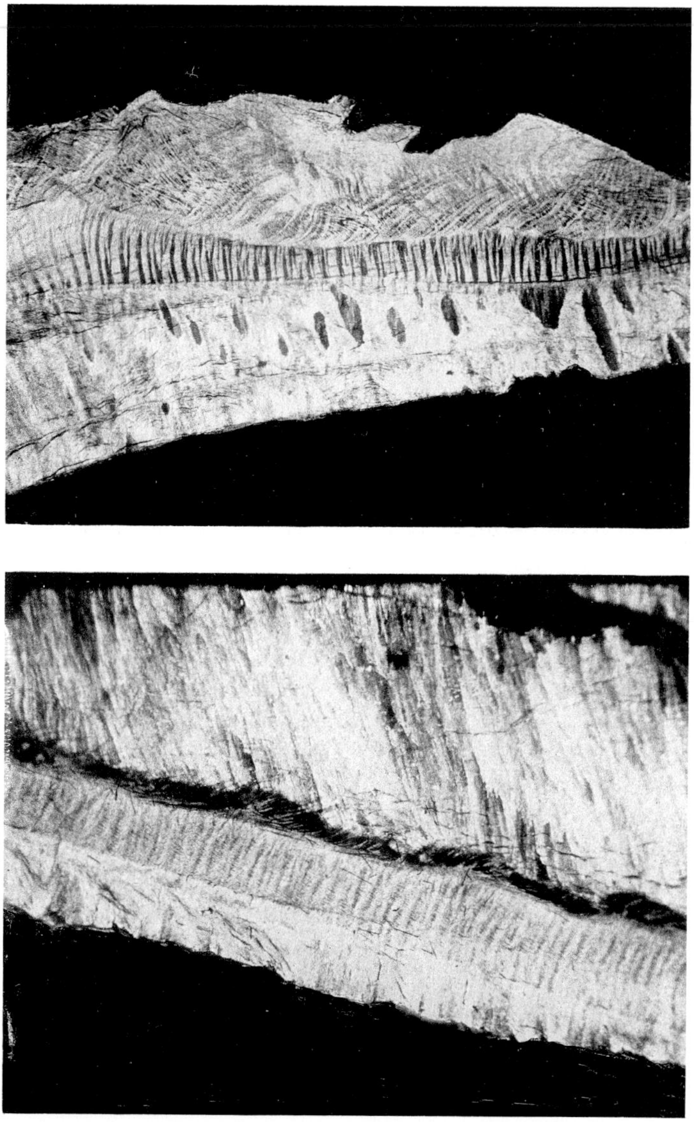

Figure 11.4. Light micrographs of thin sections along the median of (a) warm water (Quaymas, Mexico) species of *Chama sordita* showing three different layers, all of which are aragonitic. (b) Temperate water species (Iquique, Chile) of *Chama pelucida* showing three different layers, with the outer top layer composed of prismatic calcite and the lower two inner layers composed of aragonite.

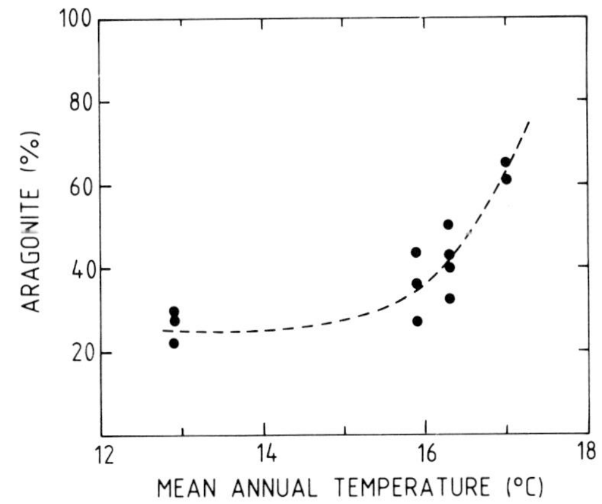

Figure 11.5. Proportions of aragonite relative to calcite (weight percent) of whole shells of the temperate water bivalve *Chama arcana* from the west coast of North America, as compared to mean annual water temperature.

some ways analogous to the marine calcareous algae described in the previous section.

11.3 Environmental Influences on Trace Element and Oxygen Isotopic Composition

The known cases of environmental influences on trace element and stable oxygen isotope contents of biogenic minerals are far more common than the effect on mineralogy, described in the previous section. Most of the currently available information on the nature of these influences, their magnitudes, and the organisms they affect is limited to the calcitic and aragonitic hard parts of eukaryotes. One obvious reason for this is that these are the most widely utilized hard part constituents of present day marine animals and plants. The ecologic factors known to affect either the trace element contents and/or the stable isotopic compositions of biogenic calcium carbonates are temperature, salinity, water chemistry, and hydrostatic pressure. It is known that the same mineral type formed by various organisms living in a particular microenvironment can have different trace element contents and isotopic compositions. This shows that at least some of these values are not in equilibrium with the environmental waters, implying, in turn, that biochemical processes can either screen or modify environmental influences on trace element content and isotopic composition. This can occur even at the species or subspecies level.

The calcite crystal lattice can more readily accommodate Mg at Ca ion sites as compared to the aragonite lattice. However, the reverse is true for Sr. Thus, crystal structure is a priori a significant controlling factor in trace element uptake potential. Generally the measured trace element content or isotopic composition is the result of several superimposed biochemical as well as environmental factors. This probably accounts for the rather slow progress in our ability to sort out the various effects. In the following section we will discuss in turn the major environmental influences, first on trace element content and then on the oxygen isotopic composition of biogenic carbonates, bearing in mind that this is combined with the biochemical effect.

11.3.1 Trace Element Contents

11.3.1.1 TEMPERATURE

The distribution coefficients for strontium coprecipitated with aragonite from seawater and with calcite from seawater decrease linearly with temperature (Kinsman and Holland 1969; Kinsman, 1969). In other words, less Sr enters into the lattice with increasing temperature. In contrast to Sr, the Mg contents of calcites formed in vitro are found to increase with increasing temperatures (Füchtbauer and Hardie 1976).

A temperature effect on Mg uptake has been identified in the calcite formed by the foraminifera and coralline algae, as well as in nearly all invertebrate phyla (Clarke and Wheeler, 1922; Chave 1954; Milliman et al. 1971). As a rule, the Mg contents increase with increasing temperature, the same trend that is found in synthetically formed calcites. However, the Mg concentrations in biogenic calcites are not the same as those found inorganically at the same temperature and disparate groups of organisms can have different partition coefficients. These differences have even been detected between genera of the same class (e.g., Pilkey and Hower 1960). In other words, genetically controlled biochemical screening is an important filter of the temperature effect. This may explain why in some rare instances, as occurs, for example, in certain mollusks, the temperature effect may be reversed or altogether suppressed (Pilkey and Goodell 1963; Lowenstam 1964b).

An excellent example illustrating these phenomena involves a comparison of the trace element contents of the calcite shells from inarticulate and articulate brachiopods with changing environmental water temperatures (Lowenstam 1961 and unpublished observations). The results are shown in Figure 11.6a and 11.6b. The absolute concentrations of Sr and Mg in the inarticulates are quite different from those of the articulates. Significantly, the trace element contents of the inarticulate brachiopods are in equilibrium with the surrounding water, whereas those of the articulate brachiopods are not. The Sr concentrations of the articulate brachiopod shells increase with increasing temperature, and those of the inarticulate brachiopods decrease. Thus, although the articulate brachiopods control their trace element contents, the influence of the environmental temperature variations is still pronounced.

Almost nothing is known about the changes of Mg concentrations with temperature in biogenic aragonite, mainly because of the analytical difficulties in

Figure 11.6. Variations in the (a) $SrCO_3$ and (b) $MgCO_3$ contents of the calcitic shells of inarticulate and articulate brachiopods in relation to environmental water temperatures. The inarticulate species used all belong to the Craniidae. See Lowenstam (1961) for the list of articulate brachiopod species used.

A

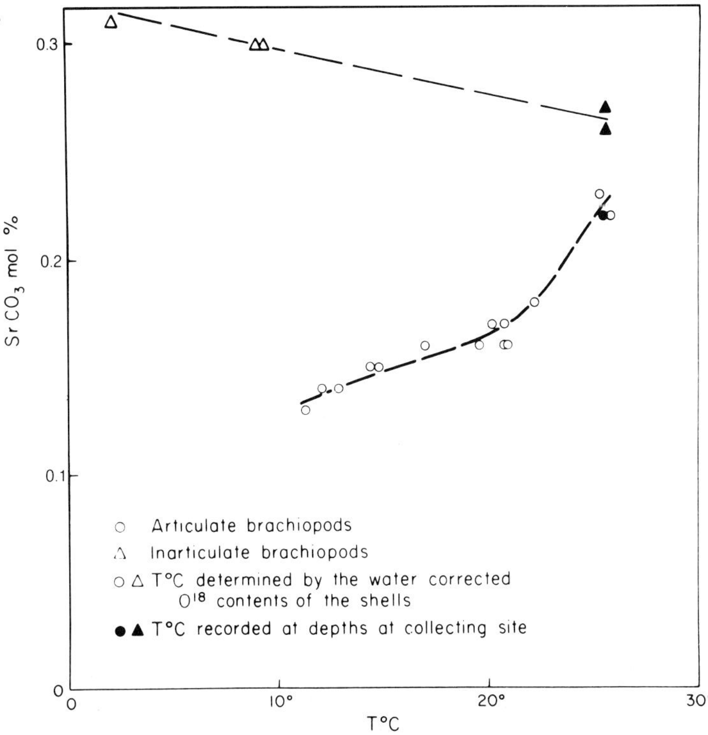
0.3
0.2
0.1
0
Sr CO3 mol %
Articulate brachiopods
Inarticulate brachiopods
T°C determined by the water corrected
O18 contents of the shells
T°C recorded at depths at collecting site
0
10°
20°
30°
T°C

B

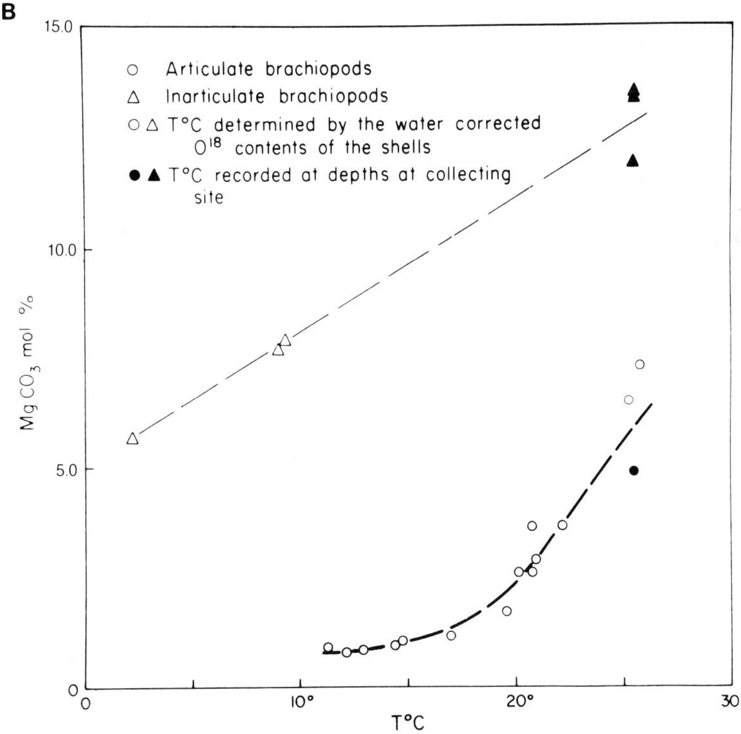
15.0
10.0
5.0
0
Mg CO3 mol %
Articulate brachiopods
Inarticulate brachiopods
T°C determined by the water corrected
O18 contents of the shells
T°C recorded at depths at collecting
site
0
10°
20°
30°
T°C

detecting the small amounts present. Dodd (1965) found a weak positive correlation between Mg content and temperature in the shells of the bivalve *Mytilus,* whereas Schifano (1982) reported a clear-cut negative correlation between Mg content and temperature in the marine gastropod, *Monodonta turbinata.*

11.3.1.2 SALINITY AND WATER CHEMISTRY

The trace element contents of populations of various euryhaline invertebrates living in seawater have been compared with those living in brackish water. These include barnacles (Gordon et al. 1970), articulate brachiopods (Lowenstam 1961), echinoids (Harris and Pilkey 1966), and various bivalves and gastropods (summarized in Dodd and Stanton 1981). In general, the observed differences are attributed to the effect of salinity. The differences, however, vary greatly from case to case.

Intuitively it is expected that salinity-induced changes in trace element concentrations should simply record the extent to which seawater is diluted to form brackish waters. This is not always the case. Certain articulate brachiopods with calcitic shells that live in brackish waters have Sr/Ca ratios greater than brachiopods living in open ocean water. The same is true for the specimens examined of a different species from a hypersaline environment (Lowenstam 1961). Studies of aragonitic bivalves living in water bodies that have the same salinity but differ in their Sr/Ca and Mg/Ca ratios show that the shells also have different trace element contents (Dodd and Crisp 1982). Experimental studies in which freshwater gastropods with aragonitic shells were grown in waters with increasing Sr/Ca ratios show that the Sr uptake into the shell increases as a function of Sr content in the water. In one extreme case a gastropod produced a shell composed entirely of strontium carbonate (strontianite) (Odum 1951; Buchardt and Fritz 1978).

Thus, it seems that the trace element contents in these skeletal hard parts are controlled by local concentrations of the trace elements under investigation and/or by salinity. We note, however, that the trace elements discussed above all substitute for calcium ions in the lattice. However, other atoms, such as boron, which do not apparently substitute for calcium, do show a linear relationship with salinity and may well provide reliable records of variations in environmental salinity (Leutwein and Waskowiak 1962; Fürst et al. 1976).

11.3.1.3 HYDROSTATIC PRESSURE

Carbonate-depositing species have been found in oceanic deep-sea trenches at depths close to 11,000 m (Wolff 1960; Beleyaev 1966), which corresponds to a hydrostatic pressure of about 1,100 atmospheres. Some eurybathial species have a wide depth range and they obviously tolerate large hydrostatic pressure differences. One of these, the benthic holothurian *Elpidia glacialis* and several of its subspecies, served as a test case to determine whether the Sr and Mg contents of their calcitic spicules are affected by hydrostatic pressure (Lowenstam 1972a). Samples were obtained from depths of 600 m to between 8,180 and 8,830 m, corresponding to 60 to about 880 atmospheres of pressure. The water temperatures in this range vary only between −1.1°C and 2.8°C. The Sr contents show a well-defined negative correlation with hydrostatic pressure for all the subspecies and even for one other related species. Significantly, the correlation with temperature is poor. In contrast to the Sr, the Mg contents show a clear-cut dichotomy between shallow and deep water *E. glacialis* subspecies. The shallow water forms show a positive correlation with temperature, whereas deep water forms correlate with depth changes rather than temperature (Lowenstam 1972a). A similar phenomenon was noted for the eurybathial holothurian *Oneirophonta mutabilis.* Interestingly, the switch from temperature to depth control in *E. glacialis* occurs at hydrostatic pressures of 500–600 atmospheres and in *Oneirophonta* at 450 atmospheres (Lowenstam 1972a). This may possibly relate to a phenomenon observed in some bacteria, in which protein synthesis was impaired at pressures exceeding 500 atmospheres (Albright 1972).

It is worth noting that the Mg content of calcitic spicules of another eurybathial holothurian *Scotoplanus globosa* that lives between 2,600 and 6,620 to 6,730 m shows no correlation with depth or with temperature, although it is possible that the two effects cancel each other out (Lowenstam 1972a).

11.3.2 The Environment and Stable Oxygen Isotopes

Calculations by Harold C. Urey (1947) on exchange reactions of the lighter elements predicted that there should be a slight dependence on temperature. Furthermore, there should also be differences in the ratios of oxygen isotopes in water and the carbonate and phosphate molecules of a mineral phase precipitated from that water. Subsequent work by Urey and co-workers (1951) showed that indeed the $^{18}O/^{16}O$ ratios of marine carbonates depend upon the temperature at which the precipitate formed, as well as on the $^{18}O/^{16}O$ ratio of the water, provided equilibrium conditions prevail. A carbonate–water isotopic temperature scale was established based on skeletal carbonates formed by marine organisms in nature, as well as in the laboratory under controlled temperature conditions (Epstein et al. 1951, 1953).

Urey originally conceived of the idea to provide, for the first time, a quantitative paleotemperature tool for tracing the changing climates through geological times. It was not coincidental that the first paleoclimatic study using oxygen isotopes involved fossils from the Cretaceous (Urey et al. 1951), as Urey was fascinated by the "sudden" disappearance of the dinosaurs. He hoped to show that this mass extinction event that occurred some 65 million years ago, of which the dinosaurs were a part, was related to some catastrophic event involving changes in temperature. (Stated in private by Urey to H. A. Lowenstam when they were sampling the famous fossil locality at Coon Creek, Tennessee in 1949.) Unfortunately, no temperature information from the critical time interval at the very end of the Cretaceous was obtained in the first study (Urey et al. 1953; Lowenstam and Epstein 1954), and the riddle of the extinction of the dinosaurs remained unanswered. At a much later date Urey (1973) proposed that the mass extinction at the end of the Cretaceous may have been caused by the impact on earth by a comet. We now know that Urey was probably close to being correct on both accounts, as there is increasing evidence that a large extraterrestrial meteoritic body impacted the earth at that time (Alvarez et al. 1980) and the scenario of events that followed includes extreme temperature fluctuations (Hsu et al. 1982).

The relationship between the environment and the isotopic composition of biogenic minerals is more easily understood for oxygen. The pioneering work of

Urey and his associates showed that the major factors that determine the $^{18}O/^{16}O$ ratio in carbonates is temperature and the $^{18}O/^{16}O$ ratio of the water. The situation is quite different for the oxygen isotopes of phosphate minerals, where the mineral and the dissolved phosphate equilibrate only when organisms make and break P–O bonds (Urey et al. 1951; Tudge 1960; Kolodny et al. 1983). Carbon isotopic compositions of biogenic carbonates are even more complicated as some of the carbon in the skeleton can be derived from the metabolism of the organism and not only from the dissolved carbon in the environmental water. For this reason, we will focus the discussion only on the oxygen isotopes.

Seawater in the well-mixed reservoir of the open oceans is rather uniform with respect to oxygen isotopic composition. The variations occur mostly in certain near-shore environments and are either due to dilution with freshwater from rivers or ice or to evaporation as a consequence of water bodies being isolated from the open oceans (Epstein and Mayeda 1953; Lloyd 1964). Thus, the isotopic composition of the carbonate shells of organisms living in these marginal seas or in freshwater will vary both as a function of temperature and as a function of water isotopic composition. Variations in the oxygen isotopic composition of those living in the open marine environment are due to temperature changes.

Some organisms can lay down their carbonate skeletons in isotopic equilibrium with the environmental water, whereas the oxygen isotopic composition of others is clearly out of equilibrium. The carbonate shells of most mollusks, articulate brachiopods, serpulid annelids, and some ectoprocts are in equilibrium with the water (e.g., Epstein and Lowenstam 1953; Epstein et al. 1953; Lowenstam and Epstein 1957; Lowenstam 1961). Scleractinian corals, echinoderms, barnacles, as well as many calcareous algae and foraminifera have shells that are not in equilibrium with the environmental water. The disequilibrium precipitates usually have lower ^{18}O concentrations than the equilibrium precipitates. This is true for most corals and echinoderms (Weber and Woodhead 1970; Weber and Raup 1966). On the other hand, some ahermatypic corals (Land et al. 1977), large foraminifera (Wefer and Berger 1980), and balanomorph barnacles (Killingley and Newman 1982) are enriched in ^{18}O relative to equilibrium values. In corals the extent of the offset varies from genus to genus, but is uniform within a genus (Weber and Woodhead 1970; 1972; Erez 1978). In contrast all balanomorph barnacles are enriched in ^{18}O to about the same extent (1.3‰ in $\delta^{18}O$) (Killingley and Newman 1982).

It is of interest to note that nonequilibrium precipitation was already recognized during the calibration of the paleotemperature scale. The isotopic compositions of samples obtained from regenerated material formed by abalones to cover up holes drilled in their shells all had small deviations from the calculated curve (Epstein et al. 1953). Apparently the abalones laid down this material so rapidly that isotopic equilibrium was not achieved. The term *vital effect* was coined to describe the disequilibrium phenomenon, and, for the most part, is really just a facade hiding our ignorance of the underlying causes of this phenomenon.

Mass spectrometric measurements of $^{18}O/^{16}O$ ratios are accurate enough to detect differences equivalent to 1°C. This is, therefore, a very powerful tool for tracing many diverse effects of temperature on skeletal mineralization, in addition to the analysis of fossils for paleotemperature reconstruction. The following are a few examples of such applications.

Oxygen isotope analyses of consecutive increments of mollusk bivalve shells usually show continuous records of the seasonal temperature range at the site at which the animal lived (e.g., Epstein et al. 1953; Epstein and Lowenstam 1953; Wefer and Killingley 1980). However, some mollusks do not form shell during the colder periods of the year and this too can be recognized from their growth increment isotopic fluctuations. An interesting case is the bivalve *Chama macerophylla* studied by Epstein and Lowenstam (1953). It lives in tropical Caribbean seas and extends into the marginal subtropical waters of Bermuda. It can develop and reproduce there, but apparently does not deposit shell carbonate during the colder parts of the year. Interesting differences in seasonal growth have been discovered in this way. For example, bivalves in Bermuda tend to primarily deposit shell material during the warm months of the year, whereas nearly all the gastropods form shells throughout the year (Epstein and Lowenstam 1953). The temperate water bivalve *Mytilus californianus* from the west coast of North America tends to form more shell during the colder parts of the year as opposed to the summer (Killingley and Berger 1979).

Oxygen isotopes have also been used to study molluscan larval development and subsequent growth rates. Thorson (1950) first noted that, as a rule, the larval shells of gastropods that developed in nurse or brood eggs on the sea bottom were large and massive, whereas those that developed floating in the water (planktotrophic larvae) were small and fragile. This became known as the "shell apex theory" and was further investigated by Shuto (1974), who used the SEM to study the larval shell surface geometry (see also Jablonski and Lutz 1980). Killingley and Rex (1985) tested this hypothesis by analyzing the oxygen isotopic compositions of larval shells of both types and compared them with the adult shell. They found that in four cases the theory was incorrect as the larval shell and the adult had the same isotopic composition even though the larva is supposed to be planktotrophic and the adult is known to be a bottom-dwelling deep sea species. In four other cases the expected differences were observed.

Another quite ingenious study involved analysis of growth increments of calcitic barnacles attached to California grey whales (Killingley 1980). A preliminary study showed that it is feasible to use this information to fairly accurately track the migration routes of the whales. Various sized barnacles, however, need to be studied to get an overlapping record covering a period of a few years. A similar approach could be used to track migrations of turtles.

We emphasize that the vast majority of oxygen isotope studies involve the determination of paleotemperatures, which in turn reflect changes in paleoclimates. There is by now a good record of oxygen isotopic fluctuations from the present all the way back to the Late Cretaceous (e.g., Savin 1977), a remarkable achievement indeed.

11.4 Environmental Influence on Skeletal Growth

Of all the different ways that the environment can and does influence skeletal structure, probably the most common effect is on the rate of growth. There is obviously no direct linkage between environment and skeletal growth, but as often happens a change in an environmental parameter affects some aspect of the physiology of the organism, which in turn influences skeletal deposition. Thus, organisms that form their skeletons by the addition of new material at the margins and do not remove or alter earlier formed skeletal structures may record many of these environmental changes in the form of a growth line. A good definition of a growth line is an "abrupt or repetitive change in the character of an accreting tissue" (Clark 1974b, p. 77). In the mineralized tissue itself it usually is thought to represent a change in the relative proportions of organic material to mineral (Lutz and Rhoads 1977), although the phenomenon is poorly understood (see end of section). Examples of mineralized tissues that do often contain a record of environmental changes are the shells and pearls of mollusks (Jones 1985), coral skeletons (Dodge and Vaisnys 1980), barnacle shells (Bourget 1980), and the otoliths of fish (Pannella 1980). (The papers cited are reviews of growth lines in each tissue.) These tissues do not record environmental changes if their formation is continuous and uninterrupted. This is thought to account for the absence of such a record in, for example, the clam *Pecten diegensis* (Clark 1975).

The environmental stimuli that can cause a growth line to be formed may be traumatic events, such as storms, predator attack, or even giving birth, or subtle events, such as gradual ambient temperature changes over a whole season or the monthly illumination cycle of the moon. Obviously the responses to different stimuli will also be different (Dillon and Clark 1980) and are often superimposed. This poses serious difficulties for the investigator trying to unravel the record while at the same time attempting to provide solid evidence for the nature of the environmental stimulus. In spite of the difficulties many interesting and diverse studies have been reported, some of which are quite ingenious. The applications are relevant to fields as diverse as geophysics, planetary sciences, paleoecology, archaeology, ecology, and population dynamics. The journals in which these reports are published are as diverse as the applications. The multiauthor book edited by Rhoads and Lutz (1980) contains a collection of excellent chapters on this subject and includes a "how-to-do-it" practical guide for the enthusiast.

In this section we will first present summaries of a selection of papers so as to give the reader the "flavor" of the subject, and then briefly focus on the nature of a growth line itself.

Corals and the Earth's Rotation. Wells (1963) published a pioneering paper, which sparked off an explosion of interest in this field. The paper, a joy to read, shows that the surfaces of a modern scleractinian coral from the West Indies contains around 360 growth ridges in a section of skeleton deposited in one year, strongly suggesting that the coral lays down one growth ridge per day. Fossil corals from the Middle Devonian (some 370 million years ago) usually have 400 growth lines per annum (range being 385–410) and from the younger Pennsylvanian (about 300 million years ago) have 385–390 growth lines per annum. These results imply that the number of days per year have decreased over the last 400 million years, or, in other words, the earth's rotation rate is decreasing. This had indeed been postulated by astronomers. Wells originally hoped that the fossils could be dated by counting diurnal growth lines, on the assumption that the earth's rotation rate has decreased uniformly over time. The variability in the data does not justify this, but the direct documentation of the slowing down of the earth's rotation obtained from the fossil record represents a landmark achievement. In 1975 a book was published on the subject (Rosenberg and Runcorn 1975).

Mollusks Can Be Superb Recorders of Environmental Fluctuations. Counting of the surface expressions of growth lines as Wells (1963) did is not as reliable as using thin sections or replicas of the internal structure after slight etching to reveal the growth lines. Barker (1964) studied four different marine bivalves using thin sections and found a complex, superimposed hierarchy of growth lines, all of which were periodic. At the largest scale he identified yearly growth cycles that are the result of annual fluctuations of temperature and salinity. The yearly cycle is interrupted twice by growth lines induced by equinoctial storms and tides. Twenty-four increments were thought by Barker to represent fortnightly tidal cycles and 365 growth lines per year to represent daily cycles. A fifth order periodicity was assigned to subdaily tidal rhythms. Barker's study demonstrated the enormous potential of this approach, even though he did not provide direct supportive evidence for the causes of the assigned periodicities, but based his assignments on numerology. In the two decades since this study, a large body of data has been collected using many different techniques that do link a known environmental stimulus with a particular set of growth lines (reviewed by Jones 1985). This is not an easy task as the direct approach of marking animals, allowing them to grow for a period of time and then examining their shell structures, suffers from the fact that the disturbance during the marking itself can cause disruptions in shell deposition. The alternative approach is to judiciously use a natural sample to relate the environmental stimulus and the growth line distribution. Certainly one of the most remarkable studies of this type is that of Evans (1972). He studied a coastal area with unusual mixed semidiurnal tides. These, he anticipated, should produce unique growth line distributions. He deliberately chose an intertidal bivalve *Clinocardium nutalli* that lives below the sand at 0.6 m above the mean tide level. This is an ideal location to record tidal fluctuations assuming, as he did, that a sharp growth line represents an interruption of growth due to exposure at low tide. By counting the growth lines back from the shell margin and knowing the date of collection he identified a 4-week period of shell deposition and compared this to the predicted tidal movements over the same period. The correspondence is remarkable.

Growth Lines and Seasonal Dating in Archaeology. The fact that some mollusk growth line patterns record seasonal changes (widely spaced in summer and closely spaced in winter) as well as individual daily growth increments makes it possible to determine the season in which the animal died. Bivalves were often important food resources of coastal dwelling prehistoric man and examination of shells collected from such middens is an indication of the season of occupation of the site. Coutts (1970) demonstrated that the tool is reliable using the living bivalve *Chione stutchbury* and then showed that the various archaeological sites studied in New Zealand were occupied during mid-summer or early autumn.

Clams Monitor Thermal Effluent of a Nuclear Power Station. The clam *Mercenaria mercenaria* is known to be a good monitor of environmental fluctuations (Pannella et al. 1968) and indeed, those located in Barnegat Bay, New Jersey, faithfully record subdaily, daily, bidaily, tidal, lunar-month, and annual fluctuations (Kennish and Olsson 1975). A nuclear power plant situated alongside a creek in the bay pumps cooling water back into the bay. The tempeature at the plant site is

10–13°C above ambient levels and at the mouth of the creek is 4–5°C above ambient level. Kennish and Olsson (1975) studied both the periodic growth line patterns and the aperiodic growth breaks in the population of *Mercenaria* at the mouth of the creek. They showed that the abrupt changes in operations of the power plant results in thermal-shock breaks in the shell. The correlation is excellent.

Mercenaria also records other "natural" shock breaks due to winter freezing, excessive summer heat, spawning, storms, and neap tides. All these shock breaks result in the deposition of a thin layer of shell with a crossed-lamellar ultrastructure instead of the normal prismatic ultrastructure. The thermal-shock break appears as a sudden break in a normal pattern of growth increment addition. The breaks correlate with rapid fluctuations in water temperatures. These thermal-shock breaks adversely affect growth and, if they overlap the spawning season, may also affect the animals reproductive cycle. Kennish and Olsson gently suggest that the scheduled shutdowns of the nuclear power station should not be made during the spawning season.

The Ultrastructure of a Growth Line. The fact that growth lines are physiological responses to different stimuli that occur in widely divergent organisms makes it highly unlikely that any common structural theme exists. Daily or subdaily growth lines in corals as seen in light micrographs of thin sections appear to correspond to the presence of thin, possibly discontinuous layers of very small calcite crystals interspersed among the prevalent large spherulitic aragonite crystals. The shock breaks that occur in the clam *Mercenaria* described by Kennish and Olsson (1975) involve the deposition of a layer of a shell with an ultrastructure different from that normally laid down. However, the most commonly attributed basis for a growth line is not a change in ultrastructure, but a different proportion of organic material to mineral as compared to the adjacent areas (Lutz and Rhoads 1977; Rhoads and Lutz 1980). One of the few studies that has addressed this question directly shows by electron probe analyses that the growth line in the bivalve *Cerastroderma edule* is actually composed of purer calcium carbonate than the surrounding increments and contains less organic material (Deith 1985). Much more detailed information at the molecular level is needed. It would be of particular interest, for example, to determine if during slow periods of growth the organization of an individual matrix layer was different from that formed during rapid growth. One possible difference, at least in mollusks, could be the relative amounts of chitin–glycine-rich protein complex (insoluble fraction or framework macromolecules) and the aspartic acid-rich proteins in each layer (soluble fraction or acidic glycoproteins). A detailed understanding of the ultrastructure of growth lines must certainly be the basis for evaluating hypotheses about their formation. Lutz and Rhoads (1977) proposed a "theory of growth line formation" for mollusks in which the growth line is formed during periods of anaerobic respiration when some mineral is removed from the shell to buffer the organic acids produced by this mode of respiration. The "theory" is probably more appropriate for intertidal or subtidal mollusks and needs to be rigorously tested, particularly at the molecular organizational level. The clam *Mercenaria mercenaria,* which has been the subject of many related studies cited by Lutz and Rhoads (1977), is not the most suitable shell to use because it has a very complex ultrastructure.

12

Evolution of Biomineralization

Biomineralization among living organisms is widespread, occurring in both prokaryotes and eukaryotes. It is diverse with some 60 or so minerals known to be formed by organisms under a wide variety of conditions. They are deposited at many different locations both inside and outside cells. Biomineralization occurs on such an enormous scale that it influences processes in the biosphere itself. It is, therefore, of interest to determine how this all came about—the evolution of biomineralization.

The evolutionary history of biomineralization is a fascinating subject in its own right, which is the primary reason for including it in this book. However, a well-substantiated understanding of this subject is also of crucial importance to the interpretation of many aspects of research into the mechanisms of biomineralization in living organisms. An example is the observation by Veis et al. (1986) that antibodies raised against the rat incisor acidic proteins, phosphophoryns, cross-react with proteins extracted from a sea urchin test. The proteins presumably share some similar molecular structures. Did they inherit them from a common ancestor or did they evolve independently from each other to fulfill similar functions? This type of question can be asked about many comparative studies in biomineralization between phyla or even within lower taxa of the same phyla. As long as we do not have answers to these questions, the powerful tool of comparative biology in biomineralization is compromised.

Studying the evolution of biomineralization has one enormous advantage over many other topics in evolutionary biology; the very material that we are interested in has the best chance of surviving the vagaries of time and being preserved in the fossil record. The fossil record at least during the last 600 million years or so is, for the most part, a documentation of remnants of the history of mineralized hard part formation by organisms. Thus, the evolution of biomineralization is one topic that can, and that should be based on the direct documentation of the fossil record. This is the way it is presented in this chapter. The corollary of this statement is also worth considering. The fossil record should be interpreted bearing in mind the evolution of biomineralization. One example discussed in this chapter is the evolution of phosphate biomineralization. The vast majority of phosphatic hard parts formed by living animals are composed of amorphous calcium phosphate, which is generally not well preserved in the fossil record. Thus, the preserved record is highly skewed toward the relatively few animal taxa that form hard parts out of crystalline phosphate minerals. There are many more examples of this type for paleontologists to deal with.

The fossil record offers another exciting possibility for the student of biomineralization that has barely been exploited. Study of the originally preserved minerals themselves can provide indirect information on the evolution of the processes used to control their formation. One example is the evolution of strontium discrimination by mollusks. Most mollusks incorporate very small quantities of strontium into their shells, with the noted exception of the Chitonidae. Careful studies of well-preserved fossils suggest that this ability evolved some 300 million years ago in the Upper Carboniferous (Lowenstam 1964a). Although as with all good scientific topics, the conclusions have been disputed (Ragland et al. 1969). Other possibilities along these lines are studies of the evolution of isotopic fractionation, the manner in which control of crystal morphology evolved, and the study of the actual preserved remnants of hard part proteins.

In this chapter we initially discuss the early evolution of biomineralization, first among prokaryotes and then eukaryotes. The evolutionary history of composite skeletal mineralization is divided into three subtopics: biogenic carbonate evolution, biogenic phosphate evolution, and biogenic opal evolution. At the end of the chapter we address one of the key questions in the field: how did the evolution of composite skeletal mineralization arise almost simultaneously (in geological terms) in so many different taxonomic groups?

12.1 The Early Evolution of Biomineralization

The fossil record is, for the most part, a documentation of the evolution of the skeletal hard parts of organisms simply because they are so much more likely to be preserved than the nonskeletal soft parts. The skeletal hard part fossil record extends back in time about 570 million years and the nonskeletal macrofossil record an additional 100 million years or so. Traditionally the period of skeletal hard part deposition from about 570 million years ago until the present is referred to as the Phanerozoic, which literally means the age of visible life. It is only since the early 1950s, starting with the work of Barghoorn and his colleagues (1954), that we have firm evidence that the fossil record extends well beyond this point in time all the way back to 3500 million years before present (3.5 Byr BP). The scientific history of this subject is reviewed by Cloud (1983). This record is not, however, one of macroscopic skeletal remains, but is primarily composed of microscopic single-celled organisms or of macroscopic structures formed by colonial organisms (stromatolites). The period prior to the Phanerozoic extending back to the formation of the earth is known as the Precambrian.

12.1.1 Biologically Induced Mineralization in the Early Precambrian

The oldest solid fossil evidence showing that organisms were inducing minerals to form is derived from chemical analyses of the stable isotopic composition of sedimentary sulfide minerals (Monster et al. 1979). During bacterial sulfate reduction sulfide ions are produced intracellularly and are then released from the cell into the surrounding environment where they combine with cations to form insoluble precipitates. These may even form at some distance from the cell. During this process the heavy sulfur isotope ^{34}S is preferentially depleted relative to the light isotope ^{32}S. The extent to which it is depleted is much greater in biologically produced sulfide minerals than in inorganically precipitated ones. Furthermore, the degree of isotopic fractionation varies considerably from one biological process to another. Thus, analyses of the isotopic compositions of sulfide minerals can differentiate between biologically and abiologically produced sulfides. Analyses of sulfide minerals from many Precambrian deposits show that prior to 2.7 Byr BP the deposits were of inorganic origin and after that they were produced by organisms (reviewed by Schidlowski et al. 1983).

The oldest actual fossils that show direct evidence of biologically produced minerals occur in deposits formed 1.6 billion years ago. These are manganese-encrusting bacteria. The manganese deposits are remarkably similar in morphology to those produced by living bacteria (Muir 1978).

Both documented cases of biomineralization in the early Precambrian are of the biologically induced type. It is, however, quite conceivable that biologically induced mineralization evolved much earlier. The evidence to date is circumstantial at best, but worth considering. Banded iron formations are unusual rocks found only in Precambrian deposits older than about 1.6 billion years. The oldest banded iron formations known were formed 3.8 billion years ago. It has been proposed that their unusual banding, which is due to the intermittent precipitation of iron oxides with quartz, is itself a by-product of biological activity (Cloud 1973). Evidence consistent with this has been obtained from the study of living obligate anaerobic bacteria that are able to extracellularly reduce amorphous ferric oxyhydroxite to magnetite in the complete absence of oxygen. The magnetite crystals formed are between 10 and 50 nm in size (Lovley et al. 1988). Hence, the possibility exists that dissimilatory iron-reducing bacteria may have been responsible for the magnetite produced in the banded iron formations. *If* this proves to be the case, biologically induced mineralization may indeed have originated very early in the evolution of life on this planet.

12.1.2 Biologically Controlled Mineralization in the Precambrian

Acquiring direct documentation of biologically induced mineral products in the fossil record is very difficult because they are morphologically indistinguishable from their inorganic counterparts. Biologically controlled mineralization products on the other hand offer more promising avenues for direct investigation, because their morphologies are often, but not always, quite distinct from their inorganic counterparts. Those that are can be unequivocally recognized as being of biological origin even though no trace of the organism that formed them remains (Lowenstam and Margulis 1980b; Kirschvink and Chang 1984).

We have a number of reasons to believe that biologically controlled mineralization originated sometime in the Precambrian. Judging from the widespread distribution among living organisms of mineralization products comprising single crystals, small crystal aggregates formed in vesicles, spicules, and so on, many of which are formed under strict biological control; it is quite conceivable that the Precambrian fossil record contains these noncomposite skeletal materials dispersed among all the other sedimentary deposits. In addition, there is no a priori reason to believe that these were formed only after the eukaryotes evolved some 1.5 to 2.0 billion years ago, as we do have documented examples of living prokaryotes that form mineralized products under controlled conditions (summarized in Lowenstam and Weiner 1983). One of these examples is the magnetotactic bacteria that form intracellular chains of magnetite crystals within organic sheaths (Blakemore 1975). The magnetite crystals usually have shapes quite different from inorganically formed magnetite. They live in sediments with low levels of oxygen. In fact, the optimal oxygen requirement for *A. magnetotacticum* to form magnetite is around 5% of present atmospheric levels (PAL) (Blakemore et al. 1985). Therefore, these bacteria are not likely to have evolved before atmospheric oxygen levels reached this point, which probably occurred around 2.0 Byr BP. Thus, it might be possible to search the sedimentary record back some 2.0 billion years for uniquely shaped magnetite crystals of a specific and uniform size for evidence of the existence of fossil magnetotactic bacteria.

Kirschvink and his colleagues have been involved in this pioneering work and have incontrovertible evidence showing the presence of biologically formed magnetite crystals as far back as 600 to 700 million years ago (Chang et al. 1987) (Fig. 12.1). They have also identified prismatic and cuboidally shaped single-domain-sized magnetite crystals in Precambrian rocks as old as 2 billion years, raising the possibility that the magnetotactic bacteria evolved around that time (Chang 1988). Another exciting development along these lines is based on the discovery that certain eukaryotic magnetotactic algae form arrowhead-shaped magnetite crystals (Torres de Araujo et al. 1986). These are similar, but hopefully not identical to some magnetite crystals formed by certain magnetotactic bacteria (Blakemore et al. 1981). If the bacterial and algal crystals can be distinguished, the possibility exists of finding the algal crystals in the fossil record and in this way obtaining an indication of when eukaryotes evolved the ability to control mineralization.

A major event in the evolution of life, as documented by the fossil record, occurred possibly as early as one billion years ago with the appearance of the first metazoans. The fossils are either in the form of tracks, trails, or burrows left by organisms of unknown affinity or the actual impressions of the body parts themselves. The latter occur only in rocks deposited between 660–680 and 560–570 millions years ago—the so-called Ediacaran fauna (Glaessner 1984). These metazoans may have had fairly stiff and rigid external body coverings, but it is significant that, with few exceptions, they did not reinforce their organic structures with mineral. The exceptions in actuality are more suggestive than proven. One comprises tubes rich in organic material that are now partially impregnated with calcite (Glaessner 1976). The calcite may be of diagenetic origin, although it is possible that it could

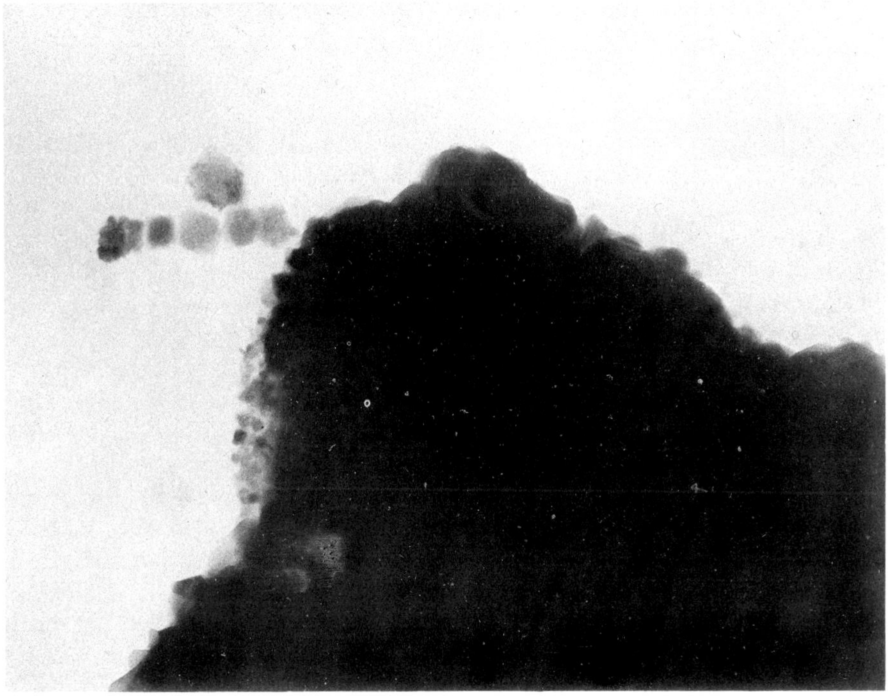

Figure 12.1. Transmission electron micrograph of a chain of fossil magnetite crystals with size and shape characteristic of those found in living magnetotactic bacteria. The crystals were obtained from rocks belonging to the Nama Group (600–700 million years old; Mooifontein Member, Kuibis subgroup). Scale bar: 100 nm. Courtesy of Dr. S. Chang.

have been induced by the metazoan that occupied the tube. The second possible exception, noted by Glaessner and Wade (1966), concerns certain elongated impressions on the surfaces of fossil colonial organisms possibly related to the Pennatulacea or sea-pens in the phylum Cnidaria. Glaessner suggested that the impressions may have been the locations of calcareous spicules that are known to be formed by living Pennatulacea and other Octocorallia (Chapter 5).

A more plausible argument for some of the Ediacaran fauna having mineralized hard parts is not based on fossil evidence, but is inferred from the fact that some of the fossils are Hydromedusae and Scyphomedusae. Cnidaria such as these float or swim in the water and many of the recent representatives have well-developed systems for orienting themselves in the earth's gravitational field. An essential part of the gravity receptor is a mineralized statolith or many small statoconia (see Chapter 10 for details), which in living Hydrozoa comprise amorphous Ca-Mg-phosphate and in the Scyphozoa gypsum or calcium sulfate (Table 5.1). It is not known whether they are preserved in the fossil record, but even if they are they would probably not have been recognized, because most paleontologists are not familiar with their appearance. As the gravity receptor is vital to the functioning of the organisms, as evidenced by the fact that it forms very early during development, it is likely that the Ediacaran medusae must also have formed these structures.

12.1.3 The Advent of Composite Skeletal Formation: The Transition Boundary Zone between the Precambrian and the Cambrian

From the geologists point of view this usually represents the transition between sedimentary rocks with macroscopic fossils and those without. From our point of view, this relates to one of the most significant events in the evolution of biomineralization: the formation of large composite mineralized skeletons. This occurred around 570 million years ago.

The first appearances of these fossilized hard parts have been studied in great detail in rocks from all over the world. Figure 12.2 is a compilation of much of these data. Not shown is the fact that the remaining Ediacaran fauna all become extinct just before or just after the boundary. Figure 12.2 shows that organisms from many different phyla belonging to the kingdoms Monera, Protoctista, Animalia, and probably others, as many of these fossils are of unknown taxonomic affinity (so-called "problematica"), all began to mineralize "instantaneously." Although it is difficult to translate the term "instantaneously" used in its geological context into absolute time, there is no denying that something or some things must have triggered this event. Furthermore, since we have no evidence of less sophisticated skeletons being formed first, we can infer that by this time many, if not all, these organisms must have evolved mechanisms for supplying the ions to be deposited in an orderly and in some cases sustained manner, for constructing organic frameworks that are already integral components of many of these early hard parts, and for producing macromolecules to control the mineralization process itself. It is also significant that the first hard parts formed by organisms at or close to the boundary already include carbonates, phosphates, and silica. It is probably no coincidence that these are also the most abundantly formed mineral types found among living organisms, and the fact that they or their diagenetic alteration products are found in the boundary zone may just be because they have the best chance of being preserved. We would not be surprised if many more biogenic mineral types were also found. Either way, the known diversity of minerals formed by these early skeletal-mineralizers supports the notion that they already possessed proficient apparatuses for forming their hard parts.

Much has been written about what may have occurred around the Precambrian–Cambrian boundary and we too will add our thoughts. We will do this, however, at the end of the chapter to give the reader the benefit of a perspective that includes events during the later Phanerozoic. The evolution of biomineralization during the Phanerozoic is reviewed in three subsections: the evolution of carbonate mineralization, phosphate mineralization, and silica (opal) mineralization.

12.2 Evolution of Carbonate Biomineralization

Of the seven different carbonate minerals formed by living organisms, only two are commonly found: calcite and aragonite (Table 2.1). Vaterite, the third polymorph

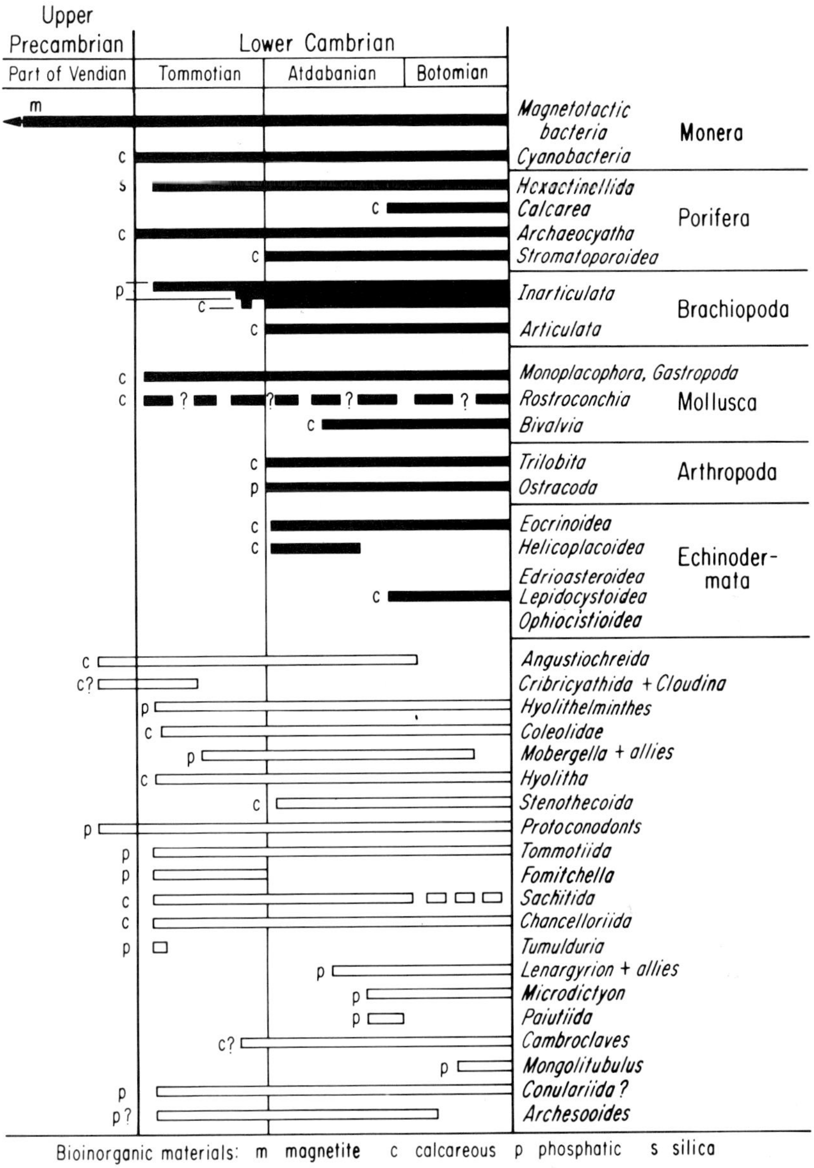

Figure 12.2. Compilation of the stratigraphic ranges of fossil taxa with mineralized hard parts from late Vendian and early Cambrian deposits. Redrawn and updated from Lowenstam and Margulis (1980b) in consultation with Dr. S. Bengtson.

of calcium carbonate, is now known to be formed by a fairly large number of organisms, but compared to calcite and aragonite it is still rather rare (Table 2.1). The other biogenic carbonate minerals (hydrocerrusite, monohydrocalcite, protodolomite, and amorphous calcium carbonate) are formed by relatively few organisms and in some cases are only transient phases of more stable end-products. Thus, from the perspective of the carbonate mineralizing activities of living organisms, the carbonate fossil record is expected to be dominated by calcitic and aragonitic hard parts, and indeed it is.

Calcite is thermodynamically more stable than aragonite and much more stable than vaterite. Thus, with time, the less stable polymorphs tend to dissolve and reprecipitate as calcite. The result is that there is little chance of finding preserved vaterite. The frequency of preserved aragonitic fossils decreases dramatically with increasing geologic age to the point that the oldest known sedimentary deposit that contains abundant aragonitic fossils is in the Carboniferous (340 million years) (Hallam and O'Hara 1962) and the oldest known preserved aragonite occurs in trace quantities in a fossil from the Ordovician (about 470 million years) (Sandberg 1975).

There is, however, a way of reliably distinguishing between a fossil with originally preserved calcite and one in which the calcite is of secondary origin as a result of the aragonite having been replaced by calcite. The secondary calcite crystals have a characteristic sparry texture and if in the same deposit sparry calcite-containing fossils occur alongside ones with originally preserved calcitic textures, it can be assumed that the former were originally aragonitic (Bøggild 1930).

The large majority of fossils with carbonate hard parts are from the Animalia or Protoctista, and, to a lesser extent, the Monera. The other two kingdoms with very few exceptions do not have preserved fossil remnants. The record is also heavily biased toward marine organisms, because marine deposits are much better preserved than continental ones, which are often destroyed by erosion. This is particularly true for the older deposits.

The first documentation of carbonate mineralization in the fossil record goes back some 1,000 million years. Lightly calcified cyanobacteria are present in rocks from this period through the Vendian times. However, in the very late Vendian, about 580 million years ago, they suddenly became heavily encrusted with calcareous mineral (Riding and Voronova 1982; 1984). Mineralized eukaryotic algae make their first appearance in late-Vendian times (Yudomian) and undergo a massive expansion in the number of mineralizing taxa at the boundary between the Vendian in the late Precambrian and Tommotian, which is the base of the Cambrian (Riding and Voronova 1984). This phenomenon repeats itself with members of many other phyla at or just after this boundary, with the result that a massive explosion of carbonate mineralizing organisms occurred. This phenomenon, of course, is not restricted to calcium carbonate mineralization (Fig. 12.2) and, in fact, carbonate-precipitating organisms constitute only about half of the first mineralizers at this time. By the end of the Cambrian (about 500 million years ago), however, they predominate.

The Lower Cambrian represents a period of experimentation in body plan design with spicular or discontinuously mineralized skeletons being rather common (Bengston 1977). The result is that a major portion of the carbonate mineralizing organisms are of unknown taxonomic affinity (problematica) (Fig. 12.2). By the end of the Cambrian most become extinct and are replaced by organisms with modern representatives. The monoplacophoran and gastropod mollusks appear at the very base of the Tommotian, as do a group of now extinct calcareous sponges (Porifera) called Archeocyathids (Debrenne and Vacelet 1984). Unlike most of the organisms with mineralized hard parts from these times that were generally small, the Archeocyathids were large and formed rigid skeletal frameworks, which constituted the first biologically precipitated reefs. The end of the Tommotian and the beginning of the Atdabanian (Fig. 12.2) is marked by the appearance of calcareous brachiopods and the well-known trilobites—arthropods with mineralized exoskeletons. The first fossil bivalves are from the mid-Atdabanian and echinoderms from the upper Atdabanian (Fig. 12.2).

It is interesting to note that almost all these organisms had a sessile benthic mode of life. Because of poor preservation, we often do not have direct documentation showing which organisms formed calcite and which aragonite. Judging both from the living representatives and the better preserved fossils, the calcareous porifera, brachiopods, arthropods, and echinoderms were all probably calcitic. The first documentation of aragonite, albeit indirect, occurs in a bivalve from the mid-Atdabanian (Runnegar and Jell 1976). We know very little about the identity of the carbonate minerals precipitated by the problematica.

During the Middle and Late Cambrian a number of important carbonate mineralizers appear in the fossil record. These include the coralline algae and the nautiloid cephalopods. The 65 or so million years following the Cambrian is called the Ordovician during which time the calcareous algae expanded significantly and the first calcitic rugose corals appeared as well as the first mineralizing foraminifera. The familiar echinoderm classes, crinoids, echinoids, and holothurians, also make their first appearance. By the end of the Ordovician most of the problematica had disappeared and all but three of the important mineralizing phyla as we know them today were all well entrenched.

It is interesting to reflect on the manner in which carbonate mineralization evolved based on the fossil record. A good example, which is probably representative of most groups, is the Echinodermata. Even the very earliest fossils display the unusual stereom structure characteristic of this phylum. In Chapter 8 we point out just how complex a mineralizing structure the echinoderm skeleton is, and yet the fossil record provides not even a hint of how it evolved. This is true for most other mineralized hard parts. In fact, the one exception that proves the rule is the Foraminiferida. The first foraminifera constructed their tests entirely out of organic material and only in the Ordovician did they begin to form biologically mineralized structures (Loeblich and Tappan 1964).

12.2.1 The Deposition of Aragonite or Calcite

The fossil record of the Cambrian carbonate mineralizers is still too poorly documented to determine whether the first mineralizers all formed calcite with aragonite a later addition. As noted, the first documented case of an aragonite-precipitating organism is that of a bivalve from the mid-Cambrian (Runnegar and Jell 1976). The impression gained by extrapolating from later times back to the Cambrian is that calcite deposition did predominate. This, however, automatically excludes the problematica as we do not know their taxonomic affinities and most of them were extinct by the end of the Cambrian.

A comparison of the biogenic carbonate skeletons formed in the Mesozoic (about 230 to 65 million years ago) and Cenozoic (65 million years ago to the present) with those formed in the earlier Paleozoic (680 to 230 million years ago) shows two major differences: more taxa form aragonite than calcite and the absolute volume of biogenic carbonate precipitated is greatly increased. To address questions as to how these changes come about, we will briefly review *some* of the pertinent data from the fossil record.

The *Mollusca* illustrate the two main pathways by which later evolving taxa came to form more aragonite than calcite. Some of the early Paleozoic monoplacophorans are known to have had shells composed entirely of calcite (Bøggild 1930). In mid-Silurian times only the outer layer of monoplacophoran shells comprised calcite; the inner layer was aragonitic (Lowenstam 1963). All known living monoplacophorans have shells composed entirely of aragonite (Lowenstam 1963). Note that there are no records of monoplacophoran fossils between the late Devonian and the Recent and, furthermore, that the living species with one exception are all deep-sea dwellers whereas the fossil ones occupied shallow shelf waters (Lowenstam 1978). A similar phenomenon occurred in a group of cephalopods called the nautiloids. Some Ordovician and Silurian nautiloid shells were composed of outer calcitic and inner aragonitic layers (Bøggild 1930), whereas from the Carboniferous (about 345 million years ago) to the present all nautiloid shells were composed entirely of aragonite (Stehli 1956; Hallam and O'Hara 1962).

The second pathway by which this is achieved is either the evolution of new taxonomic groups or the expansion of existing ones that form aragonite. There are many examples of both cases in mollusks and, in general, because the whole phylum expanded during the Phanerozoic, they contributed significantly to the overall increased production of biogenic aragonite. One important example is the nektonic pteropoda. These gastropods have shells composed of aragonite. They evolved in the early Cenozoic and became major constituents of the open oceans, where today they contribute significantly to the production of biogenic carbonate (Herman 1978).

The Cnidaria are an extremely important group in this regard both in terms of the volume of biogenic carbonates that they produce and their tendency to form large reef structures (see following). During the Paleozoic the rugose corals with their calcitic skeletons were among the major reef formers (Stehli 1956). In the Mesozoic and Cenozoic they were replaced by a taxonomically quite different group of reef-building Cnidaria, called the Scleractinia (Wells 1956); whose skeletons are composed almost entirely of aragonite (see Chapter 5). It is also of interest to note that the two "late additions" during the Cenozoic to the reef-building community from the order Hydrozoa (*Milleporina* and *Stylasterina*) were for the most part aragonitic (Lowenstam 1964b). This change from calcite to aragonite within the reef-building Cnidaria is one of the major reasons more aragonite was formed in the late Phanerozoic than in the Paleozoic.

The Protoctista contain members of two phyla that are important carbonate producers. The *Coccolithophoridae* and the planktonic *Foraminiferida* are abundant constituents of the world's open oceans today and have been through most of the late Mesozoic and the Cenozoic. They both form low magnesium calcite (Table 4.1) and, thus, significantly offset any trend toward more aragonite being formed with time in terms of the proportion of total biogenic carbonate production. It is, however, significant to note that in the mid-Mesozoic (150 million years ago) one taxon of benthonic foraminifera, the *Robertinacea,* evolved with aragonitic shells.

They represent the only exception in a long history of calcite formation by the foraminifera (Loeblich and Tappan 1964).

The *Inarticulate Brachipoda* contain taxa that from Cambrian times until the Recent form calcitic shells. One group, which evolved in the Ordovician and became extinct in the late Silurian, the *Trimerellacea,* formed aragonitic shells (Jaanusson 1966).

From this brief review we can conclude that the primary reasons for more aragonite being formed in the Cenozoic than in the Paleozoic is the replacement of the calcitic rugose corals by aragonitic scleractinian corals, the evolution of the nektonic pteropods, and the general expansion of mollusks, many of which form aragonite or aragonite-containing shells. The far more difficult question to address is why aragonite is deposited in preference to calcite or vice versa.

There is apparently no general preference by organisms for forming aragonite over calcite. Furthermore, there is clearly no specific period of time in the Phanerozoic during which groups of different carbonate mineralizers opted for one polymorph or the other. Thus, an explanation of the phenomenon involving changes in the environment such as the Mg/Ca ratio of seawater (Folk 1974; Wilkonson 1979) cannot be generally applicable, although it could of course be responsible for specific cases. Sandberg (1983) correlated periods in which inorganically formed oolites changed from calcite to aragonite or vice versa with climatic oscillations. He proposed that the most likely mechanism responsible for the mineralogical changes was the partial pressure of CO_2 in the atmosphere. The times of oolite mineralogical changes do not, however, correlate with changes from calcite to aragonite or vice versa in the fauna.

One important question is whether or not the formation of one polymorph or the other affords the organism any benefit. We do not know the answer and, unfortunately, we still have no idea how organisms form aragonite in preference to calcite or vice versa. If we knew, for example, that they do this by simply manipulating the Mg and/or Sr contents of the mother liquor from which the crystals form, then fluctuations in the seawater concentrations of these ions could well be crucial. If, on the other hand, the opposite turns out to be the case, namely that they utilize a complex process involving a battery of gene products, then factors that determine the polymorph type formed could be quite unrelated to environment. They might depend on the locations of the genes, or the nature of the neighboring genes with which they are associated, etc. It is also conceivable that an organism is selected for a particular trait that is quite unrelated to mineralogy and the actual mineral precipitated is of no direct consequence to its survival. (The phenomenon is generally known as pleiotropy.) We have no answers to these questions, but hope that this fascinating and fundamental issue will be a high priority for modern paleontology.

A thought-provoking unpublished observation by H. A. Lowenstam in this regard occurred while making tightly spaced electron probe traverses across a section of the tube of a tropical serpulid worm, *Spirobranchus gigantheus,* from Barbados. A bulk analysis of the tube showed that it is composed of aragonite, and as expected the Sr contents were high and the Mg contents very low. Every so often, however the reverse occurred with very high Mg and low Sr contents occurring in localized patches, a clear indication of the presence of high Mg calcite. It would appear that the "aragonite-producing system" was occasionally perturbed and, as a result, reverted back to forming high Mg calcite. It is also interesting to note that in Barbados two species of serpulid worms live side by side; the one forms an aragonite tube and the other a high Mg calcite tube (Lowenstam 1964b).

12.2.2 The Increase in Biogenic Carbonate Formation during the Phanerozoic

The precipitation of biogenic carbonates in the marine environment increased almost exponentially during the late Mesozoic and Cenozoic, primarily as a result of two important developments: the formation of very large biologically formed reef structures and the invasion of the open ocean by various carbonate-depositing organisms. The impact of these events on the chemistry of the present day ocean is discussed in Chapter 2. Here we briefly outline the evolutionary history of these events.

Biogenic carbonate reefs are formed today primarily by colonial scleractinian corals together with coralline algae and many other sessile skeletal-depositing organisms such as serpulid worms, sclerosponges, bryozoa, and the encrusting foraminifera *Homotrema rubrum.* Reefs can establish themselves on both consolidated and unconsolidated substrates, provided the latter are below effective wave base. The reef-building organisms construct open frameworks that support entire communities of other organisms, many of which also form carbonate skeletons. These accumulate in the spaces within the framework and with time are cemented in place to ultimately form what is called a "room and pillar" structure (Lowenstam 1950). The skeletal reef debris forms a talus slope around the reef. This, in turn, can act as a substrate for further reef expansion. Thus, reefs can grow to eventually form structures that are not only of very large dimensions and are wave resistant, but can actually build themselves out into the prevailing surf, a remarkable feat considering that most rocks are eroded by wave action. Another important quality that has contributed to their success is the ability of the scleractinian corals and the other reef builders to grow fast enough when necessary to raise themselves and with them the entire reef community from below wave base to the water surface. Furthermore, if the substrate upon which the reef resides is subsiding, they are often able to grow fast enough to keep up with the tectonic subsidence. In the Eniwetok atoll, for example, they have succeeded in maintaining contact with the surface for 60 million years even though the substrate has subsided during this time (Ladd and Schlanger 1960).

The fossil record of reef-building activity starts at the base of the Cambrian with the archeocyathids forming reefs, albeit rather modest ones. The first major reef-building period occurred some 435 million years ago in the Silurian, when structures comparable to today's Great Barrier Reef were formed. One magnificent archipelago stretched from the mid-western United States where it was anchored to the continental Ozark Mountains all the way to northwest Greenland (Lowenstam 1957). The major reef-building organisms in the Silurian were stromatoporoid sponges, tabulate and rugose corals, and coralline algae. Starting with the Triassic (about 230 million years ago) the scleractinian corals together with coralline algae become important reef builders, a trend which continued until today. They were not, however, the only reef builders during this period. In the late Paleozoic (roughly 280 million years ago) and again in the Jurassic (about 180 million years ago) reefs were formed mainly by calcisponges. The caprinid mollusks formed reefs in the early Cretaceous (100 to 120 million years ago) and related mollusks called rudistids formed reefs during the late Cretaceous (65 to 100 million years ago). The early Cenozoic, same 65 million years ago, marks the beginning of the expansion of the scleractinian coral–coralline algal reefs that are so prominent in today's oceans (see papers in Oliver et al. 1984). This last phase, in particular, reflects a marked increase in absolute volume of carbonate fixed, so much so that it is estimated that about half of all the calcium entering into present-day oceans is tied up in coral reefs (Smith 1978). It is interesting to note that during the late Pleistocene a marked change took place among the scleractinian coral fauna when the porous corals (mostly *Acroporidae*) became the dominant reef constituents rather than the massive corals. The porous corals are able to grow about 10 times faster than the massive corals and they were selected for during the rapid sea level rises that occurred during deglaciation periods (Gerth 1929).

The massive invasion of the surface waters of the open ocean by carbonate-depositing organisms is a relatively "recent" phenomenon occurring some 195 million years ago in the Jurassic. This does not imply that skeletal-bearing organisms were not able to enter this realm prior to this time. Earlier documented records exist of mobile cephalopods and trilobites with nektonic modes of life (summarized by Lowenstam 1974). In the Jurassic two important carbonate-bearing protoctist groups evolved, the planktonic Coccolithophoridae and the planktonic Foraminiferida (Chapter 4). Together with various other groups of carbonatic organisms, some of which are collectively referred to as calcareous nannoplankton, they invaded the open ocean and began to produce carbonate on a large scale. Interestingly, some of the Mesozoic planktonic constituents are not represented or are poorly represented in today's oceans. The carbonate cysts of dinoflagellates are minor constituents of today's oceans, but were common during the Jurassic and Cretaceous (in Tappan 1980). Planktonic crinoids (Echinodermata) evolved in the Triassic and became extinct in the late Cretaceous (Santonian) (Lowenstam 1974). At times they were very abundant and in the Jurassic sometimes accumulated in sufficiently large quantities to be rock forming.

The Coccolithophoridae evolved in the early Jurassic and the planktonic foraminifera in the late Jurassic. Both groups underwent adaptive radiations in the mid-Cretaceous. This together with the fact that they have short life-spans, resulted in an almost exponential increase in the volume of carbonate (in both cases low Mg calcite) formed. Thus, the Cretaceous is the first time that the major center of biogenic carbonate deposition shifted from the continental and shelf seas to the open oceans. In spite of the massive extinction that affected both these groups at the end of the Cretaceous, they recovered and again underwent adaptive radiations to regain the same niches that they had previously occupied. Furthermore, they were joined in the early Cenozoic by nektonic gastropods, the aragonitic pteropoda and heteropoda, and *Janthina,* which has a calcitic and aragonitic shell (Lowenstam 1974).

Thus, from the late Cretaceous to the present, large amounts of biogenically formed carbonates accumulated on the ocean bottoms. Their formation has affected not only the nature of the sediments, but the chemistry of the ocean water and through it the earth's atmosphere as well (Chapter 2).

Biogenic carbonate formation on land has never had a major impact on the environment except in certain local water bodies. It has its origins in the late Paleozoic (Carboniferous) with the evolution of pulmonate land snails (Solem and Yochelson 1979). Freshwater bivalves are thought to have evolved in the Mesozoic. In terms of volume of carbonate deposition in freshwater bodies, the algal taxon Charophyta is probably the most important. It has its origins in the early Devonian about 400 million years ago (Tappan 1980).

12.3 Evolution of Phosphate Mineralization

Phosphate is unlike carbonate in one important respect, it is a vital nutrient of primary producers and, of course, essential to all life forms. It is, therefore, not surprising that no phytoplankton or zooplankton species are known to form phosphatic hard parts as these organisms live in an environment in which phosphate is at a premium. A priori one would expect that the primary purpose of organisms in producing phosphatic mineral deposits would be for temporary storage of this valuable anion, rather than for constructing hard parts. A survey of the phosphatic deposits formed by living organisms shows indeed that many do form temporary storage granules composed of phosphate mineral, which is usually in the amorphous form (Brown 1982). On the other hand many others are known to construct hard parts out of phosphatic minerals and, surprisingly, with three major exceptions (vertebrate skeletons, certain inarticulate brachiopod shells, and the teeth of warm water chitons), most of them use amorphous calcium phosphate (ACP). Examples are the sternal shields of certain annelid worms *(Sternaspis)* (Lowenstam 1972b), the bulk of the gizzard plates of cephalaspidian gastropods (Lowenstam 1972b), gill supports of the bivalve *Neotrigonia* (Lowenstam 1972b), stylets of nemertian worms (Stricker and Weiner 1985), granules of polychaete worms (Gibbs and Bryan 1984), and the holothurian Molpadiidae (Lowenstam and Rossman 1975). Although we do not understand why they use the amorphous rather than the crystalline form for these structures, the fact remains that of the 22 phyla known to contain members with phosphatic minerals, 20 form amorphous minerals and only 3 form crystalline ones for hard part construction. This observation has a profound implication in terms of evaluating the fossil record of phosphatic biomineralization.

The fossil record of phosphatic hard parts is, to date, confined to the Animalia. Figure 12.3 shows the distribution of amorphous and crystlline phosphatic minerals among living and fossil animals. The amorphous form is common among living animals, whereas the fossil record is, with one exception, limited to hard parts that were probably originally deposited in a crystalline form. This is not unexpected. Most paleontologists are not familiar with the ACP-bearing hard parts of living animals and would, therefore, probably not recognize them even if they were preserved. The chances of them being preserved under normal burial conditions are not good because the spherular aggregates that generally make up the ACP structure are likely to disintegrate and be dispersed by bioturbation. Furthermore, as phos-

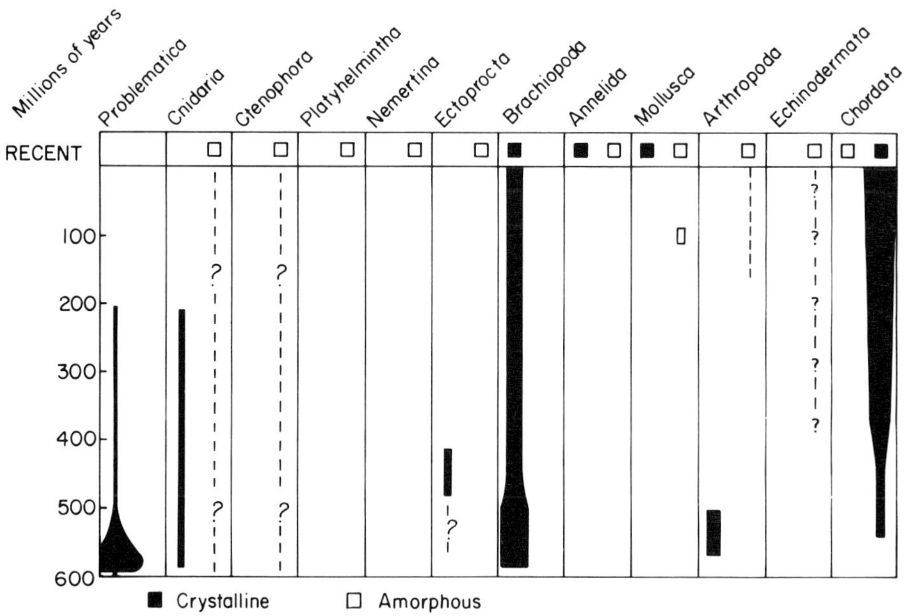

Figure 12.3. The distribution of amorphous and crystalline phosphatic mineralized hard parts among members of extant and fossil phyla. The stratigraphic ranges of the fossils are expressed in millions of years.

phate is a nutrient, there is a good chance of it being recycled through bioturbation. Recent marine sediments do appear to be largely devoid of preserved macroscopic hard parts composed of ACP (H. A. Lowenstam, personal observations).

The one known exception occurs in an Upper Cretaceous deposit (about 100 million years old) in Lebanon, in which whole fish and the impressions of "chitinous" body parts are also preserved (Lowenstam et al. 1983). An impression of a fossil nautiloid *(Somalinautilus libanoticus)* shell was found that contained piles of phosphatic uroliths bordering the region of the upper mandible. These uroliths still have the morphology characteristic of ACP, although the material itself is now crystalline carbonate apatite (dahllite). The fossil uroliths closely resemble those of living *Nautilus,* which are composed of ACP (Lowenstam et al. 1983). The important implication of this finding is that the fossil record can indeed provide information about the history of ACP-bearing hard parts, when preserved under appropriate burial conditions. However, the fossil record of phosphatic biomineralization products is currently confined, with this one exception, to fossils that were probably originally crystalline when deposited. Therefore, they almost certainly represent a minor part of the total picture.

The fossil record of phosphatic mineralized hard parts (Figs. 12.2 and 12.3) begins at the end of the Vendian, some 570 to 580 million years ago. The taxonomic affinities of most of the organisms that formed these hard parts are unknown. In fact the major part of the fossil record of phosphatic mineralizers in the Cambrian comprises these problematica, most of which become extinct by the end of the Cambrian. One wonders if there is any connection between their becoming extinct and their opting for the use of phosphate for skeletal hard part formation? It is even possible that unknown to us some survived by switching to carbonate skeletons. Their initial selection of phosphate may have been, in part, because the concentrations of phosphate in seawater toward the end of the Vendian and the beginning of the Tommotian were relatively high, as evidenced by the fact that this was a period of worldwide phosphatization (Cook and Shergold 1984). The tendency toward postmortem phosphatization introduces significant ambiguity in the interpretation of the fossil record from this period, as carbonate-bearing skeletal remains could be diagenetically replaced by phosphate. Careful studies of preserved microarchitecture are needed as one criterion for differentiating between original deposition and diagenetic alteration, particularly for the problematica. Another valuable criterion is whether the phosphatic hard part persists with the same mineralogy into an environment that is dominated by carbonate production. The evidence for original phosphate deposition by organisms of known taxonomic affinity, summarized in the following, is more easily substantiated.

Shells formed by phosphate-depositing inarticulate brachiopods are known from the early Tommotian. They are common throughout the Cambrian up to the mid-Ordovician, a period of about 100 million years. Deposits at the base of the Cambrian also contain skeletal remains of organisms thought to be conularids (Bengtson, personal communication). The Conulariida had exoskeletons with apparent tetrameral symmetry. Kinderlen (1937), therefore thought that they belong to the Scyphozoa, a class within the Cnidaria. Most investigators have subsequently concurred with this view (see Bischoff 1978). Clusters of massive, thick-walled conularids have been located firmly attached to the rocky substrate of the wave-resistant Silurian reef at Thornton, Illinois (H. A. Lowenstam and J. H. Bretz, unpublished observation). Similar massive conularids have been described from the Silurian of Bohemia where they were apparently also part of the sessile community of the reef. These are unquestionably conularids and therefore do not fit into the anatomical interpretation, taxonomic assignment, and ecologic niche proposed by Babcock and Feldmann (1986) for this group. As depicted by these authors, conularids were probably pelagic organisms that had phosphatic, flexible, four-sided bilaterally symmetrical pyramidal exoskeletons, and, therefore represent an extinct phylum (problematicum). The conularids become extinct in the late Triassic. Ostracods with phosphatic shells or carapaces evolved in the Atdabanian (Müller 1964). They formed phosphatic carapaces throughout the Cambrian, but then switched to forming calcitic exoskeletons. Living calcitic ostracods, however, may still retain vestiges of this early phase in their history (see following). Apatite mineralization has been noted in mid-Cambrian and in certain late Silurian trilobite exoskeletons, where it forms a thin surface layer. Questions still remain as to whether this layer was formed by the animal or is of diagenetic origin (Teigler and Towe 1975). Phosphatic skeletal fragments apparently of jawless armored fishes have been found in later Cambrian deposits. These would constitute the oldest known remains of the vertebrates (Bockelie and Fortey 1976), if indeed they are vertebrates (Elliot 1987). One of the phosphatic problematica may represent an early bryozoan (Ectoprocta) (Lowenstam 1972b). Some Bryozoa from younger deposits (the mid-Ordovician and the mid-Silurian) are known to have skeletons composed of an outer layer of calcite and an inner layer of a phosphate mineral, preserved as francolite (Martinsson 1964).

In spite of the difficulties in interpreting an incomplete and sparse fossil record, there can be little doubt that the proportion of organisms that formed phosphatic skeletons in the Cambrian and particularly in the early Cambrian was high compared to later times and the present. Lowenstam and Margulis (1980b) originally estimated that some two-thirds of the morphotypes formed by late Vendian and half by early Cambrian were phosphatic. The latest estimates by Bengtson (personal communication) for Tommotian animals are not any different (about half formed phosphatic skeletons). In Figure 12.2 a third of the organisms are listed as forming phosphatic skeletons. Considering that the fossil record is unlikely to include many skeletal deposits that were originally amorphous, these estimates could well be on the *low* side.

The vertebrates with their carbonate apatite skeletons are the most prominent group of phosphatic mineralizers throughout the Phanerozoic starting with the mid-Ordovician (about 470 million years ago). The jawless fishes underwent a radiation during the late Ordovician and Silurian, but by the end of the Devonian (345 million years ago) were greatly reduced in numbers. The nonteleost boney fishes make their appearance at the beginning of the Devonian, some 395 million years ago, and undergo various adaptive radiations, the last of which, starting some 65 million years ago in the teleost fishes, were by far the most impressive. The amphibians evolved in the late Devonian (350 million years ago) and were the first vertebrates to occupy niches on land. Reptiles evolved on land in the late Carboniferous (285 million years ago) and reentered the marine environment later in the Permian. The mammals were the next to evolve, with the oldest known mammalian fossils from the very late Triassic (about 200 million years ago) and finally the birds evolved in the mid-Jurassic (about 170 million years ago). For more details see Romer (1966), Olson (1977), and the beautifully illustrated book by Simpson (1983).

Thus, based on the known fossil record, we conclude that the primary period of phosphatic biomineralization, started in the late Vendian, peaked during the Cambrian, and by mid-Orodovician times was confined for the most part to the vertebrates. Many Cambrian phosphate-bearing organisms apparently became extinct, whereas others started forming carbonate minerals. Among the latter are three interesting examples in which the living descendants still seem to retain vestiges of their phosphatic ancestry. All living ostracods have calcitic carapaces and have had since the Ordovician. In the Cambrian, however, ostracods formed carapaces composed of a phosphate mineral. Rosenfeld (1979) has shown that in living freshwater and brackish water ostracods there is a layer of calcite and poorly crystalline calcium phosphate-bearing granules within the epidermis in contact with the carapace. Significantly, during the juvenile moult stages, these granules are scattered within the calcitic carapace itself. Some Ordovician and Silurian ectoprocts or bryozoa formed skeletons and calculi composed in part of a phosphate mineral (Oakley 1934; Martinsson 1964). The skeletons of certain living ectoprocts contain relatively high concentrations of phosphate (Schopf and Manheim 1967) and the infrared spectrum of the mineral from one species shows that it is ACP (Hunt 1972). The conularids formed phosphatic skeletons from the Cambrian to the Triassic. These animals were probably Scyphozoa, a class within the phylum Cnidaria (Bischoff 1978). It is interesting to note that living hydrozoan medusoids form statoliths and statoconia as part of their gravity perception apparatus, which are composed of a Ca-Mg-phosphate mineral (Chapman 1985).

The wide distribution of ACP-bearing hard parts among living organisms and their almost total absence in the fossil record imply that much still remains to be learned about the evolution of phosphate mineralization.

12.4 Evolution of Silicification

Among living organisms the formation of siliceous skeletons is common and widespread in the kingdoms Protoctista and Plantae and in the animal phylum Porifera (sponges). To date no Fungi or Monera are known to definitely form opal (also known as silica) (Table 2.1). In the Animalia, the distribution of silicified hard parts is usually limited to lower taxonomic levels, such that only species of the same genus or at most of the same family share this trait. This is probably an indication that, with the exception of the sponges, silicified deposits in the Animalia fulfill relatively specialized roles.

Present day oceans are all undersaturated with respect to opal primarily as a result of silicon being removed en masse by organisms that form siliceous skeletons (Broecker 1971). The most important are the planktonic radiolaria and diatoms. As both these protoctists are concentrated in the photic zone, the degree of undersaturation of seawater with respect to opal is greatest in the upper surface waters. This, in turn, means that most opaline skeletons deposited in continental shelf sediments dissolve. A good illustration of the rapidity of this process is that in sclerosponges with siliceous spicules the opal spicules at the base of the massive skeleton are often partially dissolved (Hartman and Goreau 1970; Land 1976). This probably occurs in tens of years and may well account for the absence of the opaline bases of patellid gastropod teeth in marine sediments, even though in some shallow water areas they are produced in very large amounts. In the open ocean fragile, lightly mineralized skeletons may be completely dissolved while settling through the water column. Siliceous skeletons on the deep-sea floor also partially or completely dissolve depending upon the rate of burial, pH of the local environment, and other chemical properties of the interstitial waters in the sediments. Only the more robust skeletons, particularly those accumulating in sediments under high productivity zones, tend to be preserved. For example, there are sediment zones in the equatorial regions that are very rich in sponge spicules (Lisitzin 1971). With time the siliceous spicules are buried and convert first to opal-CT (porcelanite) and eventually to crystalline quartz (Kastner et al. 1977; Kastner 1981).

Biogenic silica is also formed in freshwater and terrestrial environments. Diatoms, Chrysophyta, Rhizopoda, and silica sponges can all live in lakes, even when the dissolved silicon concentrations are very low (Voronkov et al. 1975). The diatoms in particular are locally able to deposit relatively large amounts of opal. On land, plants belonging to five different phyla (Table 2.1) are known to form opaline deposits. In terms of volume, the grasses and to some extent the scouring rushes of the Sphenophyta are the most important (Kaufman et al. 1973).

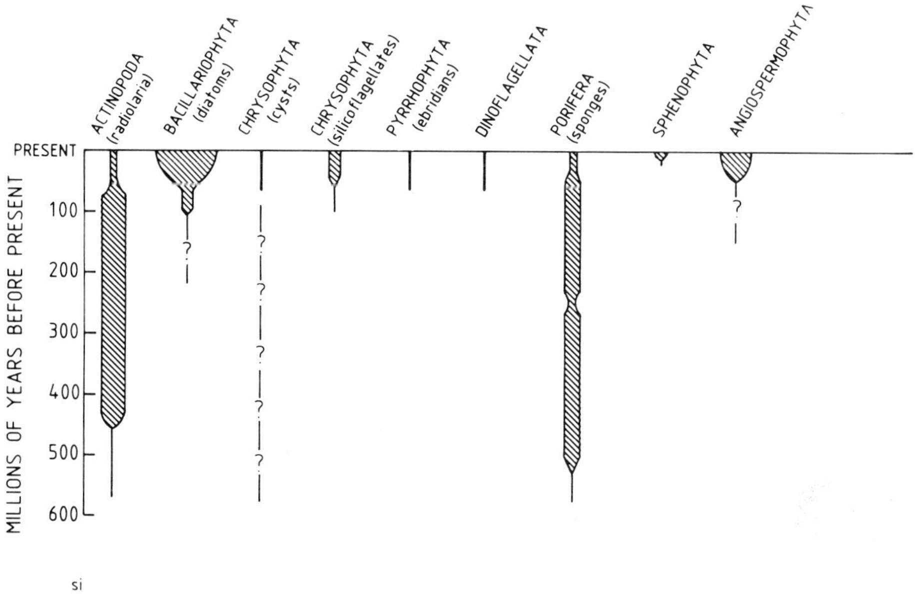

Figure 12.4. Compilation of the stratigraphic ranges of fossil taxa with siliceous mineralized hard parts. The relative abundance relationships are purely schematic.

The fossil record of biological silicification is confined to representatives of only 8 of the 22 living phyla known to form opal. These are listed in Figure 12.4. Furthermore, most of the fossil record is limited to Protoctista with two important exceptions, the sponges among the animals and the angiosperms among the plants.

The oldest known biogenic silica deposits are found at the base of the Cambrian, just above the Precambrian–Cambrian boundary (Figs. 12.2 and 12.4), where siliceous spicules of hexactinellid sponges have been found in marine deposits. These are lightly mineralized and have smooth surfaces, indicating that little or no dissolution took place (Sdzuy 1969). This may be an indication that their ability to form opal at this stage was still limited. Radiolarian skeletons are also reported to be present in Atdabanian sediments (Lipps, personal communication to H. A. Lowenstam). As a thorough search for radiolarian skeletons or siliceous skeletons of other microorganisms in Cambrian sediments has not yet been completed, it is quite conceivable that their record will be considerably extended, possibly even to older deposits.

The mid-Ordovician, some 450 million years ago, is the first time that extensive sedimentary accumulations of siliceous skeletal deposits occur. These are composed of radiolarian skeletons. This marks the beginning of mass fixation of silicon by these protoctists. Cherts derived from marine shallow water deposits of radiolarian skeletons or sponge spicules are found throughout the Paleozoic. Their robust skeletons and states of preservation, including skeletal siliceous cement, indicate that seawater was still saturated with respect to opal throughout this period (Lowenstam 1948).

An addition to the spectrum of opal-forming organisms occurred with the evolution of the diatoms in the early Jurassic about 200 million years ago (Fig. 12.4). They did not, however, become quantitatively important constituents of the marine planktonic community until about 100 million years ago in the mid-Cretaceous. During this period the silicoflagellates and also the Chrysophyta make their first appearances (Fig. 12.4). Further additions to the group of siliceous marine protoctists are the Ebridians and dinoflagellates, both of which evolved in the Paleocene, just after the massive extinction at the end of the Cretaceous (Fig. 12.4) (see Tappan 1980).

The expansion of the diatoms in the late Cretaceous and in particular the massive Cenozoic adaptive radiation during the last 65 million years had a profound effect both on the saturation levels of seawater with respect to opal and on the distribution and extent of mineralization of other siliceous organisms. Figure 12.4 shows that concurrent with the diatom expansion (number of species and absolute volume of opal formed), the radiolarians declined (Harper and Knoll 1975). This decrease occurs with respect to both the number of species and the extent to which the surviving species mineralized their skeletons. Average skeletal weight per individual radiolarian decreased during the Cenozoic (Moore 1969). The diatoms apparently had a similar effect on the siliceous sponges. A comparison of the Paleozoic and Cenozoic sponges shows that in many of the Paleozoic species the spicules are more robust and usually not cemented together with silica. These phenomena can probably be attributed to the competition with diatoms in the Cenozoic for available silicon from the seawater. Further support for this is the observation that spicule silicification of a living tetractinellid sponge from the British coast is interrupted during the period of diatom reproduction (Stone 1970). There is as yet no evidence to show that diatoms had a similar effect on silicoflagellates, Chrysophyta, and Ebridians.

Fossil silicoflagellates are generally preserved in siliceous deposits along with diatoms. In some Miocene deposits they are actually more abundant than the diatoms. This is also the time that they reach their acme in terms of species diversity. It would be interesting to determine whether the extent of silicification of their skeletons was also affected by the diatom radiation (Tappan 1980).

Sometime during the Cenozoic seawater became undersaturated with respect to opal almost certainly as a direct result of the diatom expansion. This may also account for the apparent progressive decrease in biologically derived cherts on continental shelves during the late Cenozoic. Cherts formed in shelf-sea deposits during this time were primarily derived from the diagenetic dissolution products of sponge spicules. Their decreased abundance during the Cenozoic can also be attributed to the diatoms' expansion.

Freshwater bodies do contain some opal-precipitating protoctists such as diatoms, chrysophytes, and rhyzopods as well as siliceous sponges (Table 2.1). Only the freshwater diatoms are known from the Cenozoic fossil record and not infrequently accumulated in huge quantities to form sizable diatomite deposits (Tappan 1980).

Silicification in the terrestrial environment is primarily confined to the plants. The extensive fossil record of plants goes back some 450 to 480 million years (Gray and Boucot 1977) and includes plant phyla that today are known to form silicified deposits (Fig. 12.4). Thus, the potential for a long and fairly complete record exists, but in actuality the documented record is very sparse indeed. It is limited to *Equisitum* in the phylum Sphenophyta and the angiosperms (Fig. 12.4). The oldest documented silicified plant deposits are from the beginning of the Cenozoic (Fig. 12.4). Within the angiosperms fossil opal is found in the seeds of various cereals and particularly among the grasses. The latter may have fairly heavily silicified deposits, which are often referred to as phytoliths (Jones 1964).

The grasses underwent a spectacular radiation in the Miocene, when they invaded the prairie type of environment. In the wake of this expansion, horses switched from a primarily browsing mode of life to a grazing one. The change in diet to abrasive silica-rich grasses was accommodated by extensive remodeling of the teeth to reduce their wear (Simpson 1951).

We conclude that in spite of the incompleteness of the fossil record, biological silica formation has undergone a spectacular expansion, particularly during the Cenozoic. This culminates in the wide utilization of silica for mineralization by a large number of living organisms. The diatoms are mainly responsible for the Cenozoic expansion in the marine environment, a direct consequence of which was that seawater became undersaturated with respect to opal. The expansion of terrestrial angiosperm grasses in the mid-Cenozoic resulted in huge amounts of biogenic silica being formed on land. It is curious to note that the expansions on land and in the ocean occurred more or less at the same time.

12.5 The Precambrian–Cambrian Boundary Zone: The Evolution of Composite Mineralized Skeletons

The history of biomineralization goes back 2,000 or more million years. Having reviewed this vast period of history, we have identified at least some of the important events in the evolution of mineral formation by organisms. One period undoubtedly stands out as being of exceptional interest: the events that occurred around 570 million years ago that resulted in the evolution of composite mineralized skeletons. It stands out not only because the formation of composite skeletons improved the "quality" of the fossil record enormously, but also because this event occurred during a dramatic adaptive radiation that resulted to a great extent in the establishment of the taxonomic phyla as we know them today. Why did hundreds of million years of mineralization by organisms elapse before this event took place? When it did occur, why were composite mineralized skeletons formed in so many different groups within a relatively short period of time? These are fascinating questions that have challenged many before us and will no doubt continue to do so in the future. Updated information both on the biomineralization processes of living organisms and from the fossil record does shed new light on this subject. We will broach the problem by breaking it down into four different but related subissues.

1. When did controlled biomineralization evolve?
2. What was the sequence of events in the transition zone at the end of the Precambrian and the beginning of the Cambrian?
3. What were the selection pressures that induced some organisms to form composite mineralized skeletons?
4. How did these organisms respond to the pressures?

Each of these questions will be addressed separately by drawing primarily upon the information set out in this chapter.

1. *The Evolution of Controlled Biomineralization.* We have outlined a number of lines of evidence supporting the idea that organisms evolved the ability to control mineral formation long before the end of the Precambrian. The most impressive direct evidence is that of Chang, Kirschvink, and their colleagues who have documented the ability of magnetotactic bacteria to grow magnetite crystals in an organic sheath under controlled conditions, certainly by the end of the Precambrian and probably even as early as 2.0 billion years ago. In these studies a deliberate search was made in the fossil record for particular biologically produced mineralized products. More studies of this type will, we are sure, provide support for the notion that controlled mineralization did occur well into the Precambrian. Judging from living organisms the mineralized products most likely to be found are intracellular granules used for temporary storage of mineral, and spicules or other mineralized particles that were part of discontinuously mineralized structures. Many of these have unique forms and should be recognizable in the fossil record.

We noted in this chapter and in Chapter 2 that the mineralized skeletal hard parts of many different animals and some protoctists have among their macromolecules one group that shares a number of common features. In Table 2.3 we refer to them as "acidic glycoproteins," and we distinguish them from the framework macromolecules, which are often unique to each tissue. We do not know if the wide distribution of acidic glycoproteins is a result of divergent evolution with the original genes having been derived from a common ancestor that lived some time in the Precambrian, or is a product of convergent evolution in which a presumably preadapted system was modified by many, if not all, these organisms to regulate their mineral formation processes. This is a crucial question and, as noted, is important not only in the evolutionary–historical context, but also in terms of understanding the basis for comparative biochemical studies of these macromolecules in living organisms. More information on their amino acid sequences and the organization of the gene complex from which they are derived could clarify this issue. At this stage we can only state that if it turns out that these proteins indeed evolved by divergent evolution, then their wide distribution not only shows that controlled mineral formation evolved in the Precambrian, but that the basic mechanisms used are shared by many mineralizing organisms. Another, as yet unclarified point, is whether magnetotactic bacteria use analogous glycoproteins, associated perhaps with the magnetosome membrane, for controlling their crystal formation processes. If this turns out to be the case we would know for the first time that monerans are capable of forming these macromolecules, and because of the fossil record evidence of Chang et al. (1987) and Chang (1988) they must have evolved this ability in the Precambrian. Two very important points! Although much more direct evidence from the fossil record is needed, we feel that there is good reason to believe that controlled mineralization evolved long before the advent of composite skeletal mineralization. How this ability came about and to what extent it was perfected prior to the events in the late Precambrian and early Cambrian, are not known.

2. *The Sequence of Events in the Transition Zone.* The fossil record is heavily

skewed toward events related to the advent of skeletal mineralization. There is, however, growing evidence that during the late Vendian and very early Tommotian the biosphere underwent major changes that facilitated the initiation of a massive adaptive radiation, involving mineralized as well as nonmineralized organisms (Runnegar 1982). The fossil record of nonmineralized organisms is less easily documented. Preserved tracks, trails, and burrows of organisms (so-called *trace fossils*) from the late Vendian are quite different from those of the early Tommotian. The latter become more diverse with time during the lower- and mid-Cambrian (Crimes 1987). The acritarchs are a group of nonmineralized planktonic organisms that also underwent a radiation at the base of the Cambrian (Moczydlowska and Vidal (1986). A similar phenomenon occurred among organic-walled planktonic microorganisms (Vidal and Knoll 1982). The most direct documentation of the radiation is the plethora of soft-bodied and hard-bodied organisms preserved in the Burgess Shale that formed about 540–525 million years ago (Conway Morris 1979; 1986). This unfortunately is some 30 to 50 million years after the beginning of the Cambrian. Observations such as these, together with the preserved skeletal remains, show that a massive adaptive radiation occurred during this period.

It is not possible to identify a single cause of the adaptive radiation, but there is increasing evidence to show that the biosphere underwent a series of major perturbations in the late Vendian. Organic-walled planktonic organisms underwent a minor radiation in the early Vendian (about 700 million years ago), which was followed by an extinction some 50 to 70 million years later. They recovered around 570 million years ago and participated in the Cambrian adaptive radiation (Vidal and Knoll 1982). Probably the most direct evidence of a major perturbation in the biosphere that occurred in the very late Vendian is the major shift in the carbon isotopic compositions of the carbonate rocks (Margaritz et al. 1986) that can be attributed to fluctuations in the oceans' productivity levels. It has been proposed that a major extinction occurred (Hsu et al. 1986), but it has not been demonstrated to have occurred at exactly this time.

Another possible contributing factor to the initiation of the adaptive radiation is that the atmospheric oxygen content some time prior to this event may have risen to levels that enabled organisms to perform new metabolic reactions. Towe (1970, 1982) proposed that a key event in this regard was the synthesis of two new amino acids, hydroxyproline and hydroxylysine, which play critical roles in the formation of structural glycoproteins in both plants and animals. They both require the presence of molecular oxygen for their synthesis. Their development opened the way for complex multicellularity, in that it permitted the evolution of structural proteins (in the collagen–extensin family) necessary to provide the mechanical support for increased size and morphogenetic experimentation. Thus, long vacant ecological niches could at last be occupied—in essence an adaptive radiation.

When then did skeletal mineralization evolve in relation to the onset of the adaptive radiation? The first appearance of a lineage in the fossil record does not coincide with the time it evolved. Thus, the fossil record leaves open the option of extrapolating all the first appearances that occur during the late Vendian and Tommotion, a period of 10 to 20 *million* years, to a much narrower point in time coincident with or soon after the beginning of the radiation. Alternatively, one can accept the fossil record more or less as it appears from first appearances and conclude that various phylogenetic groups evolved the ability to form mineralized skeletons at different times during the adaptive radiation. Our impression of the fossil record concurs with that of Cloud (1968) and Stanley (1976) in that the more likely scenario is that different phylogenetic groups began to mineralize over a period of millions of years, in other words at different times during the radiation rather than concurrent with its initiation.

3. *The Selection Pressure for Forming Composite Mineralized Skeletons.* Selection for composite mineralized skeletons is quite different from selection for controlling mineralization. The latter we argued evolved long before this period. Identifying *the* selection pressure that induced organisms to form composite skeletons is almost impossible given the paucity of information available. Judging simply from the types of skeletons formed by most, but not all, the organisms during the Cambrian radiation, protection from predators is probably one important reason organisms formed composite mineralized skeletons (Evans 1910). In fact, there is direct evidence from the early Cambrian of borings by predators into shells (Mathews and Missarzhevsky 1975). Stanley (1973) points out that the evolution of predators would have induced a series of self-propagating feedback systems of diversification between trophic levels and proposes that this was the reason for the massive adaptive radiation. Part of the response to predators would certainly be the production of composite mineralized skeletons, not only in animals but as noted from the fossil record in members of other kingdoms as well. This type of selection pressure is also more consistent with the staggered appearance of mineralizing skeletal biota during the late Vendian and the Cambrian.

Predation was probably just one of the factors that induced organisms to form composite skeletons. That not all the early formed skeletons were constructed for defense against predators (examples are radiolarian skeletons and the spicular skeletons of sponges) suggests that other factors were also involved. Furthermore, for many of the problematica, we have no idea of the functions for which their skeletons were designed.

4. *The Response of Organisms to the Selection Pressure to Construct Mineralized Skeletons.* The fossil record provides very limited evidence of any progressive sequences from organic skeletons to poorly mineralized skeletons and then to fully mineralized ones. Furthermore, the skeletons preserved, in some cases, seem to be comparable in complexity to those formed by living representatives of the same phylum. The fossil record is rather uninformative in this regard and we must rely more on mineralization in living organisms for guidance.

Our view of skeletal mineralization from living organisms is that it primarily involves the construction of a macromolecular framework that then fills in with mineral to varying degrees (Chapter 3). A hint that the key steps leading to skeletal framework construction may have evolved during the adaptive radiation is that the major macromolecules used for framework construction (as opposed to the acidic glycoproteins) are generally phylum specific (Table 2.3) and presumably respond to the particular mechanical requirements of each group. The framework macromolecules themselves are not unique to mineralizing systems and may well have evolved during Ediacaran times or earlier judging from the fact that some of these early metazoans probably had fairly stiff outer organic coverings (Seilacher 1983; Conway Morris 1985).

Presumably the evolution of the second stage of composite skeletal formation, namely mineralization, made use of the by then inherited ability of all organisms to manipulate ions, and in some cases to store them temporarily in the solid phase. This is at least consistent with the observation that the overwhelming majority of first mineralizers opted to form calcium minerals. Lowenstam and Margulis (1980b) point out that calcium fulfills many fundamental functions in cellular metabolism. Muscle activity is more demanding and requires the regulated release and sequestering of this ion. Thus, a high level of ability to manipulate calcium must have already evolved with the early metazoans in the late Precambrian. Calcium is thus an obvious choice as a cation for mineralization. The choice of anion, as far as we know, was mainly between carbonate and phosphate. Unlike living organisms, a fairly large proportion (about half) of first mineralizers in the Cambrian opted for phosphate. We can only point out that because calcium phosphate is more insoluble than calcium carbonate it is easier to precipitate. Furthermore, the penalty for using a vital nutrient to form a skeleton may not have been too severe during early Cambrian times as this was a period during which seawater phosphate concentrations may have been relatively high (Cook and Shergold 1984).

At few first mineralizers formed siliceous skeletons at the base of the Cambrian. Silicon is also known to be an essential nutrient in living diatoms and for that matter in normal growth and skeletal development of mammalian bone as well (Carlisle 1981). Thus, these organisms perhaps may also have utilized a preexisting system for their silica mineralization requirements.

12.6 Conclusion

Composite mineralized skeletal formation was a dramatic event in terms of the evolution of biomineralization, and it must have been preceded by a long history of hundreds, possibly even thousands of million years during which time organisms perfected their abilities to control the mineralization process. Documentation from the fossil record of these events is at present very limited. We feel confident, however, that because we are dealing with biologically formed minerals and mineralized structures that are usually different from their inorganic counterparts, they have a good chance of being preserved and being recognized in the Precambrian fossil record. The fossil record should, therefore, be able to provide much more interesting information on the evolution of biomineralization and we look forward to the unraveling of what promises to be a most interesting story in the evolution of life.

References

Abbott, D. P., and Newberry, A. T. (1980). Urochordata: The Tunicates. In *Intertidal Invertebrates of California* (eds. R. H. Morris, D. P. Abbott E. C. Haderlie), pp. 177–226. Stanford University Press, Stanford.

Addadi, L., and Weiner, S. (1985). Interactions between acidic proteins and crystals: Stereochemical requirements in biomineralization. *Proc. Natl. Acad. Sci. U.S.A.* **82,** 4110–4114.

Addadi, L., Berkovitch-Yellin, Z., Domb, N., Gati, E., Lahav, M., and Leiserowitz, L. (1982). Resolution of conglomerates by stereoselective modifications. *Nature (London)* **296,** 21–26.

Addadi, L., Berkovitch-Yellin, Z., Weissbuch, I., Van Mil, J., Shimon, L. J. W., Lahav, M., and Leiserowitz, L. (1985). Growth and dissolution of organic crystals with "tailor-made" inhibitors—implications in stereochemistry and materials science. *Angew. Chem. Int. Ed. Engl.* **24,** 466–485.

Addadi, L., Moradian, J., Shay, E., Maroudas, N. G., and Weiner, S. (1987). A chemical model for the cooperation of sulfates and carboxylates in calcite crystal nucleation. Relevance to biomineralization. *Proc. Natl. Acad. Sci. U.S.A.* **84,** 2732–2736.

Adshead, P. C. (1967). Collection and laboratory maintenance of living planktonic foraminifera. *Micropaleontology* **13,** 32–40.

Aisen, P., and Listowsky, I. (1980). Iron transport and storage proteins. *Annu. Rev. Biochem.* **49,** 357–393.

Akerston, W. A., Lowenstam, H. A., and Walker, A. (1984). "Pigmentation" of Soricine teeth composition, ultrastructure and function. *Abstr. Am. Soc. Mammalog. 64th Annu. Meet.,* 40.

Albright, L. J. (1972). Hydrostatic pressure effects upon protein synthesis by *Escherichia coli* and *Vibrio marinus.* In *Barobiology and the Experimental Biology of the Deep Sea* (ed. R. W. Brauer), pp. 115–125. North Carolina Sea Grant Program, Chapel Hill.

Alexander, G. B., Heston, W. M., and Iler, R. K. (1954). The solubility of amorphous silica in water. *J. Phys. Chem.* **58,** 435–455.

Ali, S. Y. (1983). Calcification of cartilage. in *Cartilage, Structure, Function and Biochemistry* (ed. B. K. Hall), Chap. 12, pp. 343–378. Academic Press, New York.

Ali, S. Y., Sajdera, S. W., and Anderson, H. C. (1970). Isolation and characterization of calcifying matrix vesicles from epiphyseal cartilage. *Proc. Natl. Acad. Sci. U.S.A.* **67,** 1513–1520.

Ali, S. Y., Wisby, A., Evans, L., and Craig-Gray, J. (1977). The sequence of calcium and phosphorus accumulation by matrix vesicles. *Calcif. Tissue Res.* **225,** 490–493.

Alvarez, L. W., Alvarez, W., Asaro, F., and Michel, H. V. (1980). Extraterrestrial cause for the Cretaceous-Tertiary extinction. *Science* **208,** 1095–1108.

Amprino, R., and Engström, A. (1952). Studies on X-ray absorption and diffraction of bone tissue. *Acta Anat.* **15,** 1–22.

Anderson, H. C. (1967). Electron microscopic studies of induced cartilage development and calcification. *J. Cell Biol.* **35,** 81–101.

Anderson, H. C. (1976). Matrix vesicles of cartilage and bone. In *The Biochemistry and Physiology of Bone* (ed. G. H. Bourne), pp. 135–157. Academic Press, New York.

Anderson, O. R. (1983). *Radiolaria.* Springer-Verlag, Berlin.

Anderson, O. R. (1986). Silicification in radiolaria—deposition and ontogenetic origins of form. In *Biomineralization in Lower Plants and Animals* (eds. B. S. C. Leadbeater and R. Riding), pp. 375–391. Clarendon Press, Oxford.

Angell, R. W. (1979). Calcification during chamber development in *Rosalina floridana* (Cushman). *J. Foraminiferal Res.* **9,** 341–353.

Angell, R. W. (1980). Test morphogenesis (chamber formation) in the foraminifer *Spiroloculina hyalina* Schulze. *J. Foraminiferal Res.* **10,** 89–101.

Anno, K., Otsuka, K., and Seno, N. (1974). A chitin sulfate-like polysaccharide from the test of the tunicate *Halocynthia roretzi. Biochim. Biophys. Acta* **362,** 215–219.

Arends, J., and Jongebloed, W. L. (1979). Ultrastructural studies of synthetic apatite crystals. *J. Dent. Res.* **58B,** 837–842.

Ariani, A. P., Marmo, F., Balsamo, G., and Franco, E. (1981). Vaterite in the statoliths of a mysid Crustacean *(Diamysis bahinensis). Annvar. Ist. Mus. Zool. Napoli* **24,** 69–78.

Ariani, A. P., Marmo, F., Balsamo, G., Franco, E., and Wittmann, K. (1983). The mineral composition of statoliths in relation to taxonomy and ecology in mysids. *Rapp. Comm. Int. Mer. Medit.* **28,** 333–335.

Arnott, H. J. (1973). Plant calcification. In *Biological Mineralization* (ed. Z. Zipkin), pp. 609–622. John Wiley, New York.

Arnott, H. J. (1980). Carbonates in higher plants. In *The Mechanisms of Biomineralization in Animals and Plants* (eds. M. Omori and N. Watabe), pp. 211–218. Tokai University Press, Tokyo.

Arnott, H. J., and Pautard, F. G. E. (1970). Calcification in plants. In *Biological Calcification: Cellular and Molecular Aspects* (ed. H. Schraer), pp. 441–462. Appleton Century Croft, New York.

Arnott, H. J., Pautard, F. G. E., and Steinfink, H. (1965). Structure of calcium oxalate monohydrate. *Nature (London)* **208,** 197–198.

Arny, D. C., Lindow, S. E., and Upper, C. D. (1976). Frost sensitivity of *Zea mays* increased by application of *Pseudomonas syringae. Nature (London)* **262,** 282–284.

Arsenault, A. L., and Ottensmeyer, F. P. (1983). Quantitative spatial distributions of calcium, phosphorus, and sulfur in calcifying epiphysis by high resolution electron spectroscopic imaging. *Proc. Natl. Acad. Sci. U.S.A.* **80,** 1322–1326.

Ascaso, C., Galvan, J., and Ortego, C. (1976). The pedogenic action of *Parmelia con-*

spersa, Rhizoearpon geographicum and *Umbilicaria pustulata. Lichenologist* **8,** 151–171.

Ascenzi, A., and Chiazzoto, A. (1955). Electron microscopy of the bone ground substance using the pseudo-replica technique. *Experientia* **11,** 140–142.

Ascenzi, A., Francois, C., and Steve-Bocciarelli, D. (1963). On the bone induced by estrogens in birds. *J. Ultrastruct. Res.* **8,** 491–505.

Ashton, B. A., Höhling, H. J., and Trifitt, J. T. (1976). Plasma proteins present in human cortical bone: Enrichment of the alpha-2HS-glycoprotein. *Calcif. Tissue Res.* **22,** 27–33.

Atkins, E. (1985). Conformations in polysaccharides and complex carbohydrates. *Proc. Int. Symp. Biomol. Struct. Interact. Suppl. J. Biosci.* **8,** 375–387.

Azam, F., Hemminsen, B. B., and Volcani, B. E. (1974). Role of silicon in diatom metabolism. V. Silicic acid transport and metabolism in the heterotrophic diatom *Nitzschia alba. Arch. Microbiol.* **97,** 103–114.

Babcock, L. E., and Feldman, R. M. (1986). The phylum Conulariida. In *Problematic Fossil Taxa* (eds. A. Hoffman and M. H. Nitecki), pp. 135–147. Oxford University Press, New York.

Bachmann, S., Pohla, H., and Goldschmid, A. (1980). Phagocytes in the axial complex of the sea urchin, *Sphaerechinus granularis* (Lam.). *Cell Tissue Res.* **213,** 109–120.

Bai, P., and Warshawsky, H. (1985). Morphological studies on the distribution of enamel matrix proteins using routine electron microscopy and freeze-fracture replicas in the rat incisor. *Anat. Rec.* **212,** 1–16.

Baker, R. B., Mather, J. G., and Kennaugh, J. H. (1983). Magnetic bones in human sinuses. *Nature (London)* **301,** 78–80.

Banta, W. C. (1968). The body wall of the cheilostome bryozoa. I. The ectocyst of *Watersipora nigra* (Canu and Bossler). *J. Morphol.* **125,** 497–508.

Barber, V. C. (1968). The structure of the mollusc statocysts, with particular reference to cephalopods. *Symp. Zool. Soc. London* **23,** 37–62.

Barghoorn, E. S., Tyler, S. A., and Hurley, P. (1954). Oldest plant discovered. *Traveler* **Feb 16.**

Barker, R. M. (1964). Microtextural variation in pelecypod shells. *Malacologia* **2,** 69–86.

Barnes, R. D. (1980). *Invertebrate Zoology.* Saunders College, Philadelphia.

Barrington, E. J. W. (1965). *The Biology of Hemichordata and Protochordata.* Oliver and Boyd, Edinburgh.

Barrington, E. J. W. (1974). Biochemistry of primitive deuterostomians. In *Chemical Zoology* (eds. M. Florkin and B. T. Scheer), pp. 61–95. Academic Press, New York.

Bate, R. H., and East, B. A. (1972). The structure of the ostracode carapace. *Lethaia* **5,** 177–194.

Baud, C. A., and Very, J. M. (1975). Ionic substitutions in vivo in bone and tooth apatite crystals. In *Physico-Chimie et Crystallographie des Apatites d'Interet Biologiques. Colloques Internationaux CNRS No 230,* pp. 405–410. CNRS, Paris.

Baud, C. A., and Very, J. (1982). Morphological and crystallographic analysis of bone mineral. In *The Role of Calcium in Biological Systems, I* (eds. L. J. Anghileri and A. M. Tuffet-Anghileri). CRC Press, Boca Raton, FL. 95–105.

Bauer, G. B., Fuller, M., Perry, A., Dunn, J. R., and Zoeger, J. (1985). Magnetoreception and biomineralization of magnetite in cetaceans. In *Magnetite Biomineralization and Magnetoreception in Organisms* (eds. J. L. Kirschvink, D. S. Jones, and B. J. MacFaddan), pp. 489–507. Plenum Press, New York.

Bauminger, E. R., Cohen, S. G., Dickson, D. P. E., Levi, A., Ofer, S., and Yariv, J. (1980). Mossbauer spectroscopy of *E. coli* and its iron protein. *Biochim. Biophys. Acta* **623,** 237–242.

Bayer, F. M. (1956). Octocorallia. In *Treatise on Invertebrate Paleontology Pt. F* (ed. R. C. Moore), pp. 166–232. Geological Society of America, Lawrence, Kansas.

Baylink, D., Wergedal, J., and Thompson, E. (1972). Loss of protein polysaccharides at sites where bone mineralization is initiated. *J. Histochem. Cytochem.* **20,** 279–292.

Bé, A. W. H., Hemleben, Ch., Anderson, O. R., and Spindler, M. (1979). Chamber formation in planktonic foraminifera. *Micropaleontology* **25,** 294–307.

Becher, S. (1914). Uber statische Strukturen und kristalloptische Eigentumlichkeiten des Echinodermenskeletts. *Verh. D. Zool. Ges.* ***S.307-327.***

Becker, G. L., Chen, C. H., Greenwalt, J. W., and Lehninger, A. L. (1974). Calcium phosphate granules in the hepatopancreas of the Blue Crab, *Callinectes sapidus. J. Cell Biol.* **61,** 316–326.

Beedham, G. E. (1958). Observations of the non-calcareous component of the shell of the Lamellibranchia. *Quart. J. Micr. Sci.* **99,** 341–357.

Beedham, G. E., and Truemann, E. R. (1967). The relationship of the mantle and shell of the Polyplacophora in comparison with that of other Mollusca. *J. Zool. London* **151,** 215–231.

Belcourt, A. B., Fincham, A. G., and Termine, J. D. (1982). Acid-soluble bovine fetal enamelins. *J. Dent. Res.* **61,** 1031–1032.

Beleyaev, G. M. (1966). *Bottom Fauna of the Ultra-Abyssal Depths of the World Oceans.* Akad. SSSR Inst. Okeanol., Moscow.

Bellemo, S. (1976). Wall ultramicrostructure in the foraminifer *Cibicides floridanus* (Cushman). *Micropaleontology* **22,** 352–362.

Bellemo, S. (1978). Test wall structures in some Spirillinacea (Foraminifera). *Bull. Geol. Institut. Univ. Uppsala N. S.* **8,** 77–82.

Bender, H., and Hemleben, Ch. (1988). Calcitic cement secreted by agglutinated foraminifers grown in laboratory culture. *J. Foraminiferal Res.* **18,** 42–45.

Bengtson, S. (1977). Aspects of problematic fossils in the early Paleozoic. *Acta Universit. Upsalensis Abst. Uppsla Dis. Fac. Sci.* **415,** 1–71.

Benson, S. C., Jones, E. M. E., Crise-Benson, N., and Wilt, F. (1983). Morphology of the organic matrix of the spicule of the sea urchin larva. *Exp. Cell. Res.* **148,** 249–253.

Benson, S. C., Benson, N. C., and Wilt, F. (1986). The organic matrix of the skeletal spicule of sea urchin embryos. *J. Cell. Biol.* **102,** 1878–1886.

Benson, S. C., Sucov, H., Stephens, L., Davidson, E., and Wilt, F. (1987). Lineage specific gene encoding a major matrix protein of the sea urchin embryo spicule. I. Authentication of its developmental expression. *Dev. Biol.* **120,** 499–506.

Berman, A., Addadi, L., and Weiner, S. (1988). Interactions of sea urchin skeleton macromolecules with growing calcite crystals—a study of intracrystalline proteins. *Nature (London)* **331,** 546–548.

Bernard, F. R. (1976). Living Chamidae of the eastern Pacific (Bivalvia: Heterodonta). *Contr. Sci. Nat. Hist. Mus. Los Angeles County* **278,** 1–43.

Bernard, G. W., and Pease, D. C. (1969). An electron microscopic study of initial intramembranous osteogenesis. *Am. J. Anat.* **125,** 271–290.

Berril, N. J. (1950). *The Tunicate with an Account of British Species.* Royal Society, London.

Berthet-Colominas, C., Miller, A., and White, S. W. (1979). Structural study of the calcifying collagen in turkey leg tendons. *J. Mol. Biol.* **134,** 431–445.

Bertheussen, K., and Seljelid, R. (1978). Echinoid phagocytes in vitro. *Exp. Cell. Res.* **111,** 401–412.

Berthold, W. U. (1976). Biomineralisation bei milioliden Foraminiferen und die Matrizen-Hypothese. *Naturwischenschaften* **63,** 196.

Betts, F., Blumenthal, N. C., Posner, A. S., Becker, G. L., and Lehninger, A. L. (1975). Atomic structure of intracellular amorphous calcium phosphate deposits. *Proc. Natl. Acad. Sci. U.S.A.* **72,** 2088–2092.

Bevelander, G., and Nakahara, H. (1960). Development of the skeleton of the sand dollar. In *Calcification in Biological Systems* (ed. R. F. Sognnaes), pp. 41–56. Amer. Assoc. Adv. Science, Washington, DC.

Bevelander, G., and Nakahara, H. (1969a). An electron microscope study of the formation of the ligament of *Mytilus edulis* and *Pinctada radiata. Calcif. Tissue Res.* **4,** 101–112.

Bevelander, G., and Nakahara, H. (1969b). An electron microscope study of the formation of the nacreous layer in the shell of certain bivalve molluscs. *Calcif. Tissue Res.* **3,** 84–92.

Bevelander, G., and Nakahara, H. (1980). Compartment and envelope formation in the process of biological mineralization. In *The Mechanisms of Biomineralization in Animals and Plants* (eds. M. Omori and N. Watabe), pp. 19–27. Tokai University Press, Tokyo.

Bhattacharyya, P., and Volcani, B. E. (1980). Sodium-dependent silicate transport in the apochlorotic diatom *Nitzschia alba. Proc. Natl. Acad. Sci. USA* **77,** 6386–6390.

Bhattacharyya, P., and Volcani, B. E. (1983). Isolation of silicate ionophore(s) from the apochlorotic diatom *Nitzschia alba. Biochem. Biophys. Res. Commun.* **114,** 365–372.

Birchall, J. D., and Davey, R. J. (1981). The crystallization of sodium chloride from aqueous solution in the presence of polysaccharides. *J. Cryst. Growth* **54,** 323–329.

Bischoff, G. C. O. (1978). Internal structures of conulariid tests and their functional significance with special reference to Circonulariina, n. suborder (Cnidaria, Scyphozoa) *Senckenbergiana. Lethaea* **59,** 275–327.

Black, M. (1963). The fine structure of the mineral parts of coccolithophoridae. *Proc. Linn. Soc. London* **174,** 41–46.

Blackmon, P. D., and Todd, R. (1959). Mineralogy of some foraminifera as related to their classification and ecology. *J. Paleontol.* **33,** 1–15.

Blackwell, J., and Weih, M. A. (1980). Structure of chitin-protein complexes: Ovipositor of the ichneumon fly *Megarhyssa. J. Mol. Biol* **137,** 49–60.

Blake, D. F., Peacor, D. R., and Allard, L. F. (1984). Ultrastructural and macroanalytical results from echinoderm calcite: Implications for biomineralization and diagenesis of skeletal material. *Micron Microscop. Acta* **15,** 85–90.

Blakemore, R. P. (1975). Magnetotactic bacteria. *Science* **190,** 377–379.

Blakemore, R. P., Frankel, R. B., and Kalmijn, A. J. (1981). South-seeking magnetotactic bacteria in the southern hemisphere. *Nature (London)* **286,** 384–385.

Blakemore, R. P., Short, K. A., Bazylinski, D. A., Rosenblatt, C., and Frankel, R. B. (1985). Microaerobic conditions are required for magnetite formation within *Aquaspirillum magnetotacticum. Geomicrobiol. J. **4,** 53–72.*

Blakey, P. R., Earland, C., and Stell, J. G. P. (1963). Calcification of keratin. *Nature (London)* **198,** 481.

Block, M. R. (1971). Epiphragms: Some observations. *J. Conch.* **26,** 388–409.

Blumenthal, N. C., Posner, A. S., Silverman, L. D., and Rosenberg, L. C. (1979). Effect of proteoglycans on in vitro hydroxyapatite formation. *Calcif. Tissue Int.* **27,** 75–82.

Bockelie, T., and Fortey, R. A. (1976). An early Ordovician vertebrate. *Nature (London)* **260,** 36–38.

Boedtker, H., and Doty, P. (1955). On the nature of the structural element of collagen. *J. Am. Chem. Soc.* **77,** 248–249.

Bøggild, O. B. (1930). The shell structure of the mollusks. *K. Dan. Vidensk. Selsk. Skr. Naturvidensk. Math. Afd.* **9,** 233–326.

Böhm, E. L., Futterer, D., and Kaminiski, E. (1978). Algal calcification in some Codiaceae (Chlorophyceae): Ultrastructure and location of skeletal deposits. *J. Phycol.* **14,** 486–493.

Boltovsky, E., and Wright, R. (1976). *Recent Foraminifera.* W. Junk, The Hague.

Bonar, L. C., Roufousse, A. H., Sabine, W. K., Grynpas, M. D., and Glimcher, M. J. (1983). X-Ray diffraction studies of the crystallinity of bone mineral in newly synthesized and density fractionated bone. *Calcif. Tiss. Int.* **35,** 202–209.

Bonar, L. C., Grynpas, M. D., Roberts, J. E., Griffin, R. G., and Glimcher, M. J. (1985). Physical and chemical characterization of the development and maturation of bone mineral. In *The Chemistry and Biology of Mineralized Tissues* (ed. W. T. Butler), pp. 226–233. EBSCO Media, Birmingham, AL.

Bonucci, E. (1967). Fine structure of early cartilage calcification. *J. Ultrastruct. Res.* **20,** 33–50.

Bordas, J., Woodhead-Galloway, J., and Hukins, D. W. L. (1978). Energy dispersive X-ray diffraction by collagen fibrils in costal cartilage using synchrotron radiation. *Biochem. Biophys. Res. Commun.* **84,** 627–630.

Borman, A. H., de Jong, E. W., Huizinga, M., Kok, D. J., Westbroek, P., and Bosch, L. (1982). The role of $CaCO_3$ crystallization of an acid Ca^{2+}-binding polysaccharide associated with coccoliths of *Emiliania huxleyi. Eur. J. Biochem.* **129,** 179–183.

Borman, A. H., Kok, D. J., de Jong, E. W., Bluys van Treslang, C. J., and Bosch, L. (1987). Molar mass determination of the polysaccharide associated with coccoliths of *Emiliania huxleyi. Eur. Polymer J.,* in press.

Bornfield, B. D., and Milliman, J. D. (1973). Generic and environmental control of carbonate mineralogy in serpulid (Polychaete) tubes. *J. Geol.* **81,** 363–373.

Bornstein, P., and Traub, W. (1978). The chemistry and biology of collagen. In *The Proteins* (eds. H. Neurath and R. Hill), pp. 411–632. Academic Press, New York.

Borowitzka, M. A. (1982). Morphological and cytological aspects of algal calcification. *Int. Rev. Cytol.* **74,** 127–160.

Borowitzka, M. A., Larkum, A. W. D., and Nockolds, C. E. (1974). A scanning electron microscope study of the structure and organization of the calcium carbonate deposits of algae. *Phycologia* **13,** 195–203.

Boschma, H. (1956). Milleporina and Stylasterina. In *Treatise on Invertebrate Paleontology Pt. F. Coelenterata* (ed. R. C. Moore), pp. F90–F106. Geological Society of America, Lawrence, Kansas.

Boskey, A. L. (1981). Current concepts of the physiology and biochemistry of calcification. *Clin. Orthop. Rel. Res.* **157,** 225–257.

Boskey, A. L. (1985). Overview of cellular elements and macromolecules implicated in the initiation of mineralization. In *The Chemistry and Biology of Mineralized Tissues* (ed. W. T. Butler), pp. 335–343. EBSCO Media, Birmingham, AL.

Boskey, A. L. (1986). Phospholipids and calcification: An overview. In *Cell Mediated Calcification and Matrix Vesicles* (ed. S. Y. Ali), pp. 175–179. Elsevier, Amsterdam.

Boskey, A. L., and Posner, A. S. (1977). The role of synthetic and bone-extracted Ca-phospholipid-PO_4 complexes on hydroxyapatite formation. *Calcif. Tissue Res.* **23,** 251–258.

Bouligand, Y. (1965). Sur une architecture torsadée repandue dans de nombreuses cuticules d'Arthropodes. *C.R. Hebd. Sceances Acad. Sci,* **261,** 3665–3668.

Bouligand, Y. (1972). Twisted fibrous arrangements in biological material, and cholestric mesophages. *Tissue Cell* **4,** 189–217.

Bourget, E. (1980). Barnacle shell growth and its relationship to environmental factors. In *Skeletal Growth of Aquatic Organisms* (eds. D. C. Rhoads and R. A. Lutz), pp. 469–491. Plenum Press, New York.

Bovee, E. C. (1981). Distribution and forms of siliceous structures among protozoa. In *Silicon and Siliceous Structures in Biological Systems* (eds. T. L. Simpson and B. E. Volcani), pp. 233–279. Springer-Verlag, New York.

Boyan, B. D., and Boskey, A. L. (1984). Co-isolation of proteolipids and calcium-phospholipid-phosphate complexes. *Calcif. Tissue Int.* **36,** 214–218.

Boyan, B. D., Landis, W. J., Knight, J., Dereszewski, G., and Zeagler, J. (1984). Microbial hydroxyapatite formation as a model of proteolipid membrane-mediated calcification. *Scanning Electron Micros.* **4,** 1793–1800.

Boyde, A. (1965). The structure of developing mammalian dental enamel. In *Tooth Enamel* (eds. M. V. Stack and R. W. Fearnhead), pp. 163–167. John Wright and Sons, Bristol.

Boyde, A. (1971). Comparative histology of mammalian teeth. In *Dental Morphology and Evolution* (ed. A. A. Dahlberg), pp. 81–94. Chicago Univ. Press, Chicago.

Brodsky, B., and Eikenberry, E. F. (1982). Characterization of fibrous forms of collagen. In *Methods in Enzymology* (eds. L. W. Cunningham and D. W. Frederikson), pp. 127–174. Academic Press, New York.

Broecker, W. S. (1971). A kinetic model for the chemical composition of sea water. *Quatern. Res.* **1,** 188–207.

Broecker, W. S. (1974). *Chemical Oceanography,* Harcourt Brace Jovanovich, New York.

Brown, B. E. (1982). The form and function of metal-containing "granules" in invertebrate tissues. *Biol. Rev.* **57,** 621–667.

Brown, C. H. (1952). Some structural proteins of *Mytilus edulis. Quart. J. Micr. Sci.* **93,** 487–502.

Brown, R. A., Yeh, Y., Burcham, T. S., and Feeney, R. E. (1985). Direct evidence for antifreeze glycoprotein adsorption onto ice surface. *Biopolymers* **24,** 1265–1270.

Brown, W. E. (1966). Crystal growth of bone mineral. *Clin. Orthop. Rel. Res.* **44,** 205–220.

Brown, W. E., and Chow, L. C. (1976). Chemical properties of bone mineral. *Annu. Rev. Mat. Sci.* **6,** 213–236.

Brown, W. E., Mathew, M., and Tung, M. S. (1981). Crystal chemistry of octacalcium phosphate. *Prog. Crystal Growth Charact.* **4,** 59–87.

Bryan, W. H., and Hill, D. (1941). Spherulitic crystallization as a mechanism of skeletal growth in the hexacorals. *Proc. R. Soc. Queensl.* **52,** 78–91.

Brydon, J. E., Dore, W. G., and Clark, J. S. (1963). Silicified plant asterosclerids preserved in soils. *Proc. Soil Soc. Am.* **27,** 476–477.

Buchardt, B., and Fritz, P. (1978). Strontium uptake in shell aragonite from the freshwater gastropod *Limnaea stagnalis. Science* **199,** 291–292.

Buchler, E. R., and Wasilewski, P. J. (1985). Magnetic remanence in bats. In *Magnetite Biomineralization and Magnetoreception in Organisms* (eds. J. L. Kirschvink, D. S. Jones, and B. J. MacFadden), pp. 483–487. Plenum Press, New York.

Buck, J. D., and Greenfield, L. J. (1964). Calcification in marine-occurring yeasts. *Bull. Mar. Sci. Gulf Caribbean* **14,** 239–245.

Buckwalter, J. A. (1983). Proteoglycan structure in calcifying cartilage. *Clin. Orthop. Rel. Res.* **172,** 207–232.

Buddenbroek, W. (1915). Die Statocyste von Pecten, ihre Histologie und Physiologie. *Zool. Jb. Abt. Anat. Physiol.* **35,** 302–358.

Bürger, O. (1895). Die Nemertinen des Golfes von Neapel. In *Fauna und Flora des Golfes von Neapel* Vol. 22. R. Friedlander and Sohn, Berlin.

Burnell, J. M., Teubner, E. J., and Miller, A. G. (1980). Normal maturational changes in bone matrix, mineral, and crystal size in the rat. *Calcif. Tissue Int.* **310,** 13–19.

Burton, R. F. (1972). The storage of calcium and magnesium phosphates and of calcium in the digestive glands of pulmonata (Gastropoda). *Comp. Biochem. Physiol.* **43A,** 655–663.

Buskirk, R. E., and O'Brien, W. P. (1985). Magnetic remanence and response to magnetic fields in Crustacea. In *Magnetite Biomineralization and Magnetoreception in Organisms* (eds. J. L. Kirschvink, D. S. Jones, and B. J. MacFadden), pp. 365–383. Plenum Press, New York.

Butler, W. T. (1985). *The Chemistry and Biology of Mineralized Tissues.* EBSCO Media, Birmingham, AL.

Butler, W. T., Bhown, M., Dimuzio, M. T., and Linde, A. (1981). Noncollagenous proteins of dentin. Isolation and partial characterization of rat dentin proteins and proteoglycans using a three-step preparative method. *Coll. Res.* **1,** 187–199.

Bütschli, O. (1906). Ueber die chemische Natur der Skeletsubstanz der Acantharier. *Zool. Anz.* **30,** 784.

Bütschli, O. (1908). Untersuchungen über organische Kalkgebilde nebst Bemerkungen über organische Kieselgebilde. *Abhdl. Ges. Wiss., Göttingen, Math-Physik. Kl. N.F.* **6,** 1–177.

Cain, C. J., and Heyn, A. N. J. (1964). X-Ray diffraction studies of the crystalline structure of the avian egg shell. *Biophys. J.* **4,** 23–39.

Campbell, R. D. (1972). Statocyst lacking cilia in the coelenterate *Corymorpha palma. Nature (London)* **238,** 49–51.

Campo, R. D., and Dziewiatkowski, D. (1963). Turnover of the organic matrix of cartilage and bone as visualized by autoradiography. *J. Cell Biol.* **18,** 19–29.

Campo, R. D., and Romano, J. E. (1986). Changes in cartilage proteoglycans associated with calcification. *Calcif. Tissue Int.* **39,** 175–184.

Caplan, A. I. (1984). Cartilage. *Sci. Am.* **251,** 84–94.

Caplan, A. I., and Pechak, D. G. (1987). The cellular and molecular embryology of bone formation. In *Bone and Mineral Research* (ed. W. A. Peck), pp. 117–183. Elsevier, Amsterdam.

Carlisle, E. M. (1981). Silicon in bone formation. In *Silicon and Siliceous Structures in Biological Systems* (eds. T. L. Simpson and B. E. Volcani), pp. 69–94. Springer-Verlag, Berlin.

Carlström, D. (1963). A crystallographic study of vertebrate otoliths. *Biol. Bull.* **125,** 441–463.

Carpenter, W. B. (1870). On the reparation of the spines of echinoida. *Monthly Microsc. J.* **3,** 224–228.

Carr, S. A., Hauschka, P. V., and Biemann, K. (1981). Gas chromatographic mass spectrometric sequence determination of osteocalcin, a gamma-carboxyglutamic acid-containing protein from chicken bone. *J. Biol. Chem.* **256,** 9944–9950.

Carson, D. D., Farach, M. C., Earles, D. S., Decker, G. L., and Lennarz, W. J. (1985). A monoclonal antibody inhibits calcium accumulation and skeletal formation in cultured embryonic cells of the sea urchin. *Cell* **41,** 639–648.

Carter, J. G. (1980a). Selected mineralogical data for the bivalvia. In *Skeletal Growth of Aquatic Organisms* (eds. D. C. Rhoads and R. A. Lutz), pp. 627–643. Plenum Press, New York.

Carter, J. G. (1980b). Guide to bivalve shell microstructures. In *Skeletal Growth of Aquatic Organisms* (eds. D. C. Rhoads and R. A. Lutz), pp. 645–673. Plenum Press, New York.

Carter, J. G., and Aller, R. C. (1975). Calcification in the bivalve periostracum. *Lethaia* **8,** 315–320.

Carter, J. G., and Clark, G. R. (1985). Classification and phylogenetic significance of molluscan shell microstructure. In *Mollusks, Notes for a Short Course, Univ. of Tennessee Dept. Geol. Sci. Studies in Geology* (ed. T. W. Broadhead), pp. 50–71.

Caveney, S. (1971). Cuticle reflectivity and optical activity in scarab beetles: The role of uric acid. *Proc. R. Soc. London Ser. B* **178,** 205–225.

Chalker, B. E., and Taylor, D. L. (1978). Rhythmic variations in calcification and photosynthesis associated with the coral *Acropora carvicornis* (Lamarck). *Proc. R. Soc. London Ser. B* **201,** 179–189.

Chang, S. (1988). Bacterial magnetite in sedimentary deposits and its geophysical and paleoecological implications. Ph.D. Thesis. California Institute of Technology.

Chang, S., Kirschvink, J. L., and Stolz, J. F. (1987). Bacterial magnetofossils in modern and ancient stromatolites. *Geol. Soc. Am. Abstr.* **19,** 617.

Chapman, D. M. (1985). X-Ray microanalysis of selected coelenterate statoliths. *J. Mar. Biol. Assoc. U.K.* **65,** 617–627.

Chave, K. E. (1952). A solid solution between calcite and dolomite. *J. Geol.* **60,** 190–192.

Chave, K. E. (1954). Aspects of the biogeochemistry of magnesium-I. *J. Geol.* **62,** 266–283.

Chen, C. H., Greenaway, J. W., and Lehninger, A. L. (1974). Biochemical and ultrastructural aspects of Ca^{2+} transport by mitochondria of the hepatopancreas of the blue crab *Callinectes sapidus. J. Cell Biol.* **61,** 301–305.

Choi, H. V., Tang, L. H., Johnson, T. L., Pal, S., Rosenberg, L. C., Reiner, A., and Poole, A. R. (1983). Isolation and characterization of a 35,000 molecular weight subunit fetal cartilage matrix protein. *J. Biol. Chem.* **258,** 655–661.

Christner, P. J., Lally, E. T., Miller, R. D., Leontzwich, P., Rosenbloom, J., and Herold, R. C. (1985). Monoclonal antibodies to different epitopes in amelogenins from fetal bovine teeth recognize high molecular-weight components. *Arch. Oral Biol.* **30,** 849–854.

Clark, A. H. (1911). On the inorganic constituents of two Recent crinoids. *Proc. U.S. Nat. Mus.* **39,** 487–488.

Clark, G. R. (1974a). Calcification on an unstable substrate: Marginal growth in the mollusk *Pecten diegensis. Science* **183,** 968–970.

Clark, G. R. (1974b). Growth lines in invertebrate skeletons. *Annu. Rev. Earth Planet Sci.* **2,** 77–99.

Clark, G. R. (1975). Periodic growth and biological rhythms in experimentally grown bivalves. In *Growth Rhythms and the History of the Earth's Rotation* (eds. G. D. Rosenberg and S. K. Runcorn), pp. 103–117. Wiley, London.

Clark, G. R. (1976). Shell growth in the marine environment: Approaches to the problem of marginal calcification. *Am. Zool.* **16,** 617–626.

Clarke, F. W., and Wheeler, W. C. (1922). The inorganic constituents of marine invertebrates. *U.S. Geol. Surv. Prof. Pap.* **124.**

Clarkson, E. N. K. (1973). The eyes of *Asaphus raniceps* Dalman (Trilobita). *Paleontology* **16,** 425–444.

Clarkson, E. N. K., and Levi-Setti, R. (1975). Trilobite eyes and the optics of Des Cartes and Huygens. *Nature (London)* **254,** 663–667.

Clegg, G. A., Fitton, J. E., Harrison, P. M., and Treffry, A. (1980). Ferritin: Molecular structure and iron-storage mechanisms. *Prog. Biophys. Mol. Biol.* **36,** 56–86.

Cloud, P. (1968). Premetazoan evolution and the origins of the metazoa. In *Evolution and Environment* (ed. E. T. Drake), pp. 1–72. Yale Univ. Press, New Haven, CT.

Cloud, P. (1973). Paleoecological significance of the banded iron-formation. *Econ. Geol.* **68,** 1135–1143.

Cloud, P. (1983). Early biogeologic history: The emergence of a paradigm. In *Earth's Earliest Biosphere* (ed. J. W. Schopf), pp. 14–31. Princeton Univ. Press, Princeton.

Collip, J. B. (1921). A further study of the respiratory process in *Mya arenaria* and other marine molluscs. *J. Biol. Chem.* **49,** 297–310.

Constantz, B. R. (1986). Coral skeleton construction: A physiochemically dominated process. *Palaios* **1,** 152–157.

Constantz, B. R., and Meike, A. (1988). Calcite centers of calcification in *Mussa angulosa* (Scleractinia). In *Origin, Evolution and Modern Aspects of Biomineralization in Plants and Animals* (ed. R. E. Crick). Elsevier, Amsterdam. In press.

Conway-Morris, S. (1979). The Burgess Shale (Middle Cambrian fauna). *Ann. Rev. Ecol. Syst.* **10,** 327–349.

Conway-Morris, S. (1985). The Ediacaran biota and early metazoan evolution. *Geol. Mag.* **122,** 77–81.

Conway-Morris, S. (1986). The community structure of the middle Cambrian phyllopod bed (Burgess Shale). *Palaeontology* **29,** 423–467.

Cook, P. J., and Shergold, J. H. (1984). Phosphorus, phosphorites and skeletal evolution at the Precambrian-Cambrian boundary. *Nature (London)* **308,** 231–238.

Coombs, J., and Volcani, B. E. (1968). Studies on the biochemistry and fine structure of silica-shell formation in diatoms. *Planta (Berlin)* **82,** 280–292.

Cornish, V., and Kendall, P. F. (1888). On the mineralogical constitution of calcareous organisms. *Geol. Mag.* **5,** 66–73.

Cotmore, J. M., Nichols, G., and Wuthier, R. E. (1971). Phospholipid-calcium-phosphate complex. Enhanced calcium migration in the presence of phosphate. *Science* **172,** 1339–1341.

Coutts, P. J. F. (1970). Bivalve-growth patterning as a method for seasonal dating in archeology. *Nature (London)* **226,** 874.

Cragg, S., and Nott, J. (1977). The ultrastructure of the statocysts in the pedvelliger larvae of *Pecten maximus* (L.) (Bivalvia). *J. Exp. Mar. Biol. Ecol.* **27,** 23–26.

Crawford, R. M. (1981). The siliceous components of the diatom cell wall and their morphological variation. In *Silicon and Siliceous Structures in Biological Systems* (eds. T. L. Simpson and B. E. Volcani), pp. 127–156. Springer-Verlag, Berlin.

Crawford, R. M., and Schmid, A. M. (1986). Ultrastructure of silica deposition in diatoms. In *Biomineralization in Lower Plants and Animals* (eds. B. S. C. Leadbeater and R. Riding), pp. 291–314. Clarendon Press, Oxford.

Crenshaw, M. A. (1972a). The soluble matrix of *Mercenaria mercenaria* shell. *Biomineral. Res. Rep.* **6,** 6–11.

Crenshaw, M. A. (1972b). The inorganic composition of molluscan extrapallial fluid. *Biol. Bull.* **143,** 506–512.

Crenshaw, M. A. (1980). Mechanisms of shell formation and dissolution. In *Skeletal Growth of Aquatic Organisms* (eds. D. C. Rhoads and R. A. Lutz), pp. 115–132. Plenum Press, New York.

Crenshaw, M. A. (1982). Mechanism of normal biological mineralization of calcium carbonates. In *Biological Mineralization and Demineralization* (ed. G. H. Nancollas), pp. 243–257. Springer-Verlag, Berlin.

Crenshaw, M. A., and Neff, J. M. (1969). Decalcification at the mantle-shell interface in molluscs. *Am. Zool.* **9,** 881–885.

Crenshaw, M. A., and Ristedt, H. (1976). The histochemical localization of reactive groups in septal nacre from *Nautilus pompilius* L. In *The Mechanisms of Mineralization in the Invertebrates and Plants* (eds. N. Watabe and K. M. Wilbur), pp. 355–367. Univ. South Carolina Press, Columbia.

Crimes, T. P. (1987). Trace fossils and correlation of late Precambrian and early Cambrian strata. *Geol. Mag.* **124,** 97–119.

Cuervo, L. A., Pita, J. C., and Howell, D. S. (1973). Inhibition of calcium phosphate mineral growth by proteoglycan aggregate fractions in a synthetic lymph. *Calcif. Tissue Res.* **13,** 1–10.

Cuisinier, F., Bres, E. F., Hemmerle, J., Voegel, J. C., and Frank, R. M. (1987). Transmission electron microscopy of lattice planes in human alveolar bone apatite crystals. *Calcif. Tissue Int.* **40,** 332–338.

Currey, J. (1984). *The Mechanical Adaptations of Bones.* Princeton Univ. Press, Princeton, NJ.

Daculsi, G., Kerebel, B., and Kerebel, L. M. (1979). Mechanisms of acid dissolution of biological and synthetic apatite crystals at the lattice pattern level. *Caries Res.* **13,** 277–289.

Daculsi, G., Menanteau, J., Kerebel, L. M., and Mitre, D. (1984). Length and shape of enamel crystals. *Calcif. Tissue Int.* **36,** 550–555.

Dafni, J., and Erez, J. (1982). Differential growth in *Tripneuster gratilla.* In *International Echinoderms Conference* (ed. J. M. Rotterdam), pp. 71–75. A. A. Ballcema, Rotterdam.

Danielsson, D., and Koren, J. (1882). Holothuroidea, the Norwegian North-Atlantic Expedition 1876–1878. *Zoology* **4.**

Darley, W. M., and Volcani, B. E. (1969). Role of silicon in diatom metabolism. *Exp. Cell Res.* **58,** 334–342.

Darlington, M. V., Meyer, H. J., Graf, G., and Freeman, T. P. (1983). The calcified puparium of the face fly *Musca autumnalis* (Diptera: Muscidae). *J. Insect Physiol.* **29,** 157–162.

Darwin, C. (1854). *A Monograph on the Sub-class Cirripedia.* Ray Society, London.

David, C. N., and Easterbroek, K. (1971). Ferritin in the fungus *Phycomyces. J. Cell Biol.* **48,** 15–28.

Davidson, E. H. (1986). *Gene Activity in Early Development,* 2nd. ed. Academic Press, New York.

Davies, T., Crenshaw, A., and Heathfield, M. (1972). The effect of temperature on the chemistry and structure of echinoid spine regeneration. *J. Paleontol.* **46,** 874–833.

Dawson, E. Y. (1944). The marine algae of the Gulf of California. In *Allen Hancock Pacific Expedition.* **3,** 189–453. Univ. Southern California Press, Los Angeles.

Dawson, E. Y. (1966). *Seashore Plants of Southern California.* Univ. of California Press. Berkeley.

Dawson, P. A. (1973). Observations on the structure of some forms of *Gomphonema parvulum*. *Helgol. Wiss. Meeresunters* **13,** 101–114.

Deakins, M. (1942). Changes in the ash, water, and organic content of pig enamel during calcification. *J. Dent. Res.* **21,** 429–435.

Debrenne, F., and Vacelet, J. (1984). Archaeocyatha: Is the sponge model consistent with their structural organization? *Palaeontogr. Americana* **54,** 358–369.

Deflandre, G. (1953). Ordres des Testacealobosa, Testaccafilosa, Thalamia ou Theca moebiens (Rhizopoda Testata). In *Traité de Zoologie* (ed. P. P. Grasse). Masson et Cie, Paris.

Degens, E. T., Carey, F. G., and Spencer, D. W. (1967a). Amino-acids and amino-sugars in calcified tissues of portunid crabs. *Nature (London)* **216,** 601–603.

Degens, E. T., Spencer, D. W., and Parker, R. W. (1967b). Paleobiochemistry of molluscan shell proteins. *Comp. Biochem. Physiol.* **20,** 553–579.

Deith, M. R. (1985). The composition of tidally deposited growth lines in the shell of the edible cockle, *Cerastoderma edule*. *J. Mar. Biol. Assoc. U.K.* **65,** 573–581.

de Jong, E. W., Bosch, L., and Westbroek, P. (1976). Isolation and characterization of a Ca^{2+}-binding polysaccharide associated with coccoliths of *Emiliania huxleyi* (Lohmann) Kamptner. *Eur. J. Biochem.* **70,** 611–621.

de Jong, W. F. (1926). La substance minerale dans les os. *Recl. Trav. Chim. Pays-Bas Belg.* **45,** 445–448.

Delmas, P. D., Tracey, R. P., Riggs, B. L., and Mann, K. G. (1984). Identification of the noncollagenous proteins of bone by two-dimensional gel electrophoresis. *Calcif. Tissue Int.* **36,** 308–316.

Denizot, M. (1968). *Les Algues Floridées Encroutantes à l'Exclusion des Corallinacées Laboratoire de Cryptogamie.* Museum National d'Histoire Natur., Paris.

Deutsch, D., and Pe'er, E. (1982). Development of enamel in human fetal teeth. *J. Dent. Res.* **61,** 1543–1551.

Deutsch, D., Shapira, L., Alayoff, A., Leviel, D., Yoeli, Z., and Arad, A. (1985). Protein and mineral changes during prenatal and postnatal development and mineralization of human deciduous enamel. In *Tooth Enamel* (eds. R. W. Fearnhead and S. Suga), Vol. IV, pp. 234–239. Elsevier, Amsterdam.

DeVries, A. L. (1982). Biological antifreeze agents in coldwater fishes. *Comp. Biochem. Physiol.* **73A,** 627–640.

DeVries, A. L., and Wohlschlag, D. E. (1969). Freezing resistance in some Antarctic fishes. *Science* **163,** 1073–1075.

de Vrind-de Jong, E. W., Borman, A. H., Thierry, R., Westbroek, P., Gruter, M., and Kamerling, J. P. (1986). Calcification in the coccolithophorids *Emiliania huxleyi* and *Pleurochrysis carterae.* II. Biochemical aspects. In *Biomineralization in Lower Plants and Animals* (eds. B. S. C. Leadbeater and R. Riding), pp. 205–217. Clarendon Press, Oxford.

Dick, A. B. (1928). On needles of rutile in the test of *Bathysiphon argenteus. Trans. Edinburgh Geol. Soc.* **12,** 19–21.

Digby, P. S. B. (1965). Semi-conduction and electrode processes in biological material. I. Crustacea and certain soft-bodied forms. *Proc. Roy. Soc. London Ser. B* ***161,*** *504–525.*

Digby, P. S. B. (1968). The mechanism of calcification in the molluscan shell. *Symp. Zool. Soc. London* **22,** 93–107.

Dillon, J. F., and Clark, G. R. (1980). Growth-line analysis as a test for contemporaneity in populations. In *Skeletal Growth of Aquatic Organisms* (eds. D. C. Rhoads and R. A. Lutz), Chap. 11, pp. 395–415. Plenum Press, New York.

Dimuzio, M. T., and Veis, A. (1978a). Phosphophoryns—major non-collagenous proteins of rat incisor dentin. *Calcif. Tissue Res.* **25,** 169–178.

Dimuzio, M. T., and Veis, A. (1978b). The biosynthesis of phosphophoryns and dentin collagen in the continuously erupting rat incisor. *J. Biol. Chem.* **253,** 6845–6852.

Dixon, H. H. (1900). On the structure of coccospheres and the origin of coccoliths. *Proc. Roy. Soc. (Biol.)* **68,** 305–315.

Dodd, J. R. (1963). Paleoecological implications of shell mineralogy in two pelecypod species. *J. Geol.* **71,** 1–11.

Dodd, J. R. (1964). Environmentally controlled variation in the shell structure of a pelecypod species. *J. Paleontol.* **38,** 1065–1071.

Dodd, J. R. (1965). Environmental control of strontium and magnesium in *Mytilus. Geochim. Cosmochim. Acta* **29,** 385–398.

Dodd, J. R. (1966). The influence of salinity on mollusc shell mineralogy: A discussion. *J. Geol.* **74,** 85–89.

Dodd, J. R., and Crisp, E. L. (1982). Non-linear variation with salinity of Sr/Ca and Mg/Ca ratios in water and aragonitic bivalve shells and implications for paleosalinity studies. *Palaeogeogr. Palaeoclimat. Palaeoecol.* **38,** 45–56.

Dodd, J. R., and Stanton, R. J. (1981). *Paleoecology, Concepts and Applications.* John Wiley, New York.

Dodge, R. E., and Vaisnys, J. R. (1980). Skeletal growth chronologies of recent and fossil corals. In *Skeletal Growth of Aquatic Organisms* (eds. D. C. Rhoads and R. A. Lutz), pp. 493–517. Plenum Press, New York.

Dogterom, A. A., and Doderer, A. (1981). A hormone dependent calcium-binding protein in the mantle edge of the freshwater snail *Lymnae stagnalis. Calcif. Tissue Int.* **33,** 505–508.

Dogterom, A. A., and Jentjens, T. (1980). The effect of the growth hormone of the pond snail *Lymnae stagnalis* on periostracum formation. *Comp. Biochem. Physiol.* **66A,** 687–690.

Dogterom, A. A., van Loenhout, H., and van der Schors, R. C. (1979). The effect of growth hormone of *Lymnae stagnalis* on shell calcification. *Gen. Comp. Endocrinol.* **39,** 63–68.

Doi, Y., and Eanes, E. D. (1984). Transmission electron microscopic study of calcium phosphate formation in supersaturated solutions seeded with apatite. *Calcif. Tissue Int.* **36,** 39–47.

Donnay, G., and Pawson, D. L. (1969). X-Ray diffraction studies of echinoderm plates. *Science* **166,** 1147–1150.

Dopping-Hepenstal, P. J. C., Ali, S. J., and Stamp, T. C. B. (1981). Matrix vesicles in the osteoid of human bone. In *Matrix Vesicles* (eds. A. Ascenzi, E. Bonucci, and B. de Bernard), pp. 229–234. Wichtig Editore, Milan, Italy.

Dorn, R. I., and Oberlander, T. M. (1982). Rock varnish. *Prog. Phys. Geogr.* **6,** 317–367.

Dorsat, D. L., and Ghiradella, H. (1983). Insect wax secretion: The growth of tubular crystals. *Biochim. Biophys. Acta* **760,** 136–142.

Doty, S., Robinson, R. A., and Schofield, B. (1976). Morphology of bone and histochemical staining characteristics of bone cells. In *Handbook of Physiology* (ed. G. D. Aurbach), pp. 3–23. American Physiology Soc., Washington, DC.

Doyle, L. J., Blake, N. J., Woo, C. C., and Yevich, P. (1978). Recent biogenic phosphorite: Concretions in mollusk kidneys. *Science* **199,** 1431–1433.

Drebes, G. (1977). Sexuality. In *The Biology of Diatoms* (ed. D. Werner), pp. 250–283. Blackwell, Oxford.

Drum, R. W., and Pankratz, H. S. (1964). Post mitotic fine structure of *Gomphonema pervulum*. *J. Ultrastruct. Res.* **10,** 217–223.

Dudich, E. (1931). Systematische und biologische Untersuchungen über die Kalkeinlagerungen des Crustaceanpanzers in polarisiertem Lichte. *Zoologica (Stuttgart)* **30,** 1–154.

Dugal, L. P. (1939). The use of calcareous shell to buffer the product of anaerobic glycolysis in *Venus merceneria*. *J. Cell. Comp. Physiol* **13,** 235–251.

Dunkelberger, D. G., and Watabe, N. (1974). An ultrastructural study on spicule formation in the pennatulid colony *Renilla reniformis*. *Tissue Cell* **6,** 573–586.

Dunkelberger, D. G., Dean, J. M., and Watabe, N. (1980). The ultrastructure of the otolithic membrane and otolith in the juvenile mummichoge, *Fundulus heteroclitus*. *J. Morphol.* **163,** 367–377.

Durliat, M. and Vranckx, R. (1986). Cuticular proteins in the crayfish *Astacus leptodactylus*—II. Relationships between proteins in plasma soft integument and calcified cuticle. *Comp. Biochem. Physiol.* **85B,** 267–274.

Durliat, M. D., Vranckx, R., and Denoyelle, M. (1980). Cuticular protein analysis from *Astacus leptodactylus*. II. Immunological identities of proteins from shell and hemolymph. *Comp. Biochem. Physiol.* **65B,** 65–73.

Eanes, E. D. (1986). Calcium phosphate formation in aqueous suspensions of anionic liposomes. In *Cell Mediated Calcification and Matrix Vesicles* (ed. S. Y. Ali), pp. 187–192. Elsevier, Amsterdam.

Eastoe, J. E. (1965). The chemical composition of bone and tooth. *Adv. Fluorine Res. Dent. Caries* **3,** 5–17.

Eastoe, J. E. (1979). Enamel protein chemistry—past, present and future. *J. Dent. Res.* **58B,** 753–763.

Ebert, T. A. (1967). Growth and repair of spines in the sea urchin *Strongylocentrotus purpuratus* (Stimpson). *Biol. Bull.* **133,** 141–147.

Ehrenberg, C. G. (1834). Phosphorsaures Kalk in den Zähnen und Kieselerde in dem Panzer von Infusorien. *Ann. Phys. Lpz.* **32,** 574–576.

Ehrenberg, C. G. (1836). Bemerkungen über feste mikroskopische, anorganische Formen in den erdigen und derben Mineralien. *Ber. Verhandlung. König. Preuss. Akad. Wissenschaft, Berlin 1836,* 84–85.

Ehrlich, H. L. (1981). *Geomicrobiology*. Marcel Dekker, New York.

Eisma, D. (1966). The influence of salinity on mollusc shell mineralogy: A discussion. *J. Geol.* **74,** 89–94.

Elliot, D. K. (1987). A reassessment of *Astraspis desiderata,* the oldest North American vertebrate. *Science* **237,** 190–191.

Ellis, B. F., and Messina, A. (1940). *Catalogue of Foraminifera*. American Museum of Natural History, New York.

Emlet, R. B. (1982). Echinoderm calcite: A mechanical analysis of larval spicules. *Biol. Bull.* **163,** 264–275.

Emlet, R. B. (1985). Crystal axes in recent and fossil adult echinoids indicate trophic mode in larval development. *Science* **230,** 937–940.

Ennever, J., and Summers, F. E. (1975). Calcification by *Candida albicans*. *J. Bacteriol.* **122,** 1391–1393.

Ennever, J., and Takazoe, I. (1973). Bacterial calcification. In *Biological Mineralization* (ed. I. Zipkin), pp. 629–648. Wiley-Interscience, New York.

Epstein, S., and Lowenstam, H. A. (1953). Temperature-shell-growth relations of recent interglacial Pleistocene shoal-water biota from Bermuda. *J. Geol.* **61,** 424–438.

Epstein, S., and Mayeda, T. (1953). Variations of O^{18} content of waters from natural sources. *Geochim. Cosmochim Acta* **4,** 213–224.

Epstein, S., Buchsbaum, R., Lowenstam, H. A., and Urey, H. C. (1951). Carbonate-water isotopic temperature scale. *Bull. Geol. Soc. Am.* **62,** 417–426.

Epstein, S., Buchsbaum, R., Lowenstam, H. A., and Urey, H. C. (1953). Revised carbonate-water isotopic temperature scale. *Bull. Geol. Soc. Am.* **64,** 1315–1326.

Erben, H. K. (1970). Ultrastrukturen und Mineralisation rezenter und fossiler Eischalen bei Vögeln und Reptilien. *Biomineralisation* **1,** 1–66.

Erben, H. K., Flajs, G., and Siehl, A. (1968). Über die Schalenstruktur von Monoplacophoren. *Akad. Wissensch. Literat. Mainz. Abhandl. Math. Naturwiss. Kl. Jhrg.* **1,** 3–24.

Erdman, J. A., Gaugh, L. P., and White, R. W. (1977). Calcium oxalate as source of high ash yields in the terricolous lichen *Parmelia chlorochroa. Bryologist* **80,** 334–339.

Erez, J. (1978). Vital effects on stable-isotope composition seen in foraminifera and coral skeletons. *Nature (London)* **273,** 199–202.

Erez, J. (1983). Calcification rates, photosynthesis and light in planktonic foraminifera. In *Biomineralization and Biological Metal Accumulation* (eds. P. Westbroek and E. W. de Jong), pp. 307–313. Reidel, Dordrecht.

Evans, J. W. (1910). The sudden appearance of the Cambrian fauna. *Compte Rendu. Congres. Geol. Int.* **XI,** 543–546.

Evans, J. W. (1972). Tidal growth increments in the cockle *Clinocardium nuttalli. Science* **176,** 416–417.

Fahrenbach, W. H. (1959). Studies on the histology and cytology of midgut diverticula of *Limnoria lignorum.* In *Marine Boring and Fowling Organisms* (ed. D. L. Ray), pp. 96–102. Univ. Washington Press, Seattle.

Farach, M. C., Valdizan, M., Park, H. R., Decker, G. L., and Lennarz, W. J. (1987). Developmental expression of a cell-surface protein involved in calcium uptake and skeleton formation in sea urchin embryos. *Dev. Biol.* **122,** 320–331.

Fawcett, D. W. (1942). The amedullary bones of the Florida manatee *(Trichechus latirostris). Am. J. Anat.* **71,** 271–309.

Fearnhead, R. W. (1979). Matrix-mineral relationships in enamel tissues. *J. Dent. Res.* **58B,** 909–916.

Feeney, R. E., Burcham, T. S., and Yeh, Y. (1986). Antifreeze glycoproteins from polar fish blood. *Annu. Rev. Biophys. Biochem.* **15,** 59–78.

Fell, H. B. (1925). The histogenesis of cartilage and bone in the long bones of the embryonic fowl. *J. Morphol.* **40,** 417–459.

Ferraiolo, J. A. (1982). A systematic classification of nonsilicate minerals. *Bull. Amer. Mus. Nat. Hist.* **172,** 1–237.

Fichtinger-Schepman, A. J., Kamerling, J. P., Versluis, C., and Vliegenhart, F. G. (1981). Structural studies of the methylated, acidic polysaccharide associated with *Emiliania huxleyi* (Lohmann) Kamptner. *Carbohydrate Res.* **93,** 105–123.

Filshie, B. K. (1982). Fine structure of the cuticle of insects and other arthropods. In *Insect Ultrastructure* (eds. R. C. King and H. Akai). Plenum Press, New York.

Fincham, A. G., and Belcourt, A. B. (1985). Amelogenin biochemistry: Current concepts. In *The Chemistry and Biology of Mineralized Tissues* (ed. W. T. Butler), pp. 240–247. EBSCO Media, Birmingham, AL.

Finlay, B. J., Hetherington, N. B., and Davison, W. (1983). Active biological participation in lacustrine barium chemistry. *Geochim. Comochim. Acta* **47,** 1325–1329.

Fischer, A. (1884). Ueber das Vorkommen von Gypskristallen bei den Desmidieen. *Jahrb. Wiss. Bot.* **14,** 133–184.

Fisher, L. W. (1985). The nature of the proteoglycans of bone. In *The Chemistry and*

Biology of Mineralized Tissues (ed. W. T. Butler), pp. 188–195. EBSCO Media, Birmingham, AL.

Fisher, L. W., and Termine, J. D. (1985). Noncollagenous proteins influencing the local mechanisms of calcification. *Clin. Orthop.* **280,** 362–385.

Fisher, L. W., Whitson, S. W., Avioli, L. V., and Termine, J. D. (1983). Matrix sialoprotein of developing bone. *J. Biol. Chem.* **258,** 12723–12727.

Fisher, L. W., Hawkins, G. R., Tuross, N., and Termine, J. D. (1987a). Purification and partial characterization of small proteoglycans I and II, bone sialoproteins I and II, and osteonectin from the mineral compartment of developing human bone. *J. Biol. Chem.* **262,** 9702–9708.

Fisher, L. W., Gehron-Robey, P., Tuross, N., Otsuka, A. S., Tepen, D. A., Esch, F. S., Shimasaki, S., and Termine, J. D. (1987b). The M_r 24,000 phosphoprotein from developing bone is the NH_2-terminal propeptide of the $\alpha 1$ chain of type 1 collagen. *J. Biol. Chem.* **262,** 13457–13463.

Folk, R. L. (1974). The natural history of crystalline calcium carbonate: Effect of magnesium content and salinity. *J. Sed. Petrol.* **44,** 40–53.

Follis, R. H. (1960). Calcification of cartilage. In *Calcification in Biological Systems* (ed. R. F. Sognnaes), pp. 245–280. AAAS Publication, Washington, DC.

Ford, G. C., Harrison, P. M., Rice, D. W., Smith, J. M. A., Treffrey, A., White, J. L., and Yariv, J. (1984). Ferritin: Design and formation of an iron-storage protein. *Phil. Trans. R. Soc. London Ser. B* **304,** 551–565.

Fowler, G. H. (1887). The anatomy of the Madreporia: III. *Quart. J. Microscop. Sci.* **28,** 1–17.

Franceschi, V. R., and Horner, H. T. (1980). Calcium oxalate crystals in plants. *Bot. Rev.* **46,** 361–427.

Frank, R. M., and Nalbandian, J. (1967). Ultrastructure of amelogenesis. In *Structural and Chemical Organization of Teeth* (ed. A. E. W. Miles), pp. 399–466. Academic Press, New York.

Frankel, R. B. (1984). Magnetite guidance of organisms. *Annu. Rev. Biophys. Bioeng.* **13,** 85–103.

Frankel, R. B., Blakemore, R. P., and Wolfe, R. S. (1979). Magnetite in freshwater magnetotactic bacteria. *Science* **203,** 1355–1356.

Frankel, R. B., Papaefthymiou, G. C., and Blakemore, R. P. (1985). Mossbauer spectroscopy of iron biomineralization products in magnetotactic bacteria. In *Magnetite Biomineralization and Magnetoreception in Organisms: A New Biomagnetism. Topics in Geobiology Series* (eds. J. L. Kirschvink, D. S. Jones, and B. J. MacFadden), pp. 269–287. Plenum Press, New York.

Franzen, A., and Heinegard, D. (1985a). Isolation and characterization of two sialoproteins present only in bone calcified matrix. *Biochem. J.* **232,** 715–724.

Franzen, A., and Heinegard, D. (1985b). Proteoglycans and proteins of rat bone. In *The Chemistry and Biology of Mineralized Tissues* (ed. W. T. Butler), pp. 132–141. EBSCO Media, Birmingham, AL.

Fraser, R. D. B. (1983). Molecular conformation and packaging in collagen fibrils. *J. Mol. Biol.* **167,** 497–521.

Frey-Wyssling, A. (1930). Über die Ausscheidung der Kieselsaure in der Pflanze. *Ber. Dtsch. Bot. Ges.* **48,** 179–183.

Friedmann, E. I. (1979). The genus *Geitleria* (Cyanophyceae or Cyanobacteria): Distribution of *G. calcarea* and *G. floridana* n.sp. Pl. *Syst. Evol.* **131,** 169–178.

Friedmann, E. I., Roth, H. C., Turner, J. B., and McEwan, R. S. (1972). Calcium oxalate crystals in the aragonite-producing green alga *Penicillus* and related genera. *Science* **177,** 891–893.

Fuchtbauer, H., and Hardie, L. A. (1976). Experimentally determined homogeneous distribution coefficients for precipitated magnesium calcites. *Geol. Soc. Am. Abstr.* **8**(6), 877.

Fukae, M., and Tanabe, T. (1987). Nonamelogenin components of porcine enamel in the protein fraction free from the enamel crystals. *Calcif. Tissue Int.* **40,** 286–293.

Fuller, M., Goree, W. S., and Goodman, W. L. (1985). An introduction to the use of SQUID magnetometers in biomagnetism. In *Magnetite Biomineralization and Magnetoreception in Organisms* (eds. J. L. Kirschvink, D. S. Jones, and B. J. MacFadden), pp. 103–151. Plenum Press, New York.

Furst, M., Lowenstam, H. A., and Burnett, D. S. (1976). Radiographic study of the distribution of boron in recent mollusc shells. *Geochim. Cosmochim. Acta* **40,** 1381–1386.

Galvan, J., Rodriguez, C., and Ascaso, C. (1981). The pedogenic action of lichens on metamorphic rocks. *Pedobiologia* **21,** 60–73.

Garant, P. R. (1970). An electron microscope study of the crystal-matrix relationship in the teeth of the dogfish *Squalus acanthias* L. *J. Ultrastruct. Res.* **30,** 441–449.

Garstang, W. (1928). The morphology of the Tunicata and its bearing on the phylogeny of the Chordata. *Quart. J. Microscop. Sci.* **72,** 51–187.

Gay, C. V., Schraer, H., and Hargest, T. E. (1978). Ultrastructure of matrix vesicles and mineral in unfixed embryonic bone. *Metab. Bone Dis. Rel. Res.* **1,** 105–108.

Geraerts, W. P. M. (1976). Control of growth by the neurosecretory hormone of the light green cells in the freshwater snail *Lymnae stagnalis. Gen. Comp. Endocrinol.* **29,** 97–108.

Gerth, H. (1929). The evolution of reef corals during the Cenozoic period. *Pac. Sci. Cong. 4th* pp. 333–350.

Geuze, J. (1968). Observations on the function and the structure of the statocysts of *Lymnae stagnalis* (L.). *Neth. J. Zool.* **18,** 155–204.

Gibbins, J. R., Tilney, L. G., and Porter, K. R. (1969). Microtubules in the formation and development of the primary mesenchyme in *Arbacia punctulata.* I. The distribution of microtubules. *J. Cell. Biol.* **41,** 201–226.

Gibbs, P. E., and Bryan, G. W. (1984). Calcium phosphate granules in muscle cells of *Nephtys* (Annelida, Polychaeta)— a novel skeleton?. *Nature (London)* **310,** 494–495.

Giraud, M. (1977). L'anhydrase carbonique et la calcification tegumentaire chez le crabe *Carcinus maenas* Linne. *C.R. Acad. Sci. Paris Ser. D* **284,** 453–456.

Glaessner, M. F. (1976). Early Phanerozoic annelid worms and their geological and biological significance. *J. Geol. Soc. London* **132,** 259–275.

Glaessner, M. F. (1984). *The Dawn of Animal Life, a Biohistorical Study, Cambridge Earth Science Series.* Cambridge Univ. Press, Cambridge.

Glaessner, M. F., and Wade, M. (1966). The Late Precambrian fossils from Ediacara, South Australia, *Palaeontology* **9,** 599–628.

Glas, J. E. (1962). Studies on the ultrastructure of dental enamel—VI: Crystal chemistry of shark's teeth. *Odont. Rev.* **13,** 315–326.

Glimcher, M. J. (1959). Molecular biology of mineralized tissues with particular reference to bone. *Rev. Med. Phys.* **31,** 359–393.

Glimcher, M. J. (1984). Recent studies of the mineral phase in bone and its possible linkage to the organic matrix by protein-bound phosphate bonds. *Phil. Trans. R. Soc. London Ser. B.* **304,** 479–508.

Glimcher, M. J., Brickley-Parsons, D., and Levine, P. T. (1977). Studies of enamel proteins during maturation. *Calcif. Tissue Res.* **24,** 259–270.

Glimcher, M. J., Kossiva, D., and Roufosse, A. (1979). Identification of phosphopeptides and γ-carboxyglutamic acid-contining peptides in epiphyseal growth plate cartilage. *Calcif. Tissue Int.* **27,** 187–191.

Goffinet, G., and Jeuniaux, C. (1969). Composition chimique de la fraction "nacroine" de la conchioline de nacre de *Nautilus pompilius* Lamarck. *Comp. Biochem. Physiol.* **29,** 277–282.

Goffinet, G., and Jeuniaux, C. (1979). Distribution et importance quantitative de le chitine dans les coquilles de mollusques. *Cah. Biol. Mar.* **20,** 341–349.

Goldberg, W. M., and Benayahu, Y. (1987). Spicule formation in the gorgonian coral *Pseudoplexaura flagellosa.* 1: Demonstration of intracellular and extracellular growth and the effect of ruthenium red during decalcification. *Bull. Mar. Sci.* **40,** 287–303.

Golubic, S., and Campbell, S. E. (1980). Biologically-formed aragonite concretions in marine *Rivularia.* In *Phanerozoic Stromatolites* (ed. C. L. V. Monty), pp. 209–229. Springer-Verlag, New York.

Gooday, A. J., and Nott, J. A. (1982). Intracellular barite crystals in two xenophyophores, *Aschemonella ramuliformis* and *Galatheammina* sp. (Protozoa: Rhizopoda) with comments on the taxonomy of *A. ramuliformis. J. Mar. Biol. Assoc. U.K.* **62,** 595–605.

Gorby, Y. A., Beveridge, T. J., and Blakemore, R. P. (1988). Characterization of the bacterial magnetosome membrane. *J. Bacteriol.* **170,** 834–841.

Gordon, C. M., Carr, R. A., and Larson, R. E. (1970). The infuence of environmental factors on the sodium and manganese content of barnacle shells. *Limnol. Oceanog.* **15,** 461–466.

Gordon, J., and Carriker, M. R. (1980). Sclerotized protein in the shell matrix of a bivalve mollusc. *Mar. Biol.* **57,** 251–260.

Goreau, T. F. (1959). The physiology of skeletal formation in corals. I. A method for measuring the rate of calcium deposition by corals under different conditions. *Biol. Bull.* **116,** 59–75.

Goreau, T. J. (1977). Coral skeletal chemistry: Physiological and environmental regulation of stable isotopes and trace metals in *Monastrea annularis. Proc. R. Soc. London Ser. B* **196,** 291–315.

Gould, J. L., Kirschvink, J. L., and Deffeyes, K. S. (1978). Bees have magnetic remanence. *Science* **201,** 1026–1028.

Gouranton, J. (1968). Composition, structure, et mode de formation des concrétions minerales dans l'intestin moyen des homopteres cercopides. *J. Cell Biol.* **37,** 316–328.

Graf, F., and Meyran, J. (1985). Calcium reabsorption in the posterior caeca of the midgut in a terrestrial crustacean, *Orchestia cavimana. Cell Tissue Res.* **242,** 83–95.

Graham, E. E. (1984). Protein biosynthesis during spiny dogfish *(Squalus acanthias)* enameloid formation. *Arch. Oral Biol.* **29,** 821–825.

Graham, E. E. (1985). Isolation of enamelinlike proteins from blue shark *(Prionace glauca)* enameloid. *J. Exp. Zool.* **234,** 185–191.

Grant, S. R., Farach, M. C. Decker, G. L., Woodward, H. D., Farach, H. A., and Lennarz, W. J. (1985). Developmental expression of cell-surface (glyco)proteins involved in gastrulation and spicule formation in sea urchin embryos. *Cold Spring Harbour Symp. Quant. Biol.* **50,** 91–98.

Graustein, W. L., Cromack, K., and Sollins, P. (1977). Calcium oxalate occurrence in soils and effect on nutrient and geochemical cycles. *Science* **198,** 1252–1254.

Grave, C. (1944). The larva of *Styela (Cynthia) partita:* Structure, activities and duration of life. *J. Morphol.* **75,** 173–188.

Grave, C. and Riley, G. (1935). Development of the sense organs of the larva *Botryllus schlosseri*. *J. Morphol.* **57,** 185–211.

Gray, J., and Boucot, A. J. (1977). Early vascular land plants: Proof and conjecture. *Lethaia* **10,** 145–174.

Green, H. W., Lipps, J. H., and Showers, W. J. (1980). Test ultrastructure of fusulinid Foraminifera. *Nature (London)* **283,** 853–855.

Green, J. C. (1986). Biomineralization in the algal class Prymnesiophyceae. In *Biomineralization in Lower Plants and Animals* (eds. B. S. C. Leadbeater and R. Riding), pp. 173–188. Clarendon Press, Oxford.

Green, J. C., and Course, P. A. (1983). Extracellular calcification in *Chryosotila lamellosa* (Prymnesiophyceae). *Br. Phycol. J.* **18,** 367–382.

Green, J. C., Hibberd, D. J., and Pienaar, R. N. (1982). The taxonomy of Prymnesium (Prymnesiophycea) including a description of a new cosmopolitan species *P. patellifera* sp. nov. and further observations on *P. parvum* M. Carter. *Br. Phycol. J.* **17,** 363–382.

Green, R. L., and Warren, G. J. (1985). Physical and functional repetition in a bacterial ice nucleation gene. *Nature (London)* **317,** 645–648.

Greenaway, P. (1971). Calcium regulation in the freshwater mollusc *Lymnaea stagnalis* (L.) (Gastropoda: Pulmonata). *J. Exp. Biol.* **54,** 609–620.

Greenaway, P. (1985). Calcium balance and moulting in the crustacea. *Biol. Rev.* **60,** 425–454.

Greenfield, E. M., Wilson, D. C., and Crenshaw, M. A. (1984). Ionotropic nucleation of calcium carbonate by molluscan matrix. *Am. Zool.* **24,** 925–932.

Greenfield, L. J. (1963). Metabolism and concentration of calcium and magnesium and precipitation of calcium carbonate by a marine bacterium. *Ann. N.Y. Acad. Sci.* **109,** 23–45.

Grégoire, C. (1972). Structure of the molluscan shell. In *Chemical Zoology* (eds. M. Florkin and B. T. Scheer), pp. 45–102. Academic Press, New York.

Grégoire, C. (1973). On the submicroscopic structure of the organic components of the siphon in the Nautilus shell. *Arch. Int. Physiol. Biochem.* **81,** 299–316.

Grégoire, C., Duchateau, G. H., and Florkin, M. (1955). La trame protidique des nacres et des perles. *Ann. Inst. Oceanogr. Monaco* **31,** 1–36.

Grodowitz, M. J., and Broce, A. B. (1983). Calcium storage in face fly (Diptera: Muscidae) larvae for puparium formation. *Ann. Entomol. Soc. Am.* **76,** 418–424.

Grodowitz, M. J., Broce, A. B., and Kramer, K. J. (1987). Morphology and biochemical composition of mineralized granules from the Malphigian tubules of *Musca autumnalis* De Geer larvae (Diptera: muscidae). *Insect Biochem.* **17,** 335–345.

Gron, P., and Hay, D. I. (1976). Inhibition of calcium phosphate precipitation by human salivary concretions. *Arch. Oral Biol.* **21,** 201–205.

Gross, R. (1926). *Die Kriestalline Struktur von Dentin und Zahnschmelz. Festschr. Zahnarztl.* Inst. Univ. Greifswald, Berlin.

Grynpas, M. D., Eyre, D. R., and Kirschner, D. A. (1980). Collagen type II differs from type I in native molecular packing. *Biochim. Biophys. Acta* **626,** 346–355.

Haas, W. (1972). Untersuchungen uber die Mikro-und Ultrastruktur der Polyplacophorenscale. *Biomineral. Res. Rep.* **5,** 1–52.

Haas, W. (1976). Observations on the shell and mantle of the Placophora. In *The Mechanisms of Mineralization in the Invertebrates and Plants* (eds. N. Watabe and K. M. Wilbur), pp. 389–402. Univ. of South Carolina Press, Columbia, SC.

Hackman, R. H., and Goldberg, M. (1960). Composition of the oothecae of three orthoptera. *J. Insect Physiol.* **5,** 73–78.

Haeckel, E. H. P. A. (1872). *Die Kalkschwämme.* G. Reimer, Berlin.

Hall, A., and Taylor, J. (1971). The occurrence of vaterite in gastropod eggshells. *Mineral Mag.* **38,** 521–525.

Hallam, A., and O'Hara, M. J. (1962). Aragonitic fossils in the Lower Carboniferous of Scotland. *Nature (London)* **195,** 273–274.

Hallam, A., and Price, N. B. (1968). Environmental and biochemical control of strontium in shells of *Cardium edule. Geochim. Cosmochin. Acta* **32,** 319–328.

Hallberg, R. O. (1965). Notes on biosynthesis of pyrite. *Stockholm Contr. Geol.* **13,** 35–37.

Hallberg, R. O. (1972). Iron and zinc sulfides formed in a continuous culture of sulfate reducing bacteria. *N. Jahrbuch Mineral. Monatsohefte,* 481–500.

Hallberg, R. O., and Wadsten, T. (1980). Crystal data of a new phosphate compound from microbial experiments on iron sulfide mineralization. *Am. Mineralog.* **65,** 200–204.

Hamilton, S. E., King, G., Tesch, D., Riddles, P. W., Keough, D. T., Jell, J., and Zerner, B. (1982). γ-Carboxyglutamic acid in invertebrates: Its identification in hermatypic corals. *Biochim. Biophys. Res. Commun.* **108,** 610–613.

Hansen, H. J., and Reiss, Z. (1972). Scanning electron microscopy of wall structures in some benthonic and planktonic foraminiferida. *Rev. Esp. Micropaleontol.* **4,** 169–179.

Haq, B. U. (1978). Calcareous nannoplankton. In *Introduction to Marine Micropaleontology* (eds. B. U. Haq and A. Boersma), pp. 79–107. Elsevier, New York.

Hare, P. E. (1963). Amino acids in the proteins from aragonite and calcite in the shells of *Mytilus californianus. Science* **139,** 216–217.

Hare, P. E., and Abelson, P. H. (1965). Amino acid composition of some calcified proteins. *Carnegie Inst. Wash. Year Book* **64,** 223–234.

Harper, H. E., and Knoll, A. H. (1975). Silica, diatoms and Cenozoic radiolarian evolution. *Geology* **3,** 175–177.

Harris, R. C. X., and Pilkey, O. H. (1966). Temperature and salinity control of the concentration of skeletal Na, Mn and Fe in *Dendraster excentricus. Pacif. Sci.* **20,** 235–238.

Harrison, P. M. (1986). The structure and function of ferritin. *Bichem. Ed.* **14,** 154–162.

Harrison, P. M., Fischbach, F. A., Hay, T. G., and Haggis, G. H. (1967). Ferric oxyhydroxide core of ferritin. *Nature (London)* **216,** 1188–1190.

Harrison, P. M., Treffry, A., and Lilley, T. H. (1986). Ferritin as an iron-storage protein: Mechanisms of iron uptake. *J. Inorg. Biochem.* **27,** 287–293.

Harrison, P. M., Andrews, S. C., Ford, G. C., Smith, J. M. A., Treffry, A., and White, J. L. (1987). Ferritin and bacterioferritin: Iron sequestering molecules from man to microbe. In *Iron Transport in Microbes, Plants and Animals* (eds. G. Winkelmann, D. van der Helm, and J. B. Neilands), pp. 445–475. VCH, Weinheim.

Hartman, W. D., and Goreau, T. F. (1970). Jamaican coralline sponges: Their morphology, ecology and fossil relatives. *Symp. Zool. Soc. London* **25,** 205–243.

Hascall, V. C., and Hascall, G. K. (1981). Proteoglycans. In *Cell Biology of Extracellular Matrix* (ed. E. D. Hay), Chap. 2, pp. 39–63. Plenum Press, New York.

Hass, W. (1916). Über die Struktur des Chitins bei Arthropoden. *Arch. Anat. Physiol.* **1916,** 295–338.

Hauschka, P. V., Lian, J. B., and Gallop, P. M. (1975). Direct identification of the calcium-binding amino acid, γ-carboxyglutamate in mineralized tissue. *Proc. Natl. Acad. Sci. U.S.A.* **72,** 3925–3929.

Hawkes, J. W., and Schraer, H. (1973). Mineralization during the molt cycle in *Lirceus brachyurus* (Isopoda: Crustacea). *Calcif. Tissue Res.* **12,** 125–136.

Hayasi, K. (1937). On the detection of calcium in the calicoblasts of some reef corals. *Palao Trop. Biol. Station Stud.* **8,** 169–179.

Hayes, R. L., and Goreau, N. I. (1977). Intracellular crystal-bearing vesicles in the epidermis of scleractinian corals, *Astrangia danae* (Agassiz) and *Porites porites* (Pallas). *Biol. Bull.* **152,** 26–40.

Heathfield, B. M., and Travis, D. F. (1975a). Ultrastructural studies of regenerating spines of the sea urchin *Strongylocentrotus purpuratus.* I. Cell types without spherules. *J. Morphol.* **145,** 13–27.

Heathfield, B. M., and Travis, D. F. (1975b). Ultrastructural studies of regenerating spines of the sea urchin *Strongylocentrotus purpuratus.* II. Cell types with spherules. *J. Morphol.* **145,** 51–72.

Hecky, R. E., Mopper, K., Kilham, P. and Degens, E. T. (1973). The amino acid and sugar composition of diatom cell walls. *Marine Biol.* **19,** 323–331.

Hedley, R. H. (1963). Cement and iron in the arenaceous foraminifera. *Micropaleontology* **9,** 433–441.

Hegdahl, T., Silness, J., and Gustavsen, F. (1977a). The structure and mineralization of the carapace of the crab (*Cancer pagurus* L.). 1. The endocuticle. *Zool. Scripta* **6,** 89–99.

Hegdahl, T., Gustavsen, F., and Silness, J. (1977b). The structure and mineralization of the carapace of the crab (*Cancer pagurus* L.). 2. The exocuticle. *Zool. Scripta* **6,** 101–105.

Hegdahl, T., Gustavsen, F., and Silness, J. (1977c). The structure and mineralization of the carapace of the crab (*Cancer pagurus* L.). 3. The epicuticle. *Zool. Scripta* **6,** 215–220.

Helseth, D. L., Lechner, J. H., and Veis, A. (1979). Role of the amino-terminal extrahelical region of type I collagen in directing the 4D overlap in fibrillogenesis. *Biopolymers* **18,** 3005–3014.

Hemleben, C., Bé, A. W. H., Anderson, O. R., and Tuntivate, S. (1977). Test morphology, organic layers and chamber formation of the planktonic foraminifer *Globorotalia menardii* (D'Orbigney). *J. Foraminiferal Res.* **7,** 1–25.

Hemleben, C., Anderson, O. R., Berthold, W., and Spindler, M. (1986). Calcification and chamber formation in foraminifera—a brief overview. In *Biomineralization in Lower Plants and Animals* (eds. B. S. C. Leadbeater and R. Riding), pp. 237–249. Clarendon Press, Oxford.

Herman, Y. (1978). Pteropods. In *Intoduction to Marine Micropaleontology* (eds. B. U. Haq and A. Boersma), pp. 151–159. Elsevier, Amsterdam.

Herring, G. M. (1972). The organic matrix of bone. In *The Biochemistry and Physiology of Bone* (ed. G. H. Bourne), pp. 127–189. Academic Press, New York.

Hiller, C. R., Robinson, C., and Weatherell, J. A. (1975). Variation in the composition of developing rat incisor enamel. *Calcif. Tissue Res.* **18,** 1–12.

Hillis, L. W. (1959). A revision of the genus *Halimeda* (order Siphonales). *Publ. Inst. Mar. Sci. Univ. Texas* **6,** 321–403.

Hillman, R. (1961). Formation of the periostracum in *Mercenaria mercenaria. Science* **134,** 1754–1755.

Hirschman, A., and Dziewiatkowski, D. D. (1966). Protein polysaccharide loss during endochondral ossification. Immunochemical evidence. *Science* **154,** 393–395.

Hobbs, R. N., Hazel, C. M., Smith, S. C., Carney, D. A., Howells, A. C., Littlewood, A. J., and Pennock, J. F. (1987). Metabolism of vitamin D_3 by a marine echinoderm, *Psammechinus. Chem. Scripta* **27,** 199–205.

Hodge, A. J., and Petruska, J. A. (1963). Recent studies with the electron microscope on ordered aggregates of the tropocollagen molecule. In *Aspects of Protein Structure* (ed. G. N. Ramachandran), pp. 289–300. Academic Press, New York.

Hoffman, S. (1949). Studien über das Integument der Solenogastren nebst Bemerkungen über die Verwandtschaft zwischen den Solenogastren und Placophoren. *Zool. Bidrag Uppsala* **27,** 293–427.

Hofmann, H., and Kuhn, K. (1981). Statistical analysis of collagen sequences with regard to fibril assembly and evolution. In *Structural Aspects of Recognition and Assembly in Biological Macromolecules* (eds. M. Balaban, J. L. Sussman, W. Traub, and A. Yonath), pp. 403–426. Balaban ISS, Rehovot.

Hofmann, H., Fietzek, P. P., and Kuhn, K. (1978). The role of polar and hydrophobic interactions for the molecular packing of type I collagen: A three-dimensional evaluation of the amino acid sequence. *J. Mol. Biol.* **125,** 137–165.

Höhling, H. J., Ashton, B. A., and Koster, H. D. (1974). Quantititative electron microscopic investigations of mineral nucleation in collagen. *Cell. Tissue Res.* **148,** 11–26.

Holland, N. D. (1965). An autoradiographic investigation of tooth renewal in the purple sea urchin *(Strongylocentrotus purpuratus). J. Exp. Zool.* **158,** 275–282.

Hollande, A., and Martoja, R. (1974). Identification du cristalloide des isospores de Radiolaires à un cristal de celestite ($SrSO_4$). *Protistologica* **10,** 603–609.

Holmes, J. M., Beebe, R. A., Posner, A. S., and Harper, R. A. (1970). Surface areas of synthetic calcium phosphates and bone mineral. *Proc. Soc. Exp. Biol.* **133,** 1250–1253.

Honjo, S., and Erez, J. (1978). Dissolution rates of calcium carbonate in the deep ocean; an in-situ experiment in the North Atlantic ocean. *Earth Planet Sci. Lett.* **40,** 287–300.

Hopkin. S. P., and Nott, J. A. (1979). Some observations on concentrically structured, intracellular granules in the hepatopancreas of the shore crab *Carcinus maenas* (L.). *J. Mar. Biol. Assoc. U.K.* **59,** 867–877.

Horner, H. T., Tiffany, L. M., and Cody, A. M. (1983). Formation of calcium oxalate crystals associated with apotheria of the discomycete *Dasyschypha capitata. Mycologia* **75,** 423–435.

Hottinger, L. (1986). Construction, structure and function of foraminiferal shells. In *Biomineralization in Lower Plants and Animals* (eds. B. S. C. Leadbeater and R. Riding), pp. 219–235. Clarendon Press, Oxford.

Howard, B., Mitchell, P. C. H., Ritchie, A., Simkiss, K., and Taylor, M. (1981). The composition of intracellular granules from the metal-accumulating cells of the common garden snail *(Helix aspersa). Biochem. J.* **194,** 507–511.

Hsu, K. J., McKenzie, J. A., and He, Q. X. (1982). Terminal Cretaceous environmental and evolutionary changes. In *Geological Implications of Impacts of Large Asteroids and Comets on the Earth* (eds. L. T. Silver and P. H. Schultz), pp. 317–328. Geol. Soc. Special Paper 196.

Hsu, K. J., Oberhansli, H., Gao, J. Y., Shu, S., Haihong, C., and Krahenbuhl, U. (1985). "Strangelove ocean" before the Cambrian explosion. *Nature (London)* **316,** 809–811.

Hubert, G., Rieder, N., Schmitt, G., and Send, W. (1975). Bariumanreicherung in den Müllerschen Körperchen der Loxididae (Ciliata, Holotricha). *Z. Naturforsch.* **30C,** 422–423.

Hulmes, D. J. S., Holmes, D. F., and Cummings, C. (1985). Crystalline regions in collagen fibrils. *J. Mol. Biol.* **184,** 473–477.

Humbert, W. (1978). Cytochemistry and X-ray microprobe analysis of the midgut of *Tomocerus minor* Lubbock (Insecta, Collembola) with special reference to the physiological significance. *Cell Tissue Res.* **187,** 397–416.

Humbert, W., Kirsch, R., and Simmonneaux, V. (1986). Is mucus involved in biocrystallization?. *Cell. Tissue Res.* **245,** 599–604.

Huner, J. V., Kowalczuk, J. G., and Avault, J. W. (1978). Postmolt calcification in subadult red swamp crayfish, *Procambarus clarkii* (Girard) (Decapoda, Cambridae). *Crustaceana* **34,** 275–280.

Hunt, S. (1972). Scleroprotein and chitin in the exoskeleton of the ectoproct *Flustra foliacea. Comp. Biochem. Physiol.* **43B,** 571–577.

Hunt, S. (1979). The structure and compositions of the love dart (gypsobellum) in *Helix pomatia. Tissue Cell* **11,** 51–61.

Hunt, S., and Oates, K. (1970). Fibrous protein ultrastructure of gastropod periostracum (*Buccinum undatum* L.). *Experientia* **26,** 1196–1197.

Hunt, S., and Oates, K. (1978). Fine structure and molecular organization of the periostracum in a gastropod mollusc *Buccinum undatum* L. and its relation to similar structural protein systems in other invertebrates. *Phil. Trans. R. Soc. London Ser. B* **283,** 417–463.

Hunter, G. K. (1987). An ion-exchange mechanism of cartilage calcification. *Conn. Tissue Res.* **16,** 111–120.

Huxley, T. H. (1858). Report on the examination of specimens of bottom. In *Deep-Sea Foundings in the North Atlantic Ocean between Ireland and Newfoundland made in HMS "Cyclops"* (ed. J. Dayman), pp. 63–68. Great Britain and Ireland Admiral., London.

Hyde, B. B., Hodge, A. J., Kahn, A., and Birnstiel, M. L. (1963). Studies on phytoferritin. 1. Identification and localization. *J. Ultrastruct. Res.* **9,** 248–258.

Hyman, L. H. (1951). *The Invertebrates: Platyhelminthes and Rhynchocoela: The Acoelemate Bilateria.* McGraw Hill, New York.

Hyman, L. H. (1955). *The Invertebrates. IV. Echinodermata.* McGraw Hill, New York.

Hyman, L. H. (1967). *The Invertebrates. VI Mollusca I.* McGraw Hill, New York.

Iler, R. (1979). *The Chemistry of Silica.* John Wiley, New York.

Isa, Y., and Okazaki, M. (1987). Some observations on the Ca^{2+}-binding phospholipid from scleractinian coral skeletons. *Comp. Biochem. Physiol.* **87B,** 507–512.

Isenberg, H. D., Levine, L. S., Moss, M. L., Kupferstein, D., and Lear, P. E. (1963). Calcification in a marine coccolithophorid. *Ann. N.Y. Acad. Sci.* **109,** 49–64.

Istin, M., and Girard, J. P. (1970). Carbonic anhydrase and mobilisation of calcium reserves in the mantle of lamellibranchs. *Calcif. Tissue Res.* **5,** 247–260.

Ivarson, K., and Hallberg, R. O. (1976). Formation of mackinawite by microbial reduction of jarosite and its application to tidal sediments. *Geoderma* **16,** 1–7.

Iwata, K. (1980). Mineralization and architecture of the larval shell of *Haliotis discus hannai* Ino, (Archaeogastropoda). *J. Fac. Sci., Hokkaido Univ., Ser. IV* **19,** 305–320.

Jaanusson, V. (1966). Fossil brachiopods with probable aragonitic shell. *Geol. Foren. Forhandl.* **88,** 279–281.

Jablonski, D., and Lutz, R. A. (1980). Molluscan larval shell morphology. In *Skeletal Growth of Aquatic Organisms* (eds. D. C. Rhoads and R. A. Lutz), pp. 323–377. Plenum Press, New York.

Jackson, S. A., Cartwright, A. G., and Lewis, D. (1978). The morphology of bone mineral crystals. *Calcif. Tissue Res.* **25,** 217–222.

Jackson, T. A., and Keller, W. D. (1970). A comparative study of the role of lichens and "inorganic" processes in the chemical weathering of recent Hawaiian lava flows. *Am. J. Sci.* **269,** 446–466.

James, N. P., Wray, J. L., and Ginsburg, R. N. (1988). Calcification of encrusting aragonitic algae (Peyssonneliaceae): Implications for the origin of late Paleozoic reefs and cements. *J. Sed. Pet.* **58,** 291–393.

Jesior, J. C., Miller, A., and Berthet-Colominas, C. (1980). Crystalline three-dimensional packing is a general characteristic of type I collagen fibrils. *FEBS Lett.* **113,** 238–240.

Jessen, H. (1968). Elliptical tubules as unit structure of forming enamel matrix in the rat. *Arch. Oral Biol.* **13,** 351–352.

Jeunieux, C. (1963). *Chitine et chitinolyse.* Masson, Paris.

Jodaikin, A., Weiner, S., and Traub, W. (1984). Enamel rod relations in the developing rat incisor. *J. Ultrastruct. Res.* **89,** 324–332.

Jodaikin, A., Traub, W., and Weiner, S. (1985). Rat enamel: A mineral-protein relation. In *Current Advances in Skeletogenesis* (eds. A. Ornoy, A. Harell, and J. Sela), pp. 441–445. Elsevier, Amsterdam.

Jodaikin, A., Traub, W., and Weiner, S. (1986). Protein conformation in rat tooth enamel. *Arch. Oral Biol.* **31,** 685–689.

Jodaikin, A., Perl-Treves, D., Weiner, S., Termine, J. D., and Traub, W. (1987). Developing enamel matrix proteins: A conformational study of enamelins and amelogenins. *Int. J. Biol. Macromol.* **9,** 166–168.

Jodaikin, A., Talmon, Y., Weiner, S., Grossman, E., and Traub, W. (1988). Mineral-organic matrix relations in tooth enamel. *Int. J. Biol. Macromol.* (in press).

Johansen, E., and Parks, H. F. (1960). Electron microscope observations on the 3-dimensional morphology of apatite crystallites of human dentine and bone. *J. Biophys. Biochem. Cytol.* **7,** 743–746.

Johnson, L. C. (1960). Mineralization of turkey leg tendon. I. Histology and histochemistry of mineralization. In *Calcification in Biological Systems* (ed. R. F. Sognnaes), pp. 117–128. AAAS, Washington, DC.

Johnston, I. S. (1980). The ultrastructure of skelatogenesis in hermatypic corals. *Int. Rev. Cytol.* **67,** 171–214.

Jones, D., and Wilson, M. J. (1986). Biomineralization in crustose lichens. In *Biomineralization in Lower Plants and Animals* (eds. B. S. C. Leadbeater and R. Riding), pp. 91–105. Clarendon Press, Oxford.

Jones, D., McHardy, W. J., and Wilson, M. J. (1976). Ultrastructure and chemical composition of spines in *Mucorales. Trans. Br. Mycol. Soc.* **66,** 153–157.

Jones, D., Wilson, M. J., and Tait, J. M. (1980). Weathering of a basalt by *Pertusaria corallina. Lichenologist* **12,** 277–289.

Jones, D., Wilson, M. J., and Laundon, J. R. (1982). Observations on the location and form of lead in *Stereocaulon vesuvianum. Lichenologist* **14,** 281–286.

Jones, D. S. (1985). Growth increments and geochemical variations in the molluscan shell. In *Mollusks. Notes for a Short Course* (ed. J. W. Broadhead), pp. 72–87. Univ. of Tennessee, Knoxville.

Jones, M. L. (1981). *Riftia pachyptila* Jones: Observations on the vestimentiferan worm from the Galapagos Rift. *Science* **213,** 333–336.

Jones. R. (1964). Note on the occurrence of opal phytoliths in some Cenozoic sedimentary rocks. *J. Paleontol.* **38,** 773–775.

Joosse, J., and Geraerts, W. P. M. (1983). Endocrinology. In *The Mollusca* (eds. A. S. M. Saleuddin and K. M. Wilbur), pp. 317–405. Academic Press, New York.

Jope, M. (1967). The protein of the brachiopod shell—I. Amino acid composition and implied protein taxonomy. *Comp. Biochem. Physiol.* **20,** 593–600.

Jorgensen, N. O. (1970). Ultrastructure of some ostracods. *Bull. Geol. Soc. Denmark* **20,** 79–92.

Kakuta, S., Golub, E. E., Haselgrove, J. C., Chance, B., Frasca, P., and Shapiro, I. M. (1986). Redox studies of the epiphyseal growth cartilage: Pyridine nucleotide metabolism and the development of mineralization. *J. Bone Mineral Res.* **1,** 433–440.

Kalmijn, A. J. (1978). Experimental evidence of geomagnetic orientation in elasmobranch fishes. In *Animal Migration, Navigation and Homing* (eds. K. Schmidt-Koenig and W. T. Keeton), pp. 347–353. Springer-Verlag, Berlin.

Kastner, M. (1981). Authigenic silicates in deep-sea sediments: Formation and diagenesis. In *The Sea* (ed. C. Emiliani), pp. 915–980. John Wiley, New York.

Kastner, M., Keene, J. B., and Gieskes, K. M. (1977). Diagenesis of siliceous oozes—1. Chemical controls on the rate of opal-A to opal-CT transformation—an experimental study. *Geochim. Cosmochim. Acta* **41,** 1041–1059.

Katz, E. P., and Li, S. (1973a). The intermolecular space of reconstituted collagen fibrils. *J. Mol. Biol.* **73,** 351–369.

Katz, E. P., and Li, S. (1973b). Structure and function of bone collagen fibrils. *J. Mol. Biol.* **80,** 1–15.

Katz, E. P., Wachtel, E. J., and Maroudas, A. (1986). The molecular packing in human articular cartilage collagen fibrils is regulated by the osmotic pressure in the extrafibrillar space. *Int. Workshop Calcif. Tissues 7th,* p. 49.

Kauffman, D., Wong, R., Bennick, A., and Keller, P. (1982). Basic proline-rich proteins from human parotid saliva: Complete covalent structure of protein 1B-9 and partial structure of protein 1B-6, members of a polymorphic pair. *Biochemistry* **21,** 6558–6562.

Kaufman, P. B., Bigelow, W. S., Schmid, R., and Ghosheh, N. (1971). Electron microprobe analyses of silica in epidermal cells of *Equisetum. Am. J. Bot.* **58,** 309–316.

Kaufman, P. B., La Croix, J. D., Dayanandan, P., Allard, L. F., Rosen, J. J., and Bigelow, W. C. (1973). Silicification of developing internodes in the perennial scouring rush (*Equisetum hyemale* var. affine). *Dev. Biol.* **31,** 124–135.

Kaufman, P. B., Dayanandan, P., Takeoka, Y., Bigelow, W. C., Jones, J. D., and Iler, R. (1981). Silica in sheets of higher plants. In *Silicon and Siliceous Structures in Biologic Systems* (eds. T. L. Simpson and B. E. Volcani), pp. 409–449. Springer-Verlag, New York.

Kelley, P. G., Oliver, P. T. P., and Pautard, F. G. E. (1953). The shell of *Lingula unguis.* In *Calcified Tissues* (eds. L. J. Richelle and M. J. Dallemagne), pp. 337–345. Univ. of Liège, Liège.

Kelly, A. (1901). Beiträge zur mineralogischen Kenntnis der Kalkausscheidungen im Tierreich. *Jena Z. Naturw.* **35,** 429–495.

Kemp, N. E. (1977). Banding pattern and fibrillogenesis of caratotrichia in shark fins. *J. Morphol.* **154,** 187–204.

Kemp, N. E. (1985). Ameloblastic secretion and calcification of the enamel layer in shark teeth. *J. Morphol.* **184,** 215–230.

Kemp, N. E., and Park, J. H. (1974). Ultrastructure of the enamel layer in developing teeth of the shark, *Carcharinus menissorrah. Arch. Oral Biol.* **19,** 633–644.

Kemp, N. E., and Westrin, S. K. (1979). Ultrastructure of calcified cartilage in the endoskeletal tesserae of sharks. *J. Morphol.* **160,** 75–102.

Kennedy, W. J., Morris, N. J., and Taylor, J. D. (1970). The shell structure, mineralogy and relationships of the *Chamacea. Palaeontology* **13,** 379–413.

Kennish, M. J., and Olsson, R. K. (1975). Effects of thermal discharges on the microstructural growth of *Mercenaria mercenaria. Environ. Geol.* **1,** 41–64.

Kessel, E. (1933). Über die Schale von *Viviparus viviparus* L. und *Viviparus fasciatus* Mull. *Z. Morphol. Okol. Tiere* **27,** 9–198.

Kessel, E. (1942). Über Bau und Bildung des Prosobranchier-Deckels. *Z. Morphol. Okol. Tiere* **38,** 197–250.

Kiderlen, H. (1937). Die Conularien. Über Bau und Leben der ersten Scyphozoa. *N. Jb. Mineral.* **77,** 113–169.

Killingley, J. S. (1980). Migrations of California gray whales tracked by O^{18} variations in their epizoic barnacles. *Science* **207,** 759–760.

Killingley, J. S., and Berger, W. H. (1979). Stable isotopes in a mollusk shell: Detection of upwelling events. *Science* **205,** 186–188.

Killingley, J. S., and Newman, W. A. (1982). ^{18}O fractionation in barnacle calcite: A barnacle paleotemperature equation. *J. Mar. Res.* **40,** 893–902.

Killingley, J. S., and Rex, M. A. (1985). Mode of larval development in some deep-sea gastropods indicated by oxygen-18 values of their carbonate shells. *Deep-Sea Res.* **32,** 809–818.

Kim, K., Webb, J., Macey, D. J., and Cohen, D. D. (1986). Compositional changes during biomineralization of the radula of the chiton *Clavarizona hirtosa. J. Inorg. Biochem.* **28,** 337–345.

Kinchington, D. (1980). Localization of intracellular calcium within the epidermis of a cool temperate coral. In *Development of Cellular Biology of Coelenterates* (eds. P. Tardent and R. Tardent), pp. 143–148. Elsevier, Amsterdam.

King, K. (1974). Preserved amino acids from silicified protein in fossil radiolaria. *Nature (London)* **252,** 690–692.

King, K. (1975). Amino acid composition of the silicified organic matrix in fossil polycystine radiolaria. *Micropaleontology* **21,** 215–226.

King, K. (1978). Distribution of γ-carboxyglutamic acid in calcified tissues. *Biochim. Biophys. Acta* **542,** 542–546.

King, K., and Hare, P. E. (1972). Amino acid composition of the test as a taxonomic character for living and fossil planktonic foraminifera. *Micropaleontology* **18,** 285–293.

Kingsley, R. J., and Watabe, N. (1982a). Ultrastructure of the axial region in *Leptogorgia virgulata* (Cnidaria: Gorgonacea). *Trans. Am. Microsc. Soc.* **101,** 325–339.

Kingsley, R. J., and Watabe, N. (1982b). Ultrastructural investigation of spicule formation in the gorgonian *Leptogorgia virgulata* (Lamarck) (Coelenterata: Gorgonacea). *Cell Tissue Res.* **223,** 325–334.

Kingsley, R. J., and Watabe, N. (1983). Analysis of proteinaceous components of the organic matrices of spicules from the gorgonian *Leptogorgia virgulata. Comp. Biochem. Physiol.* **76B,** 443–447.

Kingsley, R. J., and Watabe, N. (1984). Synthesis and transport of the organic matrix of the spicules in the gorgonian *Leptogorgia virgulata* (Lamarck) (Coelenterata: Gorgonacea). *Cell Tissue Res.* **235,** 533–538.

Kinsgley, R. J., and Watabe, N. (1985a). An autoradiographic study of calcium transport in spicule formation in the gorgonian *Leptogorgia virgulata* (Lamarck) (Coelenterata: Gorgonacea). *Cell Tissue Res.* **239,** 305–310.

Kingsley, R. J., and Watabe, N. (1985b). Ca-ATPase localization and inhibition in the gorgonian *Leptogorgia virgulata* (Lamarck) (Coelenterata: Gorgonacea). *J. Exp. Mar. Biol. Ecol.* **93,** 157–167.

Kingsley, R. J., and Watabe, N. (1987). Role of carbonic anhydrase in calcification in the gorgonian *Leptogorgia virgulata. J. Exp. Zool.* **241,** 171–180.

Kinne, R., and Fisher, L. W. (1987). Keratan sulfate proteoglycan in rabbit compact bone is bone sialoprotein II. *J. Biol. Chem.* **262,** 10206–10211.

Kinsman, D. J. J. (1969). Interpretation of Sr^{2+} concentrations in carbonate minerals and rocks. *J. Sed. Petrol.* **39,** 486–508.

Kinsman, D. J. J., and Holland, H. D. (1969). The co-precipitation of cations with $CaCO_3$—IV. The coprecipitation of Sr^{2+} with aragonite between 16° and 96°C. *Geochim. Cosmochim. Acta* **33,** 1017.

Kirschvink, J. L. (1980). Biogenic magnetite (Fe_3O_4): A ferromagnetic mineral in bacteria, animals and man. *Ferrites: Proc. Int. Conf. Jpn.*, pp. 135–138.

Kirschvink, J. L. (1981). Ferromagnetic crystals (magnetite?) in human tissue. *J. Exp. Biol.* **92,** 333–335.

Kirschvink, J. L. (1988). On the magnetotactic control of crystal orientation and iron accumulation in magnetosomes. In press.

Kirschvink, J. L., and Chang, S. R. (1984). Ultrafine-grained magnetite in deep sea sediments: Possible bacterial magnetofossils. *Geology* **12,** 559–562.

Kirschvink, J. L., and Lowenstam, H. A. (1979). Mineralization and magnetism of chiton teeth: Paleomagnetic, sedimentologic, and biologic implications of organic magnetite. *Earth Planet Sci. Lett.* **44,** 193–204.

Kirschvink, J. L., Walker, M. M., Chang, S. B., Dizon, A. E., and Peterson, K. A. (1985). Chains of single-domain magnetite crystals in chinook salmon, *Oncorhynchus tshawytscha. J. Comp. Physiol. A* **157,** 375–381.

Kitajima, T., and Okazaki, K. (1980). Spicule formation in vitro by the descendants of precocious micromere formed at the 8-cell stage of sea urchin embryo. *Dev. Growth Differ.* **22,** 265–279.

Kleinholz, L. H. (1959). Purines and pteridines from the reflecting pigment of the arthropod retina. *Biol. Bull. Mar. Biol. Lab. Woods Hole* **116,** 125–135.

Klement, R., and Trömel, G. (1932). Hydroxylapatit, der Hauptbestandteil der anorganischen Knochen-und Zahnsubstanz. *Hoppe-Seyler's Z. Physiol. Chem.* **213,** 263–269.

Knight, C. A., DeVries, A. L., and Oolman, L. D. (1984). Fish antifreeze protein and the freezing and recrystallization of ice. *Nature (London)* **308,** 295–296.

Kniprath, E. (1972). Formation and structure of the periostracum in *Lymnaea stagnalis. Calcif. Tissue Res.* **9,** 260–271.

Kniprath, E. (1974). Ultrastructure and growth of the sea urchin tooth. *Calcif. Tissue Res.* **14,** 211–228.

Kniprath, E. (1978). Growth of the shell-field in *Mytilus* (Bivalvia). *Zool. Scripta* **7,** 119–120.

Kniprath, E., and Lafargue, F. (1980). Spicule formation in the Didemnidae (Compound ascidians). In *The Mechanisms of Biomineralization in Animals and Plants* (eds. M. Omori and N. Watabe), pp. 31–36. Tokai University Press, Tokyo.

Kolodny, Y., Luz, B., and Navon, O. (1983). Oxygen isotope variations in phosphate of biogenic apatites, I. Fish bone apatite—rechecking the rules. *Earth Planet Sci.* **64,** 398–404.

Kovoor, J. (1978). Natural calcification of the prosomatic endosternite in the phalangiidae (Arachnidae: Opiliones). *Calcif. Tissue Res.* **26,** 267–269.

Krampitz, G., Erben, H. K., and Kriesten, K. (1972). Über aminosaurenzusammensetzung und Struktur von Eischalen. *Biomineralisation* **4,** 87–89.

Krampitz, G., Drolshagen, H., Hausle, J., and Hof-Irmscher, K. (1983a). Organic matrices of mollusc shells. In *Biomineralization and Biological Metal Accumulation* (eds. P. Westbroek and E. W. de Jong), pp. 231–247. Reidel, Dordrecht.

Krampitz, G., Drolshagen, H., and Hotta, S. (1983b). Simultaneous binding of calcium and bicarbonate by conchiolin of oyster shells. *Experientia* **39,** 1104–1105.

Kreger, D. R., and Boere, H. (1969). Some observations on barium sulphate in *Spirogyra. Acta Bot. Neerl.* **18,** 143–151.

Krumbein, W. E. (1974). On the precipitation of aragonite on the surface of marine bacteria. *Natuurwissenschaften* **61,** 167.

Krumbein, W. E. (1975). Biogenetic monohydrocalcite spherules in lake sediments of Lake Kivu (Africa) and the Solar Lake (Sinai). *Sedimentology* **22,** 631–634.

Krumbein, W. E. (1979). Photolithotrophic and chemoorganotrophic formation and degradation (Gulf of Aquaba, Sinai). *Geomicrobiol. J* **1,** 139–203.

Kuzirian, A., Alkon, D., and Harris, L. (1981). An inpaciliary network in statocyst hair cells. *J. Neurocytol.* **10,** 497–514.

Kyes, P., and Potter, T. S. (1934). Physiological marrow ossification in female pigeons. *Anat. Rec.* **60,** 377–379.

Kyle, H. M. (1927). Über die Entstehung und Bildung der Hartsubstanz bein Fischen. *Z. Mikr. Anat. Forsch* **9,** 317–384.

Ladd, H. S., and Schlanger, S. O. (1960). Drilling operations on Eniwetok atoll. *Geol. Surv. Prof. Paper* **2608,** 863–903.

Lafargue, F. (1977). Revision taxonomique des Didemnidae des cotes de France (Ascidies Composees). Synthe'se des resultats principaux. *Ann. Inst. oceanogr. Paris* **53,** 135–153.

Lafargue, F., and Kniprath, E. (1978). Formation des spicules de Didemnidae (Ascidies composées) I. L'apparition des spicules chez l'oozoide apres la métamorphose. *Mar. Biol.* **45,** 175–184.

Lahann, R. W. (1978). A chemical model for calcite crystal growth and morphology control. *J. Sed. Pet.* **48,** 337–344.

Lambert, G. (1979). Early post-metamorphic growth, budding and spicule formation in the compound ascidian, *Cystodytes lobatus. Biol. Bull.* **157,** 464–477.

Lambert, G., and Lambert, C. C. (1987). Spicule formation in the solitary ascidian, *Herdmania momus. J. Morphol.* **192,** 145–159.

Land, L. S. (1976). Early dissolution of sponge spicules from reef sediments, North Jamaica. *J. Sed. Petrol.* **46,** 967–969.

Land, L. S., Lang, J. C., and Barnes, D. J. (1977). On the stable carbon and oxygen isotopic composition of some shallow water ahermatypic scleractinian coral skeletons. *Geochim. Cosmochim. Acta* **41,** 169–172.

Landau, E. M., Levanon, M., Leiserowitz, L., Lahav, M., and Sagiv, J. (1985). Transfer of structural information from Langmuir monolayers to three-dimensional growing crystals. *Nature (London)* **318,** 353–356.

Landau, E. M., Popovitz-Biro, R., Levanon, M., Leiserowitz, L., Lahav, M., and Sagiv, J. (1986). Langmuir monolayers designed for the oriented growth of glycine and sodium chloride crystals at air/water interfaces. *Mol. Cryst. Liq. Cryst.* **134,** 323–335.

Landis, W. J. (1986). A study of calcification in the leg tendons from the domestic turkey. *J. Ultrastruct. Res.* **94,** 217–238.

Landis, W. J., and Glimcher, M. J. (1982). Electron optical and analytical observations of rat growth plate cartilage prepared by ultracryomicrotomy: The failure to detect a mineral phase in matrix vesicle and the identification of heterodispersed particles as the initial solid phase of deposition. *J. Ultrastruct. Res.* **78,** 227–268.

Lazaroff, N., Sigal, W., and Wasserman, A. (1982). Iron oxidation and precipitation of ferric hydroxysulfates by resting *Thiobacillus ferrooxidans* cells. *Appl. Environ. Microbiol.* **43,** 924–938.

Le Gros Clark, W. E. (1945). *The Tissues of the Body.* Oxford Univ. Press, Oxford.

Leadbeater, B.S.C. (1981). Ultrastructure and deposition of silica in loricate choanoflagellates. In *Silicon and Silceous Structures in Biological Systems* (eds. T. L. Simpson and B. E. Volcani), pp. 295–322. Springer-Verlag, New York.

Leadbeater, B.S.C. (1984). Silicification of "cell walls" of certain protistan flagellates. *Phil. Trans. R. Soc. London Ser. B* **304,** 529–536.

Leadbeater, B.S.C., and Riding, R. (1986). *Biomineralization in Lower Plants and Animals.* Clarendon Press, Oxford.

Ledger, P. W., and Franc, S. (1978). Calcification of the collagenous axial skeleton of *Veretillum cynomorium* Pall. (Cnidaria: Pennatulacea). *Cell Tissue Res.* **192,** 249–266.

Lee, J. J., Hutner, S. H., and Bovee, E. C. (1985). *An Illustrated Guide to the Protozoa.* Society of Protozoologists, Lawrence, Kansas.

Lees, S., Bonar, L. C., and Mook, H. A. (1984). A study of dense mineralized tissue by neutron diffraction. *Int. J. Biol. Macromol.* **6,** 321–326.

Legeros, R., and Suga, S. (1980). Crystallographic nature of fluoride in enameloids of fish. *Calcif. Tissue Int.* **32,** 169–174.

Legeros, R., Balmain, N., and Bonel, G. (1987). Age-related changes in mineral of rat and bovine cortical bone. *Calcif. Tissue Int.* **41,** 137–144.

Lehninger, A. L. (1970). Mitochondria and calcium ion transport. *Biochem. J.* **119,** 129–138.

Lehninger, A. L. (1983). The possible role of mitochondria and phosphocitrate in biological calcification. In *Biomineralization and Biological Metal Accumulation* (eds. P. Westbroek and E. W. de Jong), pp. 107–121. Reidel, Dordrecht.

Lehtovaara, I., and Kosinen, T. (1985). Vitamin D_3 and 25(OH)D_3 metabolism in the mussel *M. edulis.* In *Vitamin D. A Chemical, Biochemical and Clinical Update,* pp. 729–730. de Gruyter, Berlin.

Leleu, M. and Goni, J. (1974). Sur la formation biogeochimique de stalactites de galene. *Mineral Deposita (Berlin)* **9,** 27–32.

Leleu, M., Gugalski, T., and Goni, J. (1975). Synthese de wurtzite par voie bacterienne. *Mineral. Deposita (Berlin)* **10,** 323–329.

Lemche, H. (1957). A new living deep-sea mollusc of the Cambro-Devonian class Monoplacophora. *Nature (London)* **179,** 413–416.

Lemche, H., and Wingstrand, K. G. (1959). The anatomy of *Neopilina galatheae* Lemche, 1957. *Galathea Rep.* **3,** 9–72.

Lerman, A. (1965). Strontium and magnesium in water and in *Crassostrea* calcite. *Science* **150,** 745–751.

Leutwein, F., and Waskowiak, R. (1962). Geochemische Untersuchungen an rezenten marinen Molluskenschalen. *N. Jahrb. Miner. Abh.* **99,** 45–78.

Levetzow, K. G. (1932). Die Struktur einiger Schneckenschalen und ihre Entstehung durch typisches und atypisches Wachstum. *Jena Z. Naturwiss.* **68,** 41–108.

Levine, P. T., Glimcher, M. J., Seyer, J. M., Huddlestone, J. I., and Hein, J. W. (1966). Noncollagenous nature of the proteins of shark enamel. *Science* **154,** 1192–1194.

Levrat-Calviae, V., and Zylberberg, L.(1986). The structure of the osteoderms in the gekko: *Torentala mauritanica. Am. J. Anat.* **176,** 437–446.

Lewin, J. C. (1961). The dissolution of silica from diatom walls. *Geochim. Cosmochim. Acta* **21,** 182–189.

Li, C. W., and Volcani, B. E. (1984). Aspects of silicification in wall morphogenesis of diatoms. *Phil. Trans. R. Soc. London Ser. B* **304,** 519–528.

Li, C., and Volcani, B. E. (1985). Studies on the biochemistry and fine structure of silica shell formation in diatoms X. Morphogenesis of the labiate process in centric diatoms. *Protoplasma* **124,** 147–156.

Limeback, H. (1987). Isolation and characterization of pig enamelins. *Biochem. J.* **243,** 385–390.

Lindberg, D. R. (1985). Aplacophorans, monoplacophorans, polyplacophorans, scaphopods: the lesser classes. In *Mollusks* (eds. D. J. Bottjer, C. S. Hickman, P. D. Ward), pp. 230–47. Univ. Tennessee, Knoxville, Tenn.

Linde, A. (1984). Noncollagenous proteins and proteoglycans in dentinogenesis. In *Dentin and Dentinogenesis* (ed. A. Linde), pp. 55–93. CRC Press, Boca Raton, FL.

Lindow, S. E. (1982). Epiphytic ice nucleation-active bacteria. In *Phytopathogenic Prokaryotes* (eds. M. S. Mount and G. H. Lacey), pp. 335–362. Academic Press, New York.

Lins de Barros, H.G.P., Esquivel, D.M.S., Danon, J., and Oliviera, L.P.H. (1982). Magnetotactic algae. *Ann. Acad. Brasil Ciene* **54,** 258–259.

Lisitzin, A. P. (1971). Distribution of siliceous microfossils in suspension and bottom sediments. In *Micropaleontology of the Oceans* (eds. B. M. Funnel and W. R. Riedel), pp. 173–195. Cambridge Univ. Press, Cambridge.

Little, K. (1959). Electron microscope studies on human dental enamel. *J. R. Microsp. Soc.* **78,** 58–66.

Livne, E., Oliver, C., Leapman, R. D., Rosenberg, L. C., and Silberman, M. (1987). Age-related changes in the role of matrix vesicles in the mandibular condylar cartilage. *J. Anat.* **150,** 61–74.

Lloyd, R. M. (1964). Variations in the oxygen and carbon isotope ratios of Florida Bay mollusks and their environmental significance. *J. Geol.* **72,** 84–111.

Loeblich, A. R., and Tappan, H. (1964). Sarcodina chiefly "The camoebans" and Foraminifera. In *Treatise on Invertebrate Paleontology Pt. C. Protoctista 2* (ed. R. C. Moore), pp. C1–C900. Geol. Soc. Amer., Lawrence, Kansas.

Lohmander, S., and Hjerpe, A. (1975). Proteoglycans of mineralizing rib and epiphyseal cartilage. *Biochim. Biophys. Acta* **404,** 93–109.

Loven, S. (1892). Echinologica. *Svenska Vetensk-Akad. Handl.* **18,** 1–73.

Lovley, D. R., Stolz, J. F., Nord, G. L., and Phillips. E.J.P. (1988). Anaerobic production of magnetite by a dissimilatory iron reducing microorganism. Nature (London) **330,** 252–254.

Lowenstam, H. A. (1948). Biostratigraphic studies of the Niagaran inter-reef formations in northeast Illinois. *Illinois State Museum* **4,** 1–146.

Lowenstam, H. A. (1950). Niagaran reefs of the Great Lakes area. *J. Geol.* **58,** 430–487.

Lowenstam, H. A. (1954a). Factors affecting the aragonite:calcite ratios in carbonate secreting marine organisms. *J. Geol.* **62,** 284–322.

Lowenstam, H. A. (1954b). Systematic paleoecologic and evolutionary aspects of skeletal building materials. *Bull. Mus. Comp. Zool. Harvard College* **112,** 287–315.

Lowenstam, H. A. (1954c). Environmental relations of modification composition of certain carbonate secreting marine invertebrates. *Proc. Natl. Acad. Sci. U.S.A.* **40,** 39–48.

Lowenstam, H. A. (1955). Aragonite needles secreted by algae and some sedimentary implications. *J. Sed. Petrol.* **25,** 270–272.

Lowenstam, H. A. (1957). Niagaran reefs of the Great Lakes area. In *Treatise of Marine Ecology and Paleoecology* (ed. H. S. Ladd), Vol. 2, *Geol. Soc. Amer. Mem.* **67** pp. 215–248.

Lowenstam, H. A. (1961). Mineralogy O^{16}/O^{18} ratios, and strontium and magnesium contents of recent and fossil brachiopods and their bearing on the history of the oceans. *J. Geol.* **69,** 241–260.

Lowenstam, H. A. (1962a). Magnetite in denticle capping in recent chitons (Polyplacophora). *Bull. Geol. Soc. Am.* **13,** 435–438.

Lowenstam, H. A. (1962b). Goethite in radular teeth of recent marine gastropods. *Science* **137,** 279–280.

Lowenstam, H. A. (1963). Biologic problems relating to the composition and diagenesis of sediments. In *The Earth Sciences: Problems and Progress in Current Research* (ed. T. W. Donnelly), pp. 137–195. Univ. of Chicago Press, Chicago.

Lowenstam, H. A. (1964a). Sr/Ca ratio of skeletal aragonites from the recent marine biota at Palau and from fossil gastropods. In *Isotopic and Cosmic Chemistry* (ed. H. Craig, S. L. Miller and G. J. Wasserburg), pp. 113–132. North-Holland, Amsterdam.

Lowenstam, H. A. (1964b). Coexisting calcites and aragonites from skeletal carbonates of marine organisms and their strontium and magnesium contents. In *Recent Researches in the Fields of Hydrosphere, Atmosphere and Nuclear Geochemistry* (eds. Y. Miyake and T. Koyama), pp. 373–404. Maruzen Co. Ltd., Tokyo.

Lowenstam, H. A. (1967). Lepidocrocite, an apatite mineral, and magnetite in teeth of chitons (Polyplacophora). *Science* **156,** 1373–1375.

Lowenstam, H. A. (1968). Weddelite in a marine gastropod and in Antarctic sediments. *Science* **162,** 1129–1130.

Lowenstam, H. A. (1971). Opal precipitation of marine gastropods (Mollusca). *Science* **171,** 487–490.

Lowenstam, H. A. (1972a). Biogeochemistry of hard tissues: Their depth and possible pressure relations. In *Barobiology and the Experimental Biology of the Deep Sea* (ed. R. W. Brauer), pp. 19–32. Univ. of North Carolina, Chapel Hill, NC.

Lowenstam, H. A. (1972b). Phosphatic hard tissues of marine invertebrates: Their nature and mechanical functions, and some fossil implications. *Chem. Geol.* **9,** 153–166.

Lowenstam, H. A. (1974). Impact of life on chemical and physical processes. In *The Sea* (ed. E. Goldberg), pp. 715–796. John Wiley, New York.

Lowenstam, H. A. (1978). Recovery, behaviour and evolutionary implications of live Monoplacophora. *Nature (London)* **273,** 231–232.

Lowenstam, H. A. (1980). Bioinorganic constituents of hard parts. In *Biogeochemistry of Amino Acids* (eds. P. E. Hare, T. C. Hoering, and K. King), pp. 1–16. John Wiley, New York.

Lowenstam, H. A. (1981). Minerals formed by organisms. *Science* **211,** 1126–1131.

Lowenstam, H. A. (1986). Mineralization processes in monerans and protoctists. In *Biomineralization of Lower Plants and Animals* (eds. B.S.C. Leadbeater and R. Riding), pp. 1–17. Clarendon Press, Oxford.

Lowenstam, H. A. (1988). Spicular morphology and mineralogy in some *Pyuridae* (Ascidiacea). *Bull. Mar. Sci.* (in press).

Lowenstam, H. A., and Abbott, D. P. (1975). Vaterite: A mineralization product of the hard tissues of a marine organism (Ascidiacea). *Science* **188,** 363–365.

Lowenstam, H. A., and Epstein, S. (1954). Paleotemperatures of the post-Aptian Cretaceous as determined by the oxygen isotope method. *J. Geol.* **62,** 207–248.

Lowenstam, H. A., and Epstein, S. (1957). On the origin of sedimentary aragonite needles of the Great Bahama Bank. *J. Geol.* **65,** 364–375.

Lowenstam, H. A., and Fitch, J. E. (1978). Morphology, mineral diversity and ontogenetic mineralogic changes in extant elasmobranch otoconia. *Geol. Soc. Am. Abstr.* **10,** 114.

Lowenstam, H. A., and Fitch, J. E. (1981). Vaterite formation of higher teleost fishes. 62nd *Annu. Meet. Pacific Div. Am. Assoc. Adv. Sci.* p. 32.

Lowenstam, H. A., and Kirschvink, J. L. (1985). Iron biomineralization: A geobiological perspective. In *Magnetite Biomineralization and Magnetoreception in Organisms. A New Biomagnetism* (eds. J. L. Kirschvink, D. S. Jones, and B. J. MacFadden), pp. 3–15. Plenum Press, New York.

Lowenstam, H. A., and Margulis, L. (1980a). Calcium regulation and the appearance of calcareous skeletons in the fossil record. In *The Mechanisms of Biomineralization in Animals and Plants* (eds. M. Omori and N. Watabe), pp. 289–300. Tokai Univ. Press, Tokyo.

Lowenstam, H. A., and Margulis, L. (1980b). Evolutionary prerequisites for early Phanerozoic calcareous skeletons. *Biosystems* **12,** 27–41.

Lowenstam, H. A., and McConnell, D. (1968). Biologic precipitation of fluorite. *Science* **162,** 1496–1498.

Lowenstam, H. A., and Rossman, G. R. (1975). Amorphous, hydrous, ferric phosphatic dermal granules in *Molpadia* (Holothuroidea): Physical and chemical characterization and ecologic implications of the bioinorganic fraction. *Chem. Geol.* **15,** 15–51.

Lowenstam, H. A., and Weiner, S. (1983). Mineralization by organisms and the evolution of biomineralization. In *Biomineralization and Biological Metal Accumulation* (eds. P. Westbroek and E. W. de Jong), pp. 191–203. Reidel, Dordrecht.

Lowenstam, H. A., and Weiner, S. (1985). Transformation of amorphous calcium phosphate to crystalline dahllite in the radular teeth of chitons. *Science* **227,** 51–53.

Lowenstam, H. A., Traub, W., and Weiner, S. (1984). *Nautilus* hard parts: A study of the mineral and organic constituents. *Paleobiology* **10,** 268–279.

Lutz, R. A., and Rhoads, D. C. (1977). Anaerobiosis and a theory of growth line formation. *Science* **198,** 1222–1227.

Lutz, R. A., and Rhoads, D. C. (1980). Growth patterns within the molluscan shell: An overview. In *Skeletal Growth of Aquatic Organisms* (eds. D. C. Rhoads and R. A. Lutz), pp. 203–254. Plenum Press, New York.

Lyke, E. B. (1965). The histology of the sea pansies *Renilla reniformis* (Pallas) and *Renilla kollikeri* (Pfeffer) with a note on the fine structure of the latter species. *PhD Dissertation, University of Wisconsin,* 247 pp.

MacFadden, B. J., and Jones, D. S. (1985). Magnetic butterflies: A case study of the monarch (Lepidoptera, danaidae). In *Magnetite Biomineralization and Magnetoreception in Organisms* (eds. J. L. Kirschvink, D. L. Jones, and B. J. MacFadden), pp. 407–415. Plenum Press, New York.

Maier, G. D., Lechner, J. H., and Veis, A. (1983). The dynamics of formation of a collagen-phosphophoryn conjugate in relation to the passage of the mineralization front in rat incisor dentin. *J. Biol. Chem.* **258,** 1450–1453.

Mann, H., Mann, S., and Fyfe, W. S. (1987). Aragonite crystals in *Spyrogyra* Sp. (Chlorophyta) sp. *J. Phycol.* **23,** 506–509.

Mann, S. (1983). Mineralization in biological systems. *Struct. Bonding* **54,** 125–174.

Mann, S. (1985). Structure, morphology and crystal growth of bacterial magnetite. In *Magnetite Biomineralization and Magnetoreception in Organisms. A New Biomagnetism* (eds. J. L. Kirschvink, D. S. Jones, and B. J. MacFadden), pp. 311–332. Plenum Press, New York.

Mann, S. (1986). Biomineralization in lower plants and animals—chemical perspectives. In *Biomineralization in Lower Plants and Animals* (eds. B.S.C. Leadbeater and R. Riding), pp. 39–54. Oxford Univ. Press, Oxford.

Mann, S., Parker, S. B., Ross, M. D., Skarnulis, A. J., and Williams, R.J.P. (1983a). Ultrastructure of the calcium carbonate balance organ of the inner ear: An ultrathin resolution electron microscope study. *Proc. R. Soc. London Ser. B* **218,** 415–424.

Mann, S., Perry, C. C., Williams, R.J.P., Fyfe, C. A., Gobbi, G. C., and Kennedy, G. J. (1983b). The characterization of the nature of silica in biological systems. *J. Chem. Soc., Chem. Commun.* 168–170.

Mann, S., Hannington, J. P., and Williams, R.J.P. (1986a). Phospholipid vesicles as a model system for biomineralization. *Nature (London)* **324,** 565–567.

Mann, S., Perry, C. C., Webb, J., Luke, B., and Williams, R.J.P. (1986b). Structure, morphology, composition and organisation of biogenic minerals in limpet teeth. *Proc. R. Soc. London Ser. B* **227,** 179–190.

Mann, S., Williams J. M., Treffry, A., and Harrison, P. M. (1987). Reconstituted and native iron-cores of bacterioferritin and ferritin. *J. Mol. Biol.* **198,** 405–416.

Mann, S., Sparks, N.H.C., de Jong, E. W., and Westbroek, P. (1988a). Single crystalline nature of coccolith elements of the marine alga *Emiliania huxleyi* as determined by electron diffraction and high resolution transmission electron microscopy. Submitted for publication.

Mann, S., Sparks, N.H.C., Walker, M. M., and Kirschvink, J. L. (1988b). Ultrastructure, morphology and organization of biogenic magnetite from sockeye salmon, *Oncorhycus nerka:* Implications for magnetoreception. *J. Exp. Biol.,* in press.

Manton, I. (1986). Functional parallels between calcified and uncalcified periplasts. In *Biomineralization in Lower Plants and Animals* (eds. B.S.C. Leadbeater and R. Riding), pp. 157–172. Clarendon Press, Oxford.

Manton, I., and Leedale, G. F. (1963). Observations on the fine structure of *Prymnesium parvum* Carter. *Arch. Microbiol.* **45,** 285–303.

Manton, I., and Oates, K. (1980). *Polycrater galapagensis* gen. et sp. nov., a putative coccolithophorid from the Galapagos Islands with an unusual aragonitic periplast. *Br. phycol. J.* **15,** 95–103.

Margaritz, M., Holser, W. T., and Kirschvink, J. L. (1986). Carbon-isotope events across the Precambrian/Cambrian boundary on the Siberian platform. *Nature (London)* **320,** 258–259.

Margulis, L., and Schwartz, K. V. (1988). *Five Kingdoms. An Illustrated Guide to the Phyla of Life on Earth.* Freeman and Co., New York.

Märkel, K. (1970a). Morphologie der Seeigelzahne. IV. Die Zahne von Laganum und Clypeaster (Echinodermata, Echinoiden). *Z. Morphol. Tiere* **68,** 370–389.

Märkel, K. (1970b). The tooth skeleton of *Echinometra mathaei* (Blainville) (Echinodermata, Echinoidea). *Ann. Zool. Jpn.* **43,** 188–199.

Märkel, K. (1981). Experimental morphology of coroner growth in regular echinoids. *Zoomorphology* **97,** 31–52.

Märkel, K., and Gorny, P. (1973). Zur funktionellen Anatomie der Seeigelzahne (Echinodermata, Echinoidea). *Z. Morphol. Tiere* **75,** 223–242.

Märkel, K., and Röser, U. (1983a). The spine tissues in the echinoid *Eucidaris tribuloides. Zoomorphology* **103,** 25–41.

Märkel K., and Röser, U. (1983b). Calcite resorption in the spine of the echinoid *Eucidaris tribuloides. Zoomorphology* **103,** 43–58.

Märkel, K., Röser, U., Mackenstedt, U., and Klosterman, M. (1986). Ultrastructural investigation of matrix-mediated biomineralization in echinoids (Echinodermata, Echinoidea). *Zoomorphology* **106,** 232–243.

Marmo, F., Franco, E., and Balsamo, G. (1981). Scanning electron microscopic and X-ray diffraction studies of otoconia in the lizard *Podarcis sicula. Cell Tissue Res.* **218,** 265–270.

Marmo, F., Balsamo, G., and Franco, E. (1983). Calcite in the statoconia of amphibians: A detailed analysis in the frog *Rana esculenta. Cell Tissue Res.* **233,** 35–43.

Marsh, M. E. (1986). Biomineralization in the presence of calcium binding phosphoprotein particles. *J. Exp. Zool.* **239,** 207–220.

Marsh, M. E., and Sass, R. L. (1981). Matrix-mineral relationships in the scallop hinge ligament. *J. Ultrastruct. Res.* **76,** 57–70.

Marsh, M. E., and Sass, R. L. (1983). Calcium-binding phosphoprotein particles in the extrapallial fluid and innermost shell lamella of clams. *J. Exp. Zool.* **226,** 193–203.

Marshall, A. F., and Lawless, K. R. (1981). TEM study of the central dark line in enamel crystallites. *J. Dent. Res.* **60,** 1773–1782.

Marshall, A. T., Lewis, C. T., and Parry, G. (1974). Paraffin tubules secreted by the cuticle of an insect *Epipyrops anomala* (Epipyropidae: Lepidoptera). *J. Ultrastruct. Res.* **47,** 41–60.

Martin, A.V.W. (1954). An electron microscope study of the cartilagenous matrix in the developing tibia of the fowl. *J. Embryol. Exp. Morphol.* **2,** 38–48.

Martinsson, A. (1964). Phosphatic linings in bryozoan zooecia. *Geol. Foren. Stockholm Forhandl.* **86,** 404–408.

Massover, W. H., and Cowley, J. M. (1973). The ultrastructure of ferritin macromolecules. The lattice structure of the core crystallites. *Proc. Natl. Acad. Sci. U.S.A.* **70,** 3847–3851.

Mathews, J. L., Martin, J. H., Sampson, H. W., Kunin, A. S., and Roan, J. H. (1970). Mitochondrial granules in normal and rachitic rat epiphysis. *Calcif. Tissue Res.* **5,** 91–99.

Mathews, S. C., and Missarzhevsky, V. V. (1975). Small shelly fossils of late Precambrian and early Cambrian age. A review of recent work. *J. Geol. Soc. London* **131,** 289–304.

McConnell, D. (1938). A structural investigation of the isomorphism of the apatite group. *Am. Mineral* **23,** 1–19.

McConnell, D. (1952). The crystal chemistry of carbonate apatites and their relationship to the composition of calcified tissues. *J. Dent. Res.* **31,** 53–63.

McConnell, D. (1962). The crystal structure of bone. *Clin. Orthopaed.* **23,** 253–268.

McConnell, D. (1963). Inorganic constituents in the shell of the living brachiopod *Lingula. Geol. Soc. Am. Bull.* **74,** 363–364.

McConnell, D. (1969). The mineralogy of the apatites and their relation to biologic precipitation. *Proc. North Am. Paleontol. Convention,* pp. 1525–1535.

McConnell, D., and Ward, P. (1978). Nautiloid uroliths composed of phosphatic hydrogel. *Science* **199,** 208–209.

McGann, T.C.A., Kearney, R. D., Buchheim, W., Posner, A. S., Betts, F., and Blumenthal, N. C. (1983). Amorphous calcium phosphate in casein micelles of bovine milk. *Calcif. Tiss. Int.* **35,** 821–823.

McGrory, C. B., and Leadbeater, B.S.C. (1981). Ultrastructure and deposition of silica in Chrysophyceae. In *Silicon and Siliceous Structures in Biochemical Systems* (eds. T. L. Simpson and B. E. Volcani), pp. 201–230. Springer-Verlag, New York.

McLean, J. H. (1979). A new monoplacophoran limpet from the continental shelf off southern California. *Contrib. Science, Natl. Hist. Mus. Los Angeles County* **307,** 1–19.

Mechanic, G. L. (1965). The multicomponent re-equilibrating protein system of bovine embryonic enamelin-chromatrography in deaggregation solvents. In *Tooth Enamel* (eds. M. V. Stack and R. W. Fearnhead), pp. 88–94. John Wright, Bristol.

Mechanic, G. L., Katz, E. P., Henmi, M., Noyes, C., and Yamauchi, M. (1987). Locus of a histidine-based stable trifunctional helix to helix collagen cross-link: Stereospecific collagen structure of type I skin fibrils. *Biochemistry* **26,** 3500–3509.

Meenakshi, V. R., Hare, P. E., Watabe, N., Wilbur, K. M., and Menzies, R. J. (1970). Ultrastructure, histochemistry, and amino acid composition of the shell of *Neopilina. Anton Bruun Rpt. 2 Sci. Res. Southeast Pacific Exped.,* 3–12.

Meenakshi, V. R., Hare, P. E., and Wilbur, K. M. (1971). Amino acids of the organic matrix of neogastropod shells. *Comp. Biochem. Physiol.* **40B,** 1037–1043.

Meigen, W. (1903). Beiträge zur Kenntnis des Kohlensaurenkalks. *Ber. naturf. Ges. Freiburg i.B.* **13,** 1–55.

Meister, W. (1962). Histological structure of the long bones of penguins. *Anat. Rec.* **143,** 377–387.

Menanteau, J., Neuman, W. F., and Neuman, M. W. (1982). A study of bone proteins which can prevent hydroxyapatite formation. *Metab. Bone Dis. Rel. Res.* **4,** 157–162.

Menczel, J., Posner, A. S., and Harper, A. R. (1965). Age changes in the crystallinity of rat bone apatite. *Israel J. Med. Sci.* **1,** 251–252.

Menzies, R. J., George, R. Y., and Rowe, G. T. (1973). *Abyssal Environment and Ecology of the World Oceans.* Wiley-Interscience, New York.

Meyran, J., Graf, F., and Nicaise, G. (1984). Calcium pathway through a mineralizing epithelium in the crustacean *Orchestia* in pre-molt: Ultrastructural cytochemistry and X-ray microanalysis. *Tissue Cell* **2,** 269–286.

Meyran, J., Graf, F., and Nicaise, G. (1986). Pulse discharge of calcium through a demineralizing epithelium in the crustacean *Orchestia:* Ultrastructural cytochemistry and X-ray microanalysis. *Tissue Cell* **18,** 267–283.

Michaelson, W. (1919). Asidiae Ptychobranchiae und Diktyobranchiae des Roten Meeres. *Denkschr. Akad. Wiss. Wien. Math-Nat. Kl.* **95,** 1–120.

Miles, A.E.W. (1967). *Structural and Chemical Organization of Teeth.* Academic Press, New York.

Millar, R. H. (1953). *Ciona L.M.B.C. Memoirs, No. 35.* University Press, Liverpool.

Miller, A. (1984). Collagen: The organic matrix of bone. *Phil. Trans. R. Soc. London Ser. B* **304,** 455–477.

Miller, E. J. (1976). Biochemical characteristics and biological significance of the genetically-distinct collagens. *Mol. Cell. Biochem.* **13,** 165–192.

Miller, E. J. (1985). Recent information on the chemistry of collagens. In *The Chemistry and Biology of Mineralized Tissues* (ed. W. T. Butler), pp. 80–93. EBSCO Media, Birmingham, AL.

Milliman, J. D., Gastner, M., and Muller, J. (1971). Utilization of magnesium in coralline algae. *Bull. Geol. Soc. Am.* **82,** 573–580.

Millonig, G. (1970). A study on the formation and structure of the sea urchin spicule. *J. Submicrosc. Cytol.* **2,** 157–165.

Millot, J. (1949). Ordre des Araneides. In *Traité de Zoologie* (ed. P. P. Grasse), pp. 589–743. Masson, Paris.

Minganthi, A. (1957). Inhibition of melanogenesis in Phallusia embryos (Ascidians). *Acta Embryol. Morphol. Exp.* **1,** 37–47.

Mintz, G. R., DeFrancesco, S., and Lennarz, W. J. (1981). Spicule formation by cultured embryonic cells from the sea urchin. *J. Biol. Chem.* **256,** 13105–13111.

Misra, D. N., Bowen, R. L., and Mattamel, G. J. (1978). Surface area of dental enamel, bone, and hydroxyapatite: Chemisorption from solution. *Calcif. Tissue Res.* **26,** 139–142.

Mitterer, R. M. (1971). Comparative amino acid composition of calcified and non-calcified polychaete worm tubes. *Comp. Biochem. Physiol.* **38B,** 405–409.

Mitterer, R. M. (1978). Amino acid composition and binding capability of the skeletal protein of corals. *Bull. Mar. Sci.* **28,** 173–180.

Moczydlowska, M., and Vidal, G. (1986). Lower Cambrian acritarch zonation in southern Scandinavia and southeastern Poland. *Geol. Foren. Stockholm Forhandl.* **108,** 201–223.

Monniot, F. (1970). Les spicules chez les Tuniciers Aploustbranches. *Arch. Zool. Exp. Gén. (Ser. 3)* **3,** 205–336.

Monniot, F., and Buge, E. (1971). Les spicules d'ascidies fossiles et actuelles. *Ann. Paleontol.* **57,** 93–105.

Monster, J., Appel, P.W.U., Thode, H.G., Schidlowski, M., Carmichael, C. M., and Bridgewater, D. (1979). Sulfur isotope studies on early Archaean sediments from Isua, West Greenland. Implications for the antiquity of bacterial sulfate reduction. *Geochim. Cosmochim. Acta* **43,** 405–413.

Moore, T. C. (1969). Radiolaria: Change in skeletal weight and resistance to solution. *Geol. Soc. Am. Bull.* **80,** 2103–2108.

Morita, R. Y. (1980). Calcite precipitation by marine bacteria. *Geomicrobiol. J.* **2,** 63–82.

Morris, R. W., and Kittleman, L. R. (1967). Piezoelectric properties of otoliths. *Science* **158,** 368–370.

Morton, B. (1985). Statocyst structure in the *Anomalodesmata* (Bivalvia). *J. Zool. London A* **206,** 23–34.

Morton, J. E. (1958). *Molluscs.* Hutchinson University Library, London.

Moss, M. L. (1963). The biology of acellular teleost bone. *Ann. N.Y. Acad. Sci.* **109,** 337–350.

Moss, M. L. (1964). The phylogeny of mineralized tissues. *Int. Rev. Gen. Exp. Zool.* **1,** 297–331.

Moss, M. L. (1968). Bone, dentin and enamel and the evolution of vertebrates. In *Biology of the Mouth* (ed. P. Person), pp. 37–66. AAAS, Washington, DC.

Moss, M. L. (1969). Comparative histology of dermal sclerifications in reptiles. *Acta Anat.* **73,** 510–533.

Moss, M. L. (1970). Enamel and bone in shark teeth: With a note on fibrous enamel in fishes. *Acta Anat.* **77,** 161–187.

Moss, M. L. (1977). Skeletal tissues in sharks. *Am. Zool.* **17,** 335–342.

Moss, M. L., and Moss-Salentijn, L. (1983). Vertebrate cartilages. In *Cartilage, Structure, Function and Biochemistry* (ed. B. K. Hall), pp. 1–30. Academic Press, New York.

Moss, M. L., and Posner, A. S. (1960). X-Ray diffraction study of acellular teleost bone. *Nature (London)* **188,** 1037–1038.

Muir, M. D. (1978). Microenvironments of some modern and fossil iron- and manganese-oxidizing bacteria. In *Environmental Biogeochemistry and Geomicrobiology* (ed. W. E. Krumbein), pp. 937–944. Ann Arbor Sci. Pub., Ann Arbor, MI.

Müller, G. W. (1884). Die Ostracoden des Golfes von Neapel und der angrenzenden Meere Abschnitte. In *Fauna und Flora des Golfes von Neapel,* pp. 1–406. Friedlander and Sohn, Berlin.

Müller, J. (1858). Üeber die Thalassicollen, Polycystinene und Acanthezeiren des Mittelmeeres. *Abh. Preuss. Akad. Wiss.* **42,** 1.

Müller, K. J. (1964). Ostracoda (Bradorina) mit phosphatischen Gehäusen aus dem Oberkambrium von Schweden. *N. Jb. Geol. Palaontol. Abh.* **121,** 1–46.

Murray, G., and Blackman, V. H. (1898). On the nature of coccospheres and rhabdospheres. *Phil. Trans. R. Soc. London Ser. B* **190,** 427–441.

Mutvei, H. (1974). Ultrastructural studies on cephalopod shells. Part I. The septa and siphonal tube in *Nautilus. Bull. Geol. Inst. Univ. Uppsala* **3,** 237–261.

Mutvei, H. (1974). SEM studies on arthropod exoskeletons. Part I. Decapod crustaceans, *Homarus gammarus* (L.) and *Carcinus maenas* (L.). *Bull Geol. Inst. Univ. Upsala* **4,** 73–80.

Muzik, M., and Wainwright, S. (1977). Morphology and habitat of five Fijian sea fans. *Bull. Mar. Sci.* **27,** 308–337.

Nakahara, H. (1979). An electron microscope study of the growing surface of nacre in two gastropod species *Turbo cornutus* and *Tegula pfefferi. Jpn. J. Malac.* **38,** 205–211.

Nakahara, H. (1982). Electron microscope studies of lattice image and "central" dark line of crystallites in sound and carious human dentin. *Bull. Josai Dent. Univ.* **11,** 209–215.

Nakahara, H. (1983). Calcification of gastropod nacre. In *Biomineralization and Biological Metal Accumulation* (eds. P. Westbroek and E. W. de Jong), pp. 225–230. Reidel, Dordrecht.

Nakahara, H., and Bevelander, G. (1978). The formation of calcium carbonate crystals in *Halimeda incrassata* with special reference to the role of the organic matrix. *Jpn. J. Phycol.* **26,** 9–12.

Nakahara, H., and Bevelander, G. (1979). An electron microscope study of crystal calcium carbonate formation in the mouse otolith. *Anat. Rec.* **193,** 233–242.

Nakahara, H., and Kakei, M. (1983). The central dark line in developing enamel crystallite: An electron microscope study. *Bull. Josai Dent. Univ.* **12,** 1–7.

Nakahara, H., and Kakei, M. (1984). TEM observations on the crystallites of dentin and bone. *Bull. Josai Dent. Univ.* **13,** 259–263.

Nakahara, H., Bevelander, G., and Kakei, M. (1982). Electron microscopic and amino acid studies on the outer and inner shell layers of *Haliotis rufescens. Venus, Kyoto* **38,** 205–211.

Nanci, A., and Warshawsky, H. (1984). Relationship between the quality of fixation and presence of stippled material in newly formed enamel of the rat incisor. *Anat. Rec.* **208,** 15–31.

Nancollas, G. H. (1977). The mechanism of biological mineralization. *J. Crystal Growth* **42,** 185–193.

Nancollas, G. H. (1982). Phase transformation during precipitation of calcium salts. In *Biological Mineralization and Mineralization. Dahlem Conference* (ed. G. H. Nancollas), pp. 79–99. Springer-Verlag, Berlin.

Napp-Zinn, K. (1966). Anatomie des Blattes. 1. Blattanatomie der Gymnospermen. In *Handbuch der Pflanzenanatomie Z. Aufl. Specieller Teil,* (eds. W. Zimmerman, P. Ozenda, and H. D. Wolff), p. 129. Gebrüder Borntraeger, Berlin.

Nawrot, C. F., Campbell, D. J., Schroeder, J. K., and Van Valkenburg, M. (1976). Dental phosphoprotein-induced formation of hydroxylapatite during in vitro synthesis of amorphous calcium phosphate. *Biochemistry* **15,** 3445–3449.

Neff, J. M. (1971). Ultrastructure of calcium phosphate-containing cells in the serpulid polychaete worm. *Calcif. Tissue Res.* **7,** 191–200.

Neff, J. M. (1972). Ultrastructure of the outer epithelium of the mantle in the clam *Mercenaria mercenaria* in relation to calcification of the shell. *Tissue Cell* **4,** 591–600.

Nelson, D.G.A., Wood, G. J., and Barry, J. C. (1986). The structure of (100) defects in carbonated apatite crystallites: A high resolution electron microscope study. *Ultramicroscopy* **19,** 253–266.

Nemec, B. (1964). Georeceptors in plants. *Int. Bot. Congr. 10th Edinburgh. 194 Abstr.*

Nesson, M. H. (1968). Studies on radula tooth mineralization in the polyplacophora. *Ph.D. thesis,* California Institute of Technology, Pasadena, California.

Nesson, M. H., and Lowenstam, H. A. (1985). Biomineralization processes of the radula teeth of chitons. In *Magnetite Biomineralization and Magnetoreception in*

Organisms. A New Biomagnetism (eds. J. L. Kirschvink, D. S. Jones, and B. J. MacFadden), pp. 333–363. Plenum Press, New York.

Neuman, W. F., and Neuman, M. W. (1953). The nature of the mineral phase in bone. *Chem. Rev.* **53,** 1–45.

Neuman, W. F., and Neuman, M. W. (1958). *Chemical Dynamics of Bone Mineral.* Univ. of Chicago Press, Chicago.

Neumann, A. C., and Land, L. S. (1975). Lime mud deposition and calcareous algae in the bight of Abaco, Bahamas: A budget. *J. Sed. Petrol.* **45,** 163–186.

Neville, A. C. (1975). *Biology of the Arthropod Cuticle.* Springer-Verlag, Berlin.

Nissen, H. (1969). Crystal orientation and plate structure in echinoid skeletal units. *Science* **166,** 1150–1152.

Nylen, M. U. (1979). Matrix-mineral relationships—a morphologist's viewpoint. *J. Dent. Res.* **58B,** 922–926.

Nylen, M. U., Scott, D. B., and Mosley, V. M. (1960). Mineralization of turkey leg tendon. II. Collagen-mineral relations revealed by electron and X-ray microscopy. In *Calcification in Biological Systems* (ed. R. F. Sognnaes), pp. 129–142. AAAS, Washington, DC.

Nylen, M. U., Evans, E. D., and Omnel, K. A. (1963). Crystal growth in rat enamel. *J. Cell. Biol.* **18,** 109–123.

Oakley, K. P. (1934). Phosphatic calculi in Silurian polyzoa. *Proc. R. Soc. London Ser. B* **116,** 296–314.

O'Brien, G. W., Harris, J. R., Milnes, A. R., and Veeh, A. H. (1981). Bacterial origin of east Australian continental margin phosphates. *Nature (London)* **294,** 442–446.

Odum, H. T. (1951). Notes on the strontium content of seawater, celestite radiolaria and strontianite snail shells. *Science* **114,** 211–213.

Oelschläger, H. A. (1986). Tympanohyal bone in toothed whales and the formation of the tympano-periotic complex (Mammalia: Cetacea). *J. Morphol.* **188,** 157–165.

Ogden, C. G. and Hedley, R. H. (1980). *An Atlas of Freshwater Amoebae. British Museum (Nat. Hist.).* Oxford Univ. Press, Oxford.

Ogilvie, M. (1896). Microscopic and systematic study of Madreporian types of corals. *Phil. Trans. R. Soc. London Ser. B* **187,** 83–345.

Oikita, T. W., and Volcani, B. E. (1980a). Role of silicon in diatom metabolism. IX. Differential synthesis of DNA polymerases and DNA binding proteins during silicate starvation and recovery in *Cylindrotheca fusiformis. Biochim. Biophys. Acta* **519,** 76–86.

Oikita, T. W., and Volcani, B. E. (1980b). Role of silicon in diatom metabolism. X. Polypeptide labelling patterns during the cell cycle, silicate starvation and recovery in *Cylindrotheca fusiformis. Exp. Cell Res.* **125,** 471–481.

Okazaki, K. (1960). Skeletal formation of sea urchin larvae. II. Organic matrix of the spicule. *Embryologia* **5,** 283–320.

Okazaki, K. (1962). Skeleton formation of sea urchin larvae. IV. Correlation in shape of spicule and matrix. *Embryologia* **7,** 21–38.

Okazaki, K. (1975). Spicule formation by isolated micromeres of the sea urchin embryo. *Am. Zool.* **15,** 567–582.

Okazaki, K., McDonald, K., and Inoue, S. (1980). Sea urchin larval spicule observed with scanning electron microscope. In *Mechanisms of Biomineralization in Animals and Plants* (eds. M. Omori and N. Watabe), pp. 159–168. Tokai Univ. Press, Tokyo.

Okazaki, M., and Furuya, K. (1977). Site and nature of calcium carbonate deposits in calcareous algae. *Int. Symp. Mechanism Biomineralization Invertebrates Plants 3rd,* Kashikojima, Japan, pp. 30–31.

Oliver, W. A., Sando, W. J., Cairns, S. D., Coates, A. G., MacIntyre, I. G., Bayer, F. M., and Sorauf, J. E. (1984). Recent advances in the paleobiology and geology of the Cnidaria. *Palaeontograph. Amer.* **54,** 175–226.

Olson, E. C. (1977). The history of the vertebrates. In *Handbook of Sensory Physiology* (ed. F. Crescitelli), pp. 1–45. Springer-Verlag, Berlin.

O'Neill, P. L. (1981). Polycrystalline echinoderm calcite and its fracture mechanics. *Science* **213,** 646–648.

Onozato, H., and Watabe, N. (1979). Studies on fish scale formation and resorption. III. Fine structure and calcification of the fibrillary plates of the scales in *Carassius auratus* (Cypriniformes: Cyprinidae). *Cell Tissue Res.* **201,** 409–422.

Ørvig, T. (1967). Phylogeny of tooth tissues: Evolution of some calcified tissues in early vertebrates. In *Structural and Chemical Organization of Teeth* (ed. A.E.W. Miles), pp. 45–110. Academic Press, New York.

Outka, D. E., and Williams, D. C. (1971). Sequential coccolith morphogenesis in *Hymenomonas carterae. J. Protozool.* **18,** 285–297.

Packard, M. J. (1980). Ultrastructural morphology of the shell and shell membrane of eggs of the common snapping turtles *(Chelydra serpentina). J. Morphol.* **165,** 157–204.

Pannella, G. (1974). Otolith growth patterns: An aid in age determination in temperate and tropical fishes. In *Ageing of Fish* (ed. T. B. Begenal), pp. 28–39. Unwin Brothers, Old Woking.

Pannella, G. (1980). Growth patterns in fish sagittae. In *Skeletal Growth of Aquatic Organisms* (eds. D. C. Rhoads and R. A. Lutz), pp. 519–560. Plenum Press, New York.

Pannella, G., McClintock, C., and Thompson, M. N. (1968). Paleontological evidence of variations in length of synodic month since late Cambrian. *Science* **162,** 782–796.

Parke, M., and Adams, L.(1960). The motile (*Crystallolithus hyalinus* Gaardner & Markali) and non-motile phases in the life history of *Coccolithus pelagicus* (Wallich) Schiller. *J. Mar. Biol. Assoc. U.K.* **39,** 263–274.

Parker, K. D., and Rudall, K. M. (1955). Calcium citrate in an insect. *Biochim. Biophys. Acta* **17,** 287.

Paul, J. H., and Sharpe, J. S. (1916). Studies in calcium metabolism. I. The deposition of lime salts in the integument of decapod Crustacea. *J. Physiol.* **50,** 183–192.

Pautard, F.G.E. (1963). Mineralization of keratin and its comparison with the enamel matrix. *Nature (London)* **199,** 531–535.

Pautard, F.G.E. (1970). Calcification in unicellular organisms. In *Biological Calcification* (ed. H. Schraer), pp. 105–201. Appleton-Century-Crofts, New York.

Pearse, V. B., and Muscatine, L. (1971). Role of symbiotic algae (zooxanthellae) in coral calcification. *Biol. Bull. Woods Hole* **141,** 287–301.

Peat, A., and Banbury, G. H. (1968). Occurrence of ferritin-like particles in a fungus. *Planta* **79,** 268–270.

Pellegrino, E. D., and Biltz, R. M. (1972). Mineralization in the chick embryo. I. Monohydrogen phosphate and carbonate relationships during maturation of the bone crystal complex. *Calcif. Tissue Res.* **10,** 118–135.

Peress, N. S., Sadjera, S. W., and Anderson, C. (1974). The lipids of matrix vesicles from bovine fetal epiphyseal cartilage. *Calcif. Tissue Res.* **14,** 274–282.

Perry, A., Bauer, G. B., and Dizon, A. E. (1985). Magnetoreception and biomineralization of magnetite in amphibians and reptiles. In *Magnetite Biomineralization and Magnetoreception in Organisms. A New Biomagnetism* (eds. J. L. Kirschvink, D. S. Jones, and B. J. MacFadden), pp. 439–453. Plenum Press, New York.

Perry, C. C., and Mann, S. (1988). Aspects of biological silicification. In *Origin, Evolution and Modern Aspects of Biomineralization in Plants and Animals* (ed. R. E. Crick), In press. Elsevier, Amsterdam.

Perry, C. C., Mann, S., and Williams, R.J.P. (1984a). Structural and analytical studies of the silicified macrohairs from the lemma of the grass *Phalaris canariensis* L. *Proc. R. Soc. London Ser. B* **222,** 427–438.

Perry, C. C., Mann, S., Williams, R.J.P., Watt, F., Grime, G. W., and Takaes, J. (1984b). A scanning proton microprobe study of microhairs from the lemma of the grass *Phalaris canariensis* L. *Proc. R. Soc. London Ser. B* **222,** 439–445.

Person, P. (1983). Invertebrate cartilages. In *Cartilage, Structure, Function and Biochemistry* (ed. B. K. Hall), pp. 31–57. Academic Press, New York.

Peters, W. (1972). Occurrence of chitin in mollusca. *Comp. Biochem. Physiol.* **41B,** 541–550.

Petit, H., Davis, W. L., Jones, R. G., and Hagler, H. K. (1980). Morphological studies on the calcification processes in the fresh-water mussel *Amblema. Tissue Cell* **12,** 13–28.

Phelps, P., Giddings, T. H., Prochoda, M., and Fall, R. (1986). Release of cell-free ice nuclei by *Erwinia herbicola. J. Bacteriol.* **167,** 496–502.

Pickett-Heaps, J. D., Tippit, D. H., and Andreozzi, J. A. (1979). Cell division in the pennate diatom *Diatoma vulgare. Protoplasma* **86,** 205–242.

Pienaar, R. N. (1969). The fine structure of *Hymenomonas (Cricosphaera) carterae.* II. Observations on scale and coccolith production. *J. Phycol.* **5,** 321–331.

Pienaar, R. N. (1981). Ultrastructural studies on the cysts of *Prymnesium* (Prymnesiophyceae). *Phycologia* **20,** 112.

Pilkey, O. H., and Goodell, H. G. (1963). Trace elements in recent mollusk shells. *Limnol. Oceanogr.* **8,** 137–148.

Pilkey, O. H., and Hower, J. (1960). The effect of environment on the concentration of skeletal magnesium and strontium in *Dendraster. J. Geol.* **68,** 203–216.

Pilkington, J. B. (1969). The organization of skeletal tissues in the spines of *Echinus esculentus. J. Mar. Biol. Assoc. U.K.* **49,** 857–877.

Pobeguin, T. (1954). Contribution a l'etude des carbonates de calcium précipitation calcaire par les végétaux comparaison avec le monde animal. *Ann. Sc. Nat. Bot. (Ser. II)* **15,** 29–109.

Poluzzi, A., and Sartori, R. (1975). Report on the carbonate mineralogy of bryozoa. *Proceedings of the Third Conference International Bryozoology Association, Docum. Lab. Geol. Fac. Sci. Lyon-H.S. 3 (fasc. 1),* pp. 193–210. Universite Claude Bernard, Lyon.

Poole, A. R., and Rosenberg, L. C. (1987). Proteoglycans, chondrocalcin, and the calcification of cartilage matrix in endochondral ossification. In *Biology of Proteoglycans* (eds. T. Wright and R. Michson), pp. 187–210. Academic Press, Orlando, FL.

Poole, A. R., Pidoux, I., and Rosenberg, L. C. (1982). Role of proteoglycans in endochondral ossification: Immunofluorescent localization of link protein and proteoglycan monomer in bovine fetal epiphyseal growth plate. *J. Cell Biol.* **92,** 249–260.

Poole, A. R., Pidoux, I., Reiner, A., Choi, H., and Rosenberg, L. C. (1984). Association of an extracellular protein (chondrocalcin) with the calcification of cartilage in endochondral bone formation. *J. Cell Biol.* **98,** 54–65.

Poole, D.F.G. (1956). The fine structure of the scales and teeth of *Raia clavata. Quart. J. Microsp. Soc.* **97,** 99–107.

Poole, D.F.G. (1967). Enameloid and enamel in recent vertebrates. In *Structural and*

Chemical Organization of Teeth (ed. A.E.W. Miles), pp. 111–149. Academic Press, New York.

Poole, D.F.G. (1971). An introduction to the phylogeny of calcified tissues. In *Dental Morphology and Evolution* (ed. A. A. Dahlberg), pp. 65–79. Chicago Univ. Press, Chicago.

Poser, J. W., and Price, P. A. (1979). A method for decarboxylation of γ-carboxyglutamic acid in proteins. *J. Biol. Chem.* **254,** 431–436.

Poser, J. W., Esch, F. S., Ling, N. C., and Price, P. A. (1980). Isolation and sequence of the vitamin K-dependent protein from human bone. *J. Biol. Chem.* **255,** 8685–8689.

Posner, A. S., and Betts, F. (1981). Molecular control of tissue mineralization. In *The Chemistry and Biology of Mineralized Connective Tissues* (ed. A. Veis), pp. 257–266. Elsevier, Amsterdam.

Posner, A. S., Betts, F., and Blumenthal, N. C. (1978). Properties of nucleating systems. *Metab. Bone Dis. Rel. Res.* **1,** 179–183.

Poulicek, M., and Jeuniaux, Ch. (1981). Matrice organique la coquille et position phyletique de *Neopilina galatheae* (Mollusques, Monoplacophores). *Ann. Soc. Roy. Zool. Belg.* **111,** 143–150.

Poulicek, M., Voss-Foucart, M. F., and Jeuniaux, Ch. (1986). Chitinoproteic complexes and mineralization in mollusk skeletal structures. In *Chitin in Nature and Technology* (eds. R. Muzzarelli, Ch. Jeuniaux, and G. W. Gooday), pp. 7–12. Plenum Press, New York.

Prenant, M. (1925). Contribution a l'étude II. Sur les conditions de formation des spicules de Didemnidae. *Bull. Biol. Fr. Belg.* **59,** 403–435.

Prenant, M. (1926). Notes histoligique sur la structure et la croissance des dents d'oursin. *Arch. Zool. Exp. Gen. Note Rev.* **65,** 25–38.

Prenant, M. (1927). Les formes mineralogiques du calcaire chez les etres vivants, et le probleme de leur determinisme. *Biol. Rev.* **2,** 365–393.

Presti, D. E. (1985). Avian navigation, geomagnetic field sensitivity, and biogenic magnetite. In *Magnetite Biomineralization and Magnetoreception in Organisms* (eds. J. L. Kirschvink, D. S. Jones, and B. J. MacFadden), pp. 455–482. Plenum Press, New York.

Price, P. A., and Williamson, M. K. (1981). Effects of warfarin on bone. *J. Biol. Chem.* **256,** 12754–12759.

Price, P. A., Otsuka, A. S., Poser, J. W., Kristaponis, J., and Raman, N. (1976). Characterization of a γ-carboxyglutamic-acid containing protein from bone. *Proc. Natl. Acad. Sci. U.S.A.* **73,** 1447–1451.

Price, P. A., Lothringer, J. W., Bavkol, S. A., and Reddi, A. H. (1981). Developmental appearance of the vitamin K-dependent protein of bone during calcification. *J. Biol. Chem.* **256,** 3781–3784.

Price, P. A., Williamson, M. K., Haba, T., Dell, R. B., and Jee, W.S.S. (1982). Excessive mineralization with growth plate closure in rats on chronic warfarin treatment. *Proc. Natl. Acad. Sci. U.S.A.* **79,** 7734–7738.

Price, P. A., Urist, M. A., and Otawara, Y. (1983). Matrix Gla protein, a new γ-carboxyglutamic acid-containing protein which is associated with the organic matrix of bone. *Biochem. Biophys. Res. Commun.* **117,** 765–771.

Prien, E. L., and Frondel, C. (1947). Studies in urolithiasis: I. The composition of urinary calculi. *J. Urology* **57,** 949–994.

Prince, C. W., Oosawa, T., Butler, W. T., Tomana, M., Bhown, A. S., Bhown, M., and Schrohenloher, R. E. (1987). Isolation, characterization and biosynthesis of a phosphorylated glycoprotein from rat bone. *J. Biol. Chem.* **262,** 2900–2907.

Pringsheim, E. G. (1946). On iron flagellates. *Phil. Trans. R. Soc. London Ser. B* **232,** 311–342.

Prockop, D. J., Kivirikko, K. I., and Tuderman, L. (1979). The biosynthesis of collagen and its disorders. *N. Engl. J. Med.* **301,** 13–23.

Prostak, K., and Skobe, Z. (1985). The effects of colchicine on the ultrastructure of the dental epithelium and odontoblasts of teleost tooth enamel buds. *J. Craniofac. Genet. Dev. Biol.* **5,** 57–88.

Prostak, K., and Skobe, Z. (1986). Ultrastructure of the dental epithelium during enameloid mineralization in a teleost fish. *Arch. Oral Biol.* **31,** 73–85.

Purvis, O. W. (1984). The occurrence of copper oxalate in lichens growing on copper sulphide-bearing rocks in Scandinavia. *Lichenologist* **16,** 197–204.

Quinn, T. P. (1980). Evidence for celestial and magnetic compass orientation in lake migrating sockeye salmon smolts. *J. Comp. Physiol.* **A137,** 243–248.

Raggio, C. L. (1986). *In vivo* hydroxyapatite formation induced by lipids. *J. Bone Mineral. Res.* **1,** 409–415.

Raggio, C. L., Boskey, A., Boyan, B. D., and Urist, M. L. (1983). *In vivo* induction of hydroxyapatite formation by lipid macromolecules. *Orthopaed. Trans.* **5,** 21.

Ragland, P. C., Pilkey, O. H., and Blackwelder, B. W. (1969). Comparison of the Sr/Ca ratio of fossil and Recent mollusk shells. *Nature (London)* **224,** 1223–1224.

Ramachandran, G. N., and Kartha, G. (1955). Structure of collagen. *Nature (London)* **176,** 593–595.

Raup, D. M. (1959). Crystallography of echinoid calcite. *J. Geol.* **67,** 661–674.

Raup, D. M. (1960). Ontogenic variation in the crystallography of echinoid calcite. *J. Paleontol.* **34,** 1041–1050.

Raup, D. M. (1961). The geometry of coiling in gastropods. *Proc. Natl. Acad. Sci. U.S.A.* **47,** 602–609.

Raup, D. M. (1966). The endoskeleton. In *Physiology of Echinodermata* (ed. R. A. Boolootian), pp. 379–395. Interscience Publishers, New York.

Rawitscher, F. (1932). *Der Geotropismus der Pflanzen.* Gustav Fischer, Jena.

Register, T. C., McLean, F. M., Low, M. G., and Wuthier, R. E. (1986). Roles of alkaline phosphatase and labile internal mineral in matrix vesicle-mediated calcification. *J. Biol. Chem.* **261,** 9354–9360.

Reid, D. M. (1943). Occurence of crystals in the skin of amphipoda. *Nature (London)* **151,** 504–505.

Reid, S. A. (1986). A study of lamellar organisation in juvenile and adult human bone. *Anat. Embryol.* **174,** 329–338.

Reid, S. A., and Boyde, A. (1987). Changes in the mineral density distribution in human bone with age: Image analysis using backscattered electrons in the SEM. *J. Bone Mineral Res.* **2,** 13–22.

Reimann, B.E.F. (1964). Deposition of silica inside a diatom cell. *Exp. Cell Res.* **34,** 605–608.

Reimann, B.E.F., Lewin, J. C., and Volcani, B. E. (1966). Studies on the biochemistry and fine structure of silica shell formation in diatoms. II. The structure of the cell wall of *Navicula pelliculosa* (Breb) Hilse. *J. Phycol.* **2,** 74–84.

Reith, E. J., and Butcher, E. O. (1967). Microanatomy and histochemistry of amelogenesis. In *Structural and Chemical Organization of Teeth* (ed. A.E.W. Miles), pp. 371–397. Academic Press, New York.

Renugopalakrishnan, V., Strawich, E. S., Horowitz, P. M., and Glimcher, M. J. (1986a). Studies of the secondary structures of amelogenin from bovine tooth enamel. *Biochemistry* **25,** 4879–4887.

Renugopalakrishnan, V., Uchiyama, A., Horowitz, P. M., Rapaka, R. S., Suzuki, M.,

Lefteriou, B., and Glimcher, M. J. (1986b). Preliminary studies of the secondary structure in solution of two phosphoproteins of chicken bone matrix by circular dichroism and Fourier-transform infrared spectroscopy. *Calcif. Tissue Int.* **39,** 166–170.

Resig, J. M., Lowenstam, H. A., Echols, R. J., and Weiner, S. (1980). An extant opaline foraminifer: Test ultrastructure, mineralogy and taxonomy. *Cushman Foundation Spec. Publ.* **19,** 205–214.

Reynolds, C. S. (1986). Diatoms and the geochemical cycling of silicon. In *Biomineralization in Lower Plants and Animals* (eds. B.S.C. Leadbeater and R. Riding), pp. 269–289. Clarendon Press, Oxford.

Rhoads, D. C., and Lutz, R. A. (1980). *Skeletal Growth of Aquatic Organisms.* Plenum Press, New York.

Rice, M. E. (1969). Possible boring structures of sipunculids. *Am. Zool.* **9,** 803–812.

Richelle, L. J., and Onkelinx, C. (1969). Recent advances in the physical biology of bone and other hard tissues. In *Mineral Metabolism* (eds. C. Comar and F. Bronner), pp. 123–190. Academic Press, New York.

Riding, R., and Voronova, L. (1982). Calcified cyanophytes and the Precambrian-Cambrian transition. *Naturwissenschaften* **69,** 498–499.

Riding, R., and Voronova, L. (1984). Assemblages of calcareous algae near the Precambrian-Cambrian boundary in Siberia and Mongolia. *Geol. Mag.* **121,** 205–210.

Riquier, J. (1960). Les phytoliths de certain sols tropicaux et des podsols. *Int. Congr. Soil Sci. 7th, Madison, Wisconsin,* pp. 425–431.

Rivadeneyra, M. A., Ramos-Cormenzana, A., and Garcia-Cervigon, A. (1983). Bacterial formation of struvite. *Geomicrobiol. J.* **3,** 151–163.

Roberts, A. W., and Humpherson, P. G. (1967). Phytoferritin in plastids of the cambial zone of willow. *Planta (Berlin)* **76,** 169–188.

Robinson, C., and Kirkham, J. (1984). Enamel matrix components during development and possible interactions with the mineral phase. In *Tooth Enamel IV* (eds. R. W. Fearnhead and S. Suga), pp. 261–265. Elsevier, Amsterdam.

Robinson, C., Lowe, N. R., and Weatherell, J. A. (1977). Changes in the amino acid composition of developing rat incisor enamel. *Calcif. Tissue Res.* **23,** 19–31.

Robinson, C., Briggs, H. D., and Atkinson, P. J. (1981). Histology of enamel organ and chemical composition of adjacent enamel in rat incisors. *Calcif. Tissue Int.* **33,** 513–520.

Robinson, R. A. (1952). An electron microscope study of the crystalline inorganic component of bone and its relationship to the organic matrix. *J. Bone Joint Surg.* **34A,** 389–434.

Robinson, R. A., and Cameron, D. A. (1956). Electron microscopy of cartilage and bone matrix at the distal epiphyseal line of the femur in the newborn infant. *J. Biophys. Biochem.* **2** *(Suppl.),* 253–260.

Robinson, R. A., and Watson, M. L.(1952), Collagen-crystal relationships in bone as seen in the electron microscope. *Anat. Rec.* **114,** 383–410.

Robison, R. (1923). The possible significance of hexosophosphoric esters in ossification. IX. Calcification in vitro. *Biochem. J.* **24,** 1927–1941.

Roer, R., and Dilliman, R. (1984). The structure and calcification of the crustacean cuticle. *Am. Zool.* **24,** 893–909.

Romer, A. S. (1966). *Vertebrate Paleontology,* 3rd ed.Univ. Chicago Press, Chicago.

Rönnholm, E. (1962). The amelogenesis of human teeth as revealed by electron microscopy. II. The development of the enamel crystallites. *J. Ultrastruct. Res.* **6,** 249–303.

Rose, R. D., and Stokes, D. R. (1981). A crustacean statocyst with only three hairs: Light and scanning electron microscopy. *J. Morphol.* **169,** 21–28.

Rosenberg, G. D., and Runcorn, S. K. (1975). *Growth Rhythms and the History of the Earth's Rotation.* John Wiley, London.

Rosenberg, L. C., and Buckwalter, J. A. (1986). Cartilage proteoglycans. In *Articular Cartilage Biochemistry* (ed. K. Kuettner), pp. 39–57. Raven Press, New York.

Rosenfeld, A. (1979). Structure and secretion of the carapace in some living ostracodes. *Lethaia* **12,** 353–360.

Ross, M. D. (1983). Gravity and the cells of gravity receptors in mammals. *Adv. Space Res.* **3,** 179–190.

Ross, M. D. (1984). The influence of gravity on structure and function of animals. *Adv. Space Res.* **4,** 305–314.

Roth, G., and Calmes, R. (1981). *Oral Biology.* C. V. Mosby Company, St. Louis.

Round, F. E. (1981). Morphology and phyletic relationships of the silicified algae and the architypal diatom-monopoly or polyphyly. In *Silicon and Siliceous Structures in Biological Systems* (eds. T. L. Simpson and B. E. Volcani), pp. 97–128, Springer-Verlag, New York.

Rowson, J. D., Leadbeater, B.S.C., and Green, J. C. (1986). Calcium carbonate deposition in the motile *(Crystallolithus)* phase of *Coccolithus pelagicus* (Wallich) J. Schiller, *Br. Phycol. J.,* **21,** 359–370.

Rucker, J. B,. and Carver, R. E. (1969). A survey of the carbonate mineralogy of cheilostome bryozoa. *J. Paleontol.* **43,** 791–799.

Runham, N. W., Thornton, P. R., Shaw, D. A., and Wayte, R. C. (1969). The mineralization and hardness of the radular teeth of the limpet *Patella vulgata* L. *Z. Zellforsch.* **99,** 608–626.

Runnegar, B. (1982). The Cambrian explosion: Animals or fossils?. *J. Geol. Soc. Aust.* **29,** 395–411.

Runnegar, B. (1984). Crystallography of the foliated calcite shell layers of bivalve molluscs. *Alcheringa* **8,** 273–290.

Runnegar, B., and Jell, P. A. (1976). Australian middle-Cambrian molluscs and their bearing on early molluscan evolution. *Alcheringa* **1,** 109–138.

Saffo, M. B., and Lowenstam, H. A. (1978). Calcareous deposits in the renal sac of a molgulid tunicate. *Science* **200,** 1166–1168.

Salamat, M. D., Ross, M., and Peacor, D. (1980). Otoconia formation in the fetal rat. *Ann. Otol. Rhinol. Laryngol.* **89,** 229–238.

Saleuddin, A.S.M. (1970). Electron microscopic study of the mantle of normal and regenerating *Helix. Can. J. Zool.* **48,** 409–416.

Saleuddin, A.S.M., and Petit, H. P. (1983). The mode of formation and the structure of the periostracum. In *The Mollusca,* Vol. 4. (ed. K. M. Wilbur). 199–234. Academic Press, New York.

Samata, S. P., Cazaux, C., Hamm, M., Engels, J., and Krampitz, G. (1980). Biochemical studies on components of mollusc shell. In *The Mechanisms of Biomineralization in Animals and Plants* (eds. M. Omori and N. Watabe), pp. 37–47. Tokai Univ. Press, Tokyo.

Sand, O. (1974). Directional sensitivity of microphonic potentials from the perch ear. *J. Exp. Biol.* **60,** 881–899.

Sandberg, P. A. (1975). Bryozoan diagenesis: Bearing on the nature of the original skeleton of rugose corals. *J. Paleontol.* **49,** 587–606.

Sandberg, P. A. (1983). An oscillating trend in Phanerozoic non-skeletal carbonate mineralogy. *Nature (London)* **305,** 19–22.

Sangster, A. G., and Parry, D. W. (1981). Ultrastructure of silica deposits in higher plants. In *Silicon and Siliceous Structures in Biological Systems* (eds. T. L. Simpson and B. E. Volcani), pp. 383–407. Springer-Verlag, New York.

Sasaki, T., and Higashi, S. (1983). Scanning and transmission electron microscopy of developing enamel surfaces of the kitten tooth germs. *J. Electron Microsc.* **32,** 163–171.

Sato, S., Rahemtulla, F., Prince, C. W., Tomana, M., and Butler, W. T. (1985a). Acidic glycoproteins from bovine compact bone. *Conn. Tissue Res.* **14,** 51–64.

Sato, S., Rohemtulla, F., Prince, C. W., Tomana, M., and Butler, W. T. (1985b). Proteoglycans of adult bovine compact bone. *Conn. Tissue Res.* **14,** 65–75.

Sauk, J. J., Cocking-Johnson, D., Cervenka, V. A., and Van Kampen, C. L. (1984). Non-collagenous protein derived from teleostean fish-scales. *Biochim. Biophys. Acta* **798,** 199–203.

Savin, S. M. (1977). The history of the earth surface temperature during the past 100 million years. *Annu. Rev. Earth Planet. Sci.* **5,** 319–355.

Schaffer, J. (1930). Die Stützgewebe. In *Handbuch der Mikroskopischen Anatomie des Menschen* (ed. W. von Mollendorf), pp. 1–390. Springer, Berlin.

Schidlowski, M., Hayes, K. M., and Kaplan, I. R. (1983). Isotopic inferences of ancient biochemistries: Carbon, sulfur, hydrogen and nitrogen. In *Earth's Earliest Biosphere* (ed. J. W. Schopf), pp. 149–186. Princeton Univ. Press, Princeton.

Schifano, G. (1982). Temperature-magnesium relations in the shell carbonate of some modern marine gastropods. *Chem. Geol.* **35,** 321–332.

Schmid, A. M., and Schulz, D. (1979). Wall morphogenesis in diatoms: deposition of silica by cytoplasmic vesicles. *Protoplasma* **100,** 267–288.

Schmidt, W. J. (1924). *Die Bausteine des Tierkorpers in Polarisiertem Lichte.* F. Cohen Verlag, Bonn.

Schmidt, W. J. (1932). Bestimmung der Lage der optischen Achse in Biokristallen. *Abderhalden's Handbuch biol. Arbeits. Abt. 5.2* **289,** 1357–1400.

Schmidt, W. J. (1936). Uber die Kristallorientierung im Zahnschmelz. *Naturwissenschaften* **24,** 361.

Schmidt, W. J. (1958). Die gesunden Zahngewebe. In *Die gesunden und die erkrankten Zahngewebe des Menschen und der Wirbeltiere in Polarisationsmikroskop* (ed. W. J. Schmidt). Carl Hanser, Munich.

Schmidt, W. J. (1959). Bemerkungen zur Schalenstruktur von *Neopilina galatheae. Galathea Rpt.* **3,** 13–17.

Schnell, R. C., and Vali, G. (1972). Atmospheric ice nuclei from decomposing vegetation. *Nature (London)* **236,** 163–164.

Schoffeniels, E. (1951). Utilisation du ^{45}Ca pour l'étude de l'état physicochimique du calcium sangiun echange au niveau des tissues, chez l'Anodonte. *Arch. Int. Physiol.* **59,** 137–138.

Schönbörner, A. A., Boivin, G., and Baud, C. A. (1979). The mineralization process in teleost fish scales. *Cell Tissue Res.* **202,** 203–212.

Schönleber, K. (1936). *Scytonema Julianum.* Beiträge zur normalen und pathologischen Cytologie und Cytogenese der Blaualgen. *Arch. Protistenk* **88,** 36–68.

Schopf, T.J.M., and Manheim, F. T. (1967). Chemical composition of ectoprocta (Bryozoa). *J. Paleontol.* **41,** 1197–1225.

Schraer, H., and Hunter, S. J. (1985). The development of medullary bone: A model for osteogenesis. *Comp. Biochem. Physiol.* **82A,** 13–17.

Schreiber, B., Cavalea, L., and Boltazzi-Massera, E. (1959). Ecologia degli Acantari e circolazione dello Sr. nel mare. *Boll. Zool.* **24,** 213.

Schroder, O. (1901). Neue radiolarien. *Deutsch. Südpolar-Expedition, 1901–1903. (Zool.)* **9**(H4), 209–223.

Schroeder, J. H., Dwornik, E. J., and Papike, J. J. (1969). Primary protodolomite in echinoid skeletons. *Bull. Geol. Soc. Am.* **80,** 1613–1616.

Schröter, K., Lachli, A., and Sievers, A. (1975). Mikroanalytische Identifikation von Bariumsulfat-Kristallen in den Statolithen der Rhizoide von *Chara fragilis,* Desv. *Planta* **122,** 213–225.

Schulz-Baldes, M., and Lewin, R. A. (1975). Manganese encrustation of zygospores of a *Chlamydomonas* (Chlorophyta: Volvocales). *Science* **188,** 1119–1120.

Schwab, D. and Plapp, R. (1983). Quantitative chemical analysis of the shell of the monothalamous foraminifer *Allogromia laticollaris* Arnold. *J. Foraminif. Res.* **13,** 69–71.

Schwabl, M. (1963). Solenogaster mollusks from southern California. *Pacific Sci.* **17,** 261–281.

Scott, B. L., and Pease, D. C. (1956). Electron microscopy of the epiphyseal apparatus. *Anat. Rec.* **126,** 465–495.

Scurfield, G., Michell, A. J., and Silva, S. R. (1973). Crystals in woody stems. *Bot. J. Linn. Soc.* **66,** 277–289.

Sdzuy, K. (1969). Unter- und mittelkambrische Porifera (Chancelloriida und Hexactinellida). *Palaeontol. Z.* **43,** 115–147.

Seilacher, A. (1983). Precambrian metazoan extinctions. *Geol. Soc. Am. Abst.* **15,** 683.

Sela, J., Amir, D., Schwartz, Z., and Weinberg, H. (1987). Changes in the distribution of extracellular matrix vesicles during healing of rat tibial bone (computerized morphometry and electron microscopy). *Bone* **8,** 245–250.

Selvig, K. A. (1970). Periodic lattice images of hydroxyapatite crystals in human bone and dental hard tissues. *Calcif. Tissue Res.* **6,** 227–238.

Selvig, K. A. (1972). The crystal structure of hydroxyapatite in dental enamel as seen with electron microscope. *J. Ultrastruct. Res.* **41,** 369–375.

Shapiro, I. M., and Boyde, A. (1984). Microdissection—elemental analysis of the mineralizing growth cartilage of the normal and rachitic chick. *Metab. Bone Dis.* **5,** 317–326.

Shapiro, I. M., and Greenspan, J. S. (1969). Are mitochondria directly involved in biological mineralisation? *Calif. Tissue Res.* **3,** 100–102.

Shellis, R. P., and Miles, A.E.W. (1976). Observations with the electron microscope on enameloid formation in the common eel (*Anguilla anguilla;* Teleostei). *Proc. R. Soc. London Ser. B* **194,** 253–269.

Shimuzu, M., and Yamada, J. (1980). Sclerocytes and crystal growth in the regeneration of sea urchin test and spines. In *The Mechanisms of Biomineralization in Animals and Plants* (eds. M. Omori and N. Watabe), pp. 169–178. Tokai Univ. Press, Tokyo.

Shore, R. E. (1972). Axial filament of siliceous sponge spicules, its organic components and synthesis. *Biol. Bull.* **143,** 689–698.

Shrivastava, S. C. (1970). Cuticular components of common Indian arachnids and myriopods. *Experimentia* **26,** 1028–1029.

Shuto, T. (1974). Larval ecology of prosobranch gastropods and its bearing on biogeography and paleontology. *Lethaia* **7,** 239–256.

Silverman, H., Steffens, W. L., and Dietz, T. H. (1983). Calcium concretions in the gills of a freshwater mussel serve as a calcium reservoir during periods of hypoxia. *J. Exp. Zool.* **227,** 177–189.

Silyn-Roberts, H., and Sharp, R. M. (1986). Crystal growth and the role of the organic network in eggshell biomineralization. *Proc. R. Soc. London Ser. B* **227,** 303–324.

Simkiss, K. (1965). The organic matrix of the oyster shell. *Comp. Biochem. Physiol.* **16,** 427–435.

Simkiss, K. (1968). The structure and formation of the shell and shell membranes. In *Egg Quality: A Study of the Hen's Egg* (ed. T. C. Carter), pp. 3–25. Oliver and Boyd Ltd., Edinburgh.

Simkiss, K. (1976). Intracellular and extracellular routes in biomineralization. *Symp. Soc. Exp. Biol.* **30,** 423–444.

Simkiss, K. (1986). The processes of biomineralization in lower plants and animals—an overview. In *Biomineralization in Lower Plants and Animals* (eds. B.S.C. Leadbeater and R. Riding), pp. 19–37. Clarendon Press, Oxford.

Simpson, G. G. (1951). *Horses.* Oxford Univ. Press, New York.

Simpson, G. G. (1983). *Fossils and the History of Life.* Scientific American Books Inc., New York.

Simpson, J. W., and Awapara, J. (1966). The pathway of glucose degradation in some invertebrates. *Comp. Biochem. Physiol.* **18,** 537–548.

Simpson, T. L., and Volcani, B. E. (1981). *Silicon and Siliceous Structures in Biological Systems.* Springer-Verlag, New York.

Sissons, H. A. (1956). The growth of bone. In *The Biochemistry and Physiology of Bone* (ed. G. H. Bourne), pp. 443–474. Academic Press, New York.

Slavkin, H. C., Zeichner-David, M., Snead, M., Graham, E. E., Samuel, N., and Ferguson, M.W.J. (1984). Amelogenesis in reptilia: Evolutionary aspects of enamel gene products. *Symp. Zool. Soc. London* **52,** 275–304.

Smith, B. N., Hoops, G. K., Land, L.S.D., and Folk, R. L. (1971). $^{13}C/^{12}C$ ratios of aragonite from fruit of *Celtis laevigata willd. Naturwissenschaften* **7,** 365–366.

Smith, S. V. (1978). Coral-reef area and the contributions of reefs to processes and resources of the world's oceans. *Nature (London)* **273,** 225–226.

Smith, S. V., Buddenmeier, R. W., Redalie, R. C., and Houck, J. E. (1979). Strontium-calcium thermometry in coral skeletons. *Science* **204,** 404–407.

Smout, A. H. (1954). Lower Tertiary Foraminifera of the Qatar Peninsula. *Br. Mus. (Nat. Hist.)* **3,** 1096.

Snead, M. L., Law, E. C., Zeichner-David, M., Fincham, A. G., Woo, S. L., and Slavkin, H. C. (1985). DNA sequence for cloned cDNA for murine amelogenin reveal the amino acid sequence for enamel-specific protein. *Biochem. Biophys. Res. Commun.* **129,** 812–818.

Solem, A., and Yochelson, E. L. (1979). North American Paleozoic land snails, with a summary of other Paleozoic nonmarine snails. *U.S. Geol. Surv. Prof. Paper* **1072,** 1–42.

Sorby, H. C. (1879). Anniversary address of the president: Proc. Geol. Soc. London, 1878–1879. *Quart. J. Geol. Soc. London* **35,** 56–83.

Sorby, H. C. (1897). Structure and origin of limestones. *Proc. Geol. Soc. London* **35,** 56–95.

Spangenberg, D. B. (1976). Intracellular statolith synthesis in *Auralia aurita.* In *The Mechanism of Biomineralization in the Invertebrates and Plants* (eds. N. Watabe and K. M. Wilbur), pp. 231–248. Univ. of South Carolina, Columbia, SC.

Spangenberg, D. B., and Beck, C. W. (1968). Calcium sulfate dihydrate statoliths in *Auralia. Trans. Am. Microsc. Soc.* **87,** 329–335.

Sparkes, S., and Greenaway, P. (1984). The haemolymph as a storage site for cuticular ions during premoult in the freshwater/land crab *Holthusiana transversa. J. Exp. Biol.* **113,** 43–54.

Spear, D. G., Barr, J. K., and Barr, C. E. (1969). Localization of hydrogen ion and chloride ion fluxes in *Nitella. J. Gen. Physiol.* **54,** 397–414.

Spero, H. J., and Parker, S. L. (1985). Photosynthesis in the symbiotic planktonic foraminifer *Orbulina universa,* and its potential contribution to oceanic primary-productivity. *J. Foraminif. Res.* **15,** 273–281.

Spindler, M., and Röttger, R. (1973). Der Kammerbauvorgang der Grossforaminifere *Heterostegina depressa* (Nummulitidae). *Mar. Biol.* **18,** 146–159.

Spindler, M., Hemleben, Ch., Salomons, J. B,. and Smit, L. P. (1984). Feeding behavior of some planktonic foraminifers in laboratory cultures. *J. Foraminif. Res.* **14,** 237–249.

Spiro, B. F. (1971). Ultrastructure and chemistry of the skeleton of *Tubipora musica* Linne. *Bull. Geol. Soc. Denmark* **20,** 279–284.

Sprang, M. L., and Linlow, S. E.(1981). Subcellular localization and partial characterization of ice nucleating activity of *Pseudomonas syringae* and *Erwinia herbicola. Phytopathology* **72,** 111–115.

St. Pierre, T. G., Bell, S. H., Dickson, D.P.E., Mann, S., Webb, J., Moore, G. R., and Williams, R.J.P. (1986). Mossbauer spectroscopic studies of the cores of human, limpet and bacterial ferritins. *Biochim. Biophys. Acta* **870,** 127–134.

St. Aubin, D. J., Stinson, R. H., and Geraci, J. R. (1984). Aspects of the structure and composition of baleen and some effects of exposure to petroleum hydrocarbons. *Can. J. Zool.* **62,** 193–198.

Stanley, S. M. (1973). An ecological theory for the sudden origin of multicellular life in the Late Precambrian. *Proc. Natl. Acad. Sci. U.S.A.* **70,** 1486–1489.

Stanley, S. M. (1976). Fossil data and the Precambrian-Cambrian evolutionary transition. *Am. J. Sci.* **276,** 56–76.

Stapleton, R. P. (1973). Ultrastructure of tests of some recent benthic hyaline foraminifera. *Palaeontographica Part A* **142,** 10–49.

Stasek, C. R. (1972). The molluscan framework. In *Chemical Zoology. VII. Mollusca* (eds. M. Florkin and B. T. Scheer), pp. 1–43. Academic Press, New York.

Stay, B., King, A., and Roth, L. M. (1960). Calcium oxalate in the oothecae of cockroaches. *Ann. Ent. Soc. Am.* **53,** 79–86.

Steele, V. J., and Oshel, P. E. (1985). Mineral content of the organ of Bellonci in the marine amphipod *Gammarus setosus. J. Morphol.* **185,** 51–58.

Stehli, F. G. (1956). Shell mineralogy in Paleozoic invertebrates. *Science* **123,** 1031–1032.

Stenzel, H. B. (1962). Aragonite in the resilium of oysters. *Science* **136,** 1121–1122.

Stenzel, H. B. (1963). Aragonite and calcite as constituents of adult oyster shells. *Science* **142,** 232–233.

Stenzel, H. B. (1964). Oysters: Composition of the larval shell. *Science* **145,** 155–156.

Stetler-Stevenson, W. G., and Veis, A. (1986). Type I collagen shows a specific binding affinity for bovine dentin phosphophoryn. *Calcif. Tissue Int.* **38,** 135–141.

Stetler-Stevenson, W. G., and Veis, A. (1987). Bovine dentin phosphophoryn: calcium ion binding properties of a high molecular weight preparation. *Calcif. Tissue Int.* **40,** 97–102.

Steve-Bocciarelli, D. S. (1970). Morphology of crystallites in bone. *Calcif. Tissue Res.* **5,** 261–269.

Stiefel, E. I., and Watt, G. D. (1979). Azotobacter cytochrome b557.5 is a bacterioferritin. *Nature (London)* **279,** 81–83.

Stockman, K. W., Ginsburg, R. N., and Shinn, E. A. (1967). The production of lime mud by algae in South Florida. *J. Sed. Petrol.* **37,** 633–648.

Stoermer, E. F., Pankratz, H. S., and Bowen, C. C. (1965). Fine structure of the diatom *Amphipleura pellucida.* II. Cytoplasmic fine structure and frustule formation. *Am. J. Bot.* **52,** 1067–1078.

Stolkowski, J. (1951). Essai sur le détermination des formes mineralogiques du calcaire chez les êtres vivants (calcaires coquilliers). *Ann. Inst. Oceanogr.* **26,** 1–115.

Stone, A. R. (1970). Seasonal variation of spicule size in *Hymeniacidon perleve. J. Mar. Biol. Assoc. U.K.* **50,** 343–348.

Stricker, S. A., and Weiner, S. (1985). Amorphous calcium phosphate in the stylets produced by a marine worm (Nemertea). *Experientia* **41,** 1557–1559.

Stuhler, R. (1937). Über den Feinbau des Knochens. *Fortscht. Geb. Röntgenstrahlen* **57,** 231–264.

Sucov, H. M., Benson, C. S., Robinson, J. J., Britten, J. R., Wilt, F., and Davidson, E. H. (1987). A lineage-specific gene encoding a major matrix protein of the sea urchin embryo spicule: II. Structure of the gene and derived sequence of the protein. *Dev. Biol.* **120,** 507–519.

Suga, S., Wada, K., and Ogawa, M. (1980). Fluoride concentration in the enameloid of fishes. In *The Mechanisms of Biomineralization in Animals and Plants* (eds. M. Omori and N. Watabe), pp. 229–240. Tokai Univ. Press, Tokyo.

Suga, S., Wada, K., and Ogawa, M. (1981). Fluoride concentration in teeth of tetraodontiform fishes and its phylogenetic significance. *Jpn. J. Ichthyol.* **28,** 304–312.

Sullivan, B. K., Miller, C. B., Peterson, W. T., and Soeldner, A. H. (1975). A scanning electron microscope study of the mandibular morphology of boreal copepods. *Marine Biol.* **30,** 175–182.

Sullivan, C. W. (1979). Diatom mineralization of silicic acid. IV. Kinetics of soluble Si pool formation in exponentially growing synchronized *Navicula pelliculosa. J. Phycol.* **15,** 210–216.

Sullivan, C. W. (1986). Silicification by diatoms. In *Silicon Biochemistry. Ciba Foundation Symposium 121,* pp. 59–89. Wiley, Chichester.

Sullivan, C. W., and Volcani, B. E. (1973). Role of silicon in diatom metabolism. III. The effects of silicic acid on DNA polymerase, TMP kinase and DNA synthesis in *Cylindrotheca fusiformis. Biochim. Biophys. Acta* **308,** 212–229.

Swift, D. M., Sikes, C. S., and Wheeler, A. P. (1986). Analysis and function of organic matrix from sea urchin tests. *J. Exp. Zool.* **240,** 65–73.

Swineford, A., and Franks, P. C. (1959). Opal in the Ogallala formation in Kansas. In *Silica in Sediments. Soc. Econ. Paleont. Mineral Special Publication No. 7,* pp. 111–120.

Takagi, M., Parmley, R. T., Denys, F. R., Yagasaki, H., and Toda, Y. (1984). Ultrastructural cytochemistry of proteoglycans associated with calcification of shark cartilage. *Anat. Rec.* **208,** 149–158.

Takagi, T., Suzuki, M., Baba, T., Minegishi, K., and Sasaki, S. (1984). Complete amino acid sequence of amelogenin in developing bovine enamel. *Biochem. Biophys. Res. Commun.* **121,** 592–597.

Takagi, Y., and Veis, A. (1981). Matrix protein difference between human normal and dentinogenesis imperfecta dentin. In *The Chemistry and Biology of Mineralized Connective Tissues* (ed. A. Veis), pp. 233–243. Elsevier, Amsterdam.

Takuma, S., Tohda, H., Watanabe, K., and Yama, S. (1986). Size increase of dentin crystals in the intertubular matrix due to caries. *J. Electron Microsc.* **35,** 60–65.

Tappan, H. (1980). *The Paleobiology of Plant Protists.* W. H. Freeman, San Francisco.

Taylor, D. L. (1974). Symbiotic marine algae: Taxonomy and biological fitness. In *Symbiosis in the Sea* (ed. W. B. Vernberg), pp. 245–262. Univ. South Carolina Press, Columbia.

Taylor, J. D., and Kennedy, W. J. (1969). The shell structure and mineralogy of *Chama pellucida* Broderip. *Veliger* **II,** 391–398.

Taylor, J. D., Kennedy, W. J., and Hall, A. (1969). The shell structure and mineralogy of the bivalvia. Introduction. Nuculacea-Trigonacea. *Bull. Br. Mus. Nat. Hist. (Zool.) Suppl.* **3,** 1–125.

Taylor, J. D., Kennedy, W. J., and Hall, A. (1973). The shell structure and mineralogy of the Bivalvia. II. Lucinacea-Clavagellaceae. Conclusions. *Bull. Br. Mus. Nat. Hist. (Zool.)* **22,** 256–293.

Teigler, D. J., and Arnott, H. J. (1972). X-Ray diffraction and fine structural studies of crystals in the malphigian tubules of silkworms. *Nature (London)* **235,** 166–167.

Teigler, D. J., and Towe, K. M. (1975). Microstructure and composition of the trilobite exoskeleton. *Fossils Strata* **4,** 137–149.

Tendal, O. S. (1972). A monograph of the Xenophyophoria (Rhizopodea, Protozoa). In *Galathea Report,* ed. T. Wolff **12,** pp. 1–103. Danish Science Press, Copenhagen.

ter Kuile, B., and Erez, J. (1987). Uptake of inorganic carbon and internal carbon cycling in symbiont-bearing benthonic foraminifera. *Marine Biol.* **94,** 499–509.

ter Kuile, B., and Erez, J. (1988). Mechanisms for the uptake of inorganic carbon by two species of symbiont-bearing foraminifera. Submitted for publication.

Termine, J. D. (1986). Bone proteins and mineralization. *Rheumatology* **10,** 184–196.

Termine, J. D., and Posner, A. S. (1966). Infrared analysis of rat bone: Age dependency of amorphous and crystalline mineral fractions. *Science* **133,** 1523–1525.

Termine, J. D., and Posner, A. S. (1967). Amorphous/crystalline interrelationships in bone mineral. *Calcif. Tissue Res.* **1,** 8–23.

Termine, J. D., Belcourt, A. B., Christner, P. J., Conn, K. M., and Nylen, M. U. (1980a). Properties of dissociative extracted fetal bovine tooth matrix proteins. I. Principal molecular species in developing bovine enamel. *J. Biol. Chem.* **225,** 9760–9768.

Termine, J. D., Eanes, E. D., and Conn, K. M. (1980b). Phosphoprotein modulation of apatite crystallization. *Calcif. Tissue Int.* **31,** 247–251.

Termine, J. D., Belcourt, A. B., Conn, K. M., and Kleinman, H. K. (1981). Mineral and collagen-binding proteins of fetal calf bone. *J. Biol. Chem.* **256,** 10403–10408.

Termine, J. D., Robey, P. G., Fisher, L. W., Shimukawa, H., Drum, M. A., Conn, K. M., Hawkins, G. R., and Cruz, J. B. (1984). Osteonectin, bone proteoglycan, and phosphophoryn defects in a form of osteogenesis imperfecta. *Proc. Natl. Acad. Sci. U.S.A.* **81,** 2213–2217.

Terpstra, R. A., Bennema, P., Hartman, P., Woensdregt, C. F., Perdok. W. G., and Senechal, M. L. (1986). F faces of apatite and its morphology: Theory and observation. *J. Crystal Growth* **78,** 468–478.

Tesch, F. W. (1974). Influence of geomagnetism and salinity on the directional choice of eels. *Helgol. Wiss. Meeresunters* **26,** 382–395.

Théel, H. (1876). Mémoire sur l'Elpidia. *Kongl. Svenska Vetensk. Akad. Handl.* **14**(8).

Thompson, D. W. (1942). *On Growth and Form.* Cambridge Univ. Press, Cambridge.

Thorson, G. (1950). Reproductive and larval ecology of marine bottom invertebrates. *Biol. Rev.* **25,** 1–45.

Thoulet, M. J. (1884). Sur les spicules siliceux d'eponges vivantes. *C.R. Acad. Sci. Paris* **98,** 1000–1001.

Tomes, C. S. (1898). Upon the structure and development of the enamel of elasmobranch fishes. *Phil. Trans.* **190B,** 443–464.

Tompa, A. S. (1976). A comparative study of the ultrastructure and mineralogy of calcified land snail eggs (Pulmonata: Stylommatophora). *J. Morphol.* **150,** 861–888.

Tompa, A. S. (1980). The ultrastructure and mineralogy of the dart from *Philomycus carolinianus* (Pulmonata: Gastropoda) with a brief survey of the occurrence of darts in land snails. *The Veliger* **23,** 35–42.

Tompa, A. S., and Watabe, N. (1976a). Calcified arteries in gastropods. *Calcif. Tissue Res.* **22,** 159–172.

Tompa, A. S., and Watabe, N. (1976b). Ultrastructural investigation of the mechanism of muscle attachment to the gastropod shell. *J. Morphol.* **149,** 339–352.

Torres de Araujo, F. F., Pires, M. A., Frankel, R. B., and Bicudo, C.E.M. (1986). Magnetite and magnetotaxis in algae. *Biophys. J.* **50,** 375–378.

Towe, K. M. (1967). Echinoderm calcite: Single crystal or polycrystalline aggregate. *Science* **157,** 1048–1050.

Towe, K. M. (1970). Oxygen-collagen priority and the early metazoan fossil record. *Proc. Natl. Acad. Sci. U.S.A.* **65,** 781–788.

Towe, K. M. (1973). Tribolite eyes: Calcified lenses in vivo. *Science* **179,** 1007–1009.

Towe, K. M. (1982). Biochemical keys to the emergence of complex life. In *Life in the Universe* (ed. J. Bollingham), pp. 297–306. MIT Press, Cambridge.

Towe, K. M., and Bradley, W. F. (1967). Mineralogical constitution of colloidal "hydrous ferric oxides." *J. Colloid Interface Sci.* **24,** 384–392.

Towe, K. M., and Cifelli, R. (1967). Wall ultrastructure in the calcareous foraminifera: Crystallographic aspects and a model for calcification. *J. Paleontol.* **41,** 742–762.

Towe, K. M., and Lowenstam, H. A. (1967). Ultrastructure and development of iron mineralization in the radular teeth of *Cryptochiton stelleri* (Mollusca). *J. Ultrastruct. Res.* **17,** 1–13.

Towe, K. M., Berthold, W. U., and Appleman, D. E. (1977). The crystallography of *Patellina corrugata* Williamson: a-axis preferred orientation. *J. Foraminif. Res.* **7,** 58–61.

Towe, K. M., and Rutzler, K. (1968). Lepidocrocite iron mineralization in keratose sponge granules. *Science* **162,** 268–269.

Towne, W. F., and Gould, J. L. (1985). Magnetic field sensitivity in honeybees. In *Magnetite Biomineralization and Magnetoreception in Organisms* (eds. J. L. Kirschvink, D. S. Jones, and B. J. MacFadden), pp. 385–406. Plenum Press, New York.

Tracy, R. P., Stenner, D. D., Romberg, R. W., Mintz, K. P., Riggs, B. L., and Mann, K. G. (1987). In *Development and Diseases of Cartilage and Bone Matrix* (eds. A. Sen and T. Thornhill) pp. 127–136. Alan R. Liss, New York.

Trautz, D. R., Klein, E., and Addleston, H. K. (1952). Variation in the X-ray diffractograms of dental enamel of man and shark. *J. Dent. Res.* **31,** 472–473.

Travis, D. F. (1963). Structural features of mineralization from tissue to macromolecular levels of organization in the decapod Crustacea. *Ann. N.Y. Acad. Sci.* **109,** 177–245.

Travis, D. F., and Glimcher, M. J. (1963). The structure and organization of and the relationship between the organic matrix and the inorganic crystals of embryonic bovine enamel. *J. Cell Biol.* **18,** 109–123.

Treffry, A., and Harrison, P. M. (1984). Spectroscopic studies on the binding of iron, terbium, and zinc by apoferritin. *J. Inorg. Biochem.* **21,** 9–20.

Treffry, A., Lee, P. J., and Harrison, P. M. (1984). Functional studies on rat liver isoferritin. *Biochem. Biophys. Acta* **785,** 22–29.

Treffry, A., Harrison, P. M., Cleton, M. I., de Bruijn, W. C., and Mann, S. (1987). A note on the composition and properties of ferritin iron cores. *J. Inorg. Biochem.* **28.**

Trench, R. K., and Blank, R. J. (1987). *Symbiodinium microadriatacum* Freudenthal, *S. goreauii* Sp. Nov., *S. kawagutii* Sp. Nov. and *S. pilosum* Sp. Nov.: Gymnodinioid dinoflagellate symbionts of marine invertebrates. *J. Phycol.* **23,** 469–481.

Trueman, E. R. (1975). *The Locomotion of Soft-Bodied Animals.* Edward Arnold, London.

Tschachotin, S. (1908). Die Statocyste der Heteropoden. *Z. Wiss. Zool.* **90,** 343–422.

Tsirulis, T. (1974). The fine structure of the statocyst of the univalve mollusk *Clione limacina. J. Evol. Biochem. Physiol.* **10,** 158–165.

Tsugita, A., and Yariv, J. (1985). Preliminary results for the primary structure of bacterioferritin of *Escherichia coli. Biochem. J.* **231,** 209–212.

Tudge, A. P. (1960). A method of analysis of oxygen isotopes in orthophosphates—its use in measurements of paleotemperatures. *Geochim. Cosmochim. Acta* **18,** 81–93.

Tung, M. S., and Brown, W. E. (1983). An intermediate state in hydrolysis of amorphous calcium phosphate. *Calcif. Tissue Int.* **35,** 783–790.

Tuross, N., and Fisher, L. W. (1988). The proteins in the shell of *Lingula.* In *Origin, Evolution and Modern Aspects of Biomineralization in Plants and Animals* (ed. R. E. Crick). Elsevier, Amsterdam. (in press)

Uchiyama, A., Lefteriou, B., and Glimcher, M. J. (1985). Phosphoproteins of chicken bone. In *The Chemistry and Biology of Mineralized Tissues* (ed. W. T. Butler), pp. 182–184. EBSCO Media, Birmingham, AL.

Uozomi, S., Iwata, K., and Togo, Y. (1972). The ultrastructure of the mineral in and the construction of the crossed-lamellar layer in molluscan shell. *J. Fac. Sci. Hokkaido Univ. Ser. IV* **15,** 447–478.

Urbanus, J.F.L.M., van den Ende, H., and Koch, B. (1978). Calcium oxalate crystals in the wall of *Mucor mecede. Mycologia* **70,** 829–842.

Urey, H. C. (1947). The thermodynamic properties of isotopic substances. *J. Chem. Soc.* 562–581.

Urey, H. C. (1973). Cometary collisions and geological periods. *Nature (London)* **242,** 32–33.

Urey, H. C., Lowenstam, H. A., Epstein, S., and McKinney, C. (1951). Measurement of paleotemperatures and temperature of the Upper Cretaceous of England, Denmark and the southeastern United States. *Geol. Soc. Am. Bull.* **62,** 399–416.

van der Rest, M., Rosenberg, L. C., Olsen, B. R., and Poole, A. R. (1986). Chondrocalcin is identical with the C-propeptide of type II procollagen. *Biochem. J.* **237,** 923–925.

van Emburg, P. R., de Jong, E. W., and Daems, W.T. (1986). Immunochemical localization of a polysaccharide from biomineral structures (coccoliths) of *Emiliania huxleyi. J. Ultrastruct. Mol. Struct. Res.* **94,** 246–259.

Vandermeulen, J. H., and Watabe, N. (1973). Studies on reef corals. I. Skeleton formation by newly settled planula larva of *Pocillopora damicornis. Marine Biol.* **23,** 47–57.

van Name, W. G. (1945). The North and South American ascidians. *Bull. Am. Mus. Nat. Hist.* **84,** 1–476.

Veis, A. (1984). Bones and teeth. In *Extracellular Matrix Biochemistry* (eds. K. A. Piez, and A. H. Reddi), pp. 329–374. Elsevier, Amsterdam.

Veis, A., and Perry, A. (1967). The phosphoprotein of the dentin matrix. *Biochemistry* **6,** 2409–2416.

Veis, D. J., Albinder, T. M., Clohisy, J., Rahima, M., Sabsay, B., and Veis, A. (1986). Matrix proteins of the teeth of the sea urchin *Lytechinus variegatus. J. Exp. Zool.* **240,** 35–46.

Vidal, G., and Knoll, A. H. (1982). Radiations and extinctions of plankton in the late Proterozoic and early Cambrian. *Nature (London)* **297,** 57–59.

Vinogradov, A. P. (1953). *The Elementary Chemical Composition of Marine Organisms. Memoir. Sears Foundation for Marine Research II.* Yale University, New Haven.

Vogel, J. J., and Boyan-Salyers, B. D. (1976). Acidic lipids associated with the local mechanism of calcification. *Clin. Orthop. Rel. Res.* **118,** 230–241.

Vogel, J. J., and Smith, W. N. (1976). Calcification of membranes isolated from *Bacterionema matruchotii. J. Dent. Res.* **55,** 1080–1083.

Volcani, B. E. (1981). Cell wall formation in diatoms: Morphogenesis and biochemistry. In *Silicon and Siliceous Structures in Biological Systems* (eds. T. L. Simpson and B. E. Volcani), pp. 155–200. Springer-Verlag, Berlin.

von Brand, T., Scott, D. B., Nylen, M. U., and Pugh, M. H. (1965). Variations in the mineralogical composition of cestode calcareous corpuscles. *Exp. Parasitol.* **16,** 382–391.

von Graff, L. (1875). Anatomie des *Chaetoderma nitidulum. Z. Wissensch. Zool.* **26,** 166–192.

von Heider, S. (1882). Die Gattung *Cladocora* Ehrenberg. *Sitzber. Akad. Wiss., Wien. Math.naturwiss.* **84,** 634–667,

Voronkov, M. G., Zelchan, G. I., and Lukevitz, E. (1975). *Silizium und Leben.* Akademie-Verlag, Berlin.

Vranckx, R,. and Durliat, M. (1980). Cuticular protein analysis from *Astacus leptodactylus*—I. Electrophoretical studies of the so!uble cuticular proteins. *Comp. Biochem. Physiol.* **65B,** 55–64.

Wada, K. (1961). Crystal growth of molluscan shells. *Bull. Nat. Pearl. Res. Lab.* **7,** 703–785.

Wada, K. (1980). Initiation of mineralization of bivalve molluscs. In *The Mechanisms of Biomineralization in Animals and Plants* (eds. M. Omori and N. Watabe), pp. 79–92. Tokai Univ. Press, Tokyo.

Wada, K., and Fukinuki, T. (1968). Biomineralization in bivalve molluscs with emphasis on the chemical composition of the extrapallial fluid. In *The Mechanisms of Mineralization in the Invertebrates and Plants* (eds. N. Watabe and K. M. Wilbur), pp. 175–190. Univ. South Carolina Press, Columbia.

Wadsten, T., and Moberg, R. (1985). Calcium oxalate hydrates on the surface of lichens. *Lichenologist* **17,** 239–245.

Wagge, L. E. (1952). Quantitative studies of calcium metabolism in *Helix aspersa. J. Exp. Zool.* **120,** 311–342.

Wainwright, S. A. (1963). Skeletal organisation in the coral, *Pocillopora damicornis. Quart. J. Microsc. Sci.* **104,** 169–183.

Wainwright, S. A. (1964). Studies of the mineral phase of coral skeleton. *Exp. Cell Res.* **34,** 213–230.

Wainwright, S. A., Biggs, W. D., Currey, J. D., and Gosline, J. M. (1976). *Mechanical Design in Organisms.* Princeton Univ. Press, Princeton, NJ.

Waite, J. H. (1983). Quinone-tanned scleroproteins. In *The Mollusca* (ed. K. M. Wilbur), pp. 467–504. Academic Press, New York.

Waite, J. H., Saleuddin, A.S.M., and Anderson, S. O. (1979). Periostracin—a soluble precursor of sclerotized periostracum in *Mytilus edulis* L. *J. Comp. Physiol.* **130,** 301–307.

Walcott, C., Gould, J. L., and Kirschvink, J. L. (1979). Pigeons have magnets. *Science* **205,** 1027–1029.

Walker, G., Rainbow, P. S., Foster, P., and Holland, D. L. (1975). Zinc phosphate granules in tissue surrounding the midgut gland of the barnacle *Balanus balanoides. Marine Biol.* **33,** 161–166.

Walker, M. M., Kirschvink, J. L., and Dizon, A. E. (1985). Magnetoreception and biomineralization of magnetite: Fish. In *Magnetite Biomineralization and Magne-*

toreception in Organisms (eds. J. L. Kirschvink, D. S. Jones, and B. J. McFadden), pp. 417–437. Plenum Press, New York.

Wall, D., Guillard, R.R.L., Dale, B., Swift, E. J., and Watabe, N. (1970). Calcitic resting cysts in *Peridinium trochoideum.* (Stein) Lemmermann. An autotropic marine dinoflagellate. *Phycologia* **9,** 151–156.

Waller, T. R. (1980). Scanning electron microscopy of shell and mantle in the order Arcoida (Mollusca: Bivalvia). *Smithsonian Contrib. Biol.* **313,** 1–58.

Waller, T. R. (1983). Dahllite in the periostracum of *Lithophaga nigra* (Mollusca, Bivalvia) and its taxonomic and functional implications. *Am. Malacolog. Bull.* **1,** 101.

Wallich, G. C. (1860). On the development and structure of the diatom valve. *Trans. Microsc. Soc. London N.S.* **8,** 129–145.

Wallner, J. (1933). *Oocardium stratum* Naeg. eine wichtige tuffbildende Algae Sübayerns. *Planta* **20,** 287–293.

Warshawsky, H. (1985). Ultrastructural studies on amelogenesis. In *The Chemistry and Biology of Mineralized Tissues* (ed. W. T. Butler), pp. 33–45. EBSCO Media, Birmingham, AL.

Warshawsky, H., and Nanci, A. (1982). Stereo electron microscopy of enamel crystallites. *J. Dental Res.* **61,** 1504–1514.

Watabe, N. (1965). Studies on shell formation. XI. Crystal-matrix relationships in the inner layers of mollusk shells. *J. Ultrastruct. Res.* **12,** 351–370.

Watabe, N. (1967). Crystallographic analysis of the coccolith of *Coccolithus huxleyi. Calcif. Tissue Res.* **1,** 114–121.

Watabe, N. (1984). Shell. In *Biology of the Integument* (eds. J. Bereiter-Hahn and A. G. Matolsky), Chap. 25, pp. 448–485, Springer-Verlag, Berlin.

Watabe, N., Meenakshi, V. R., Blackwelder, P. L., Kurtz, E. M., and Dunkelberger, D. G. (1976). Calcareous spherules in the gastropod *Pomacea paludosa.* In *Mechanisms of Biomineralization in the Invertebrates and Plants* (eds. N. Watabe and K. M. Wilbur), pp. 283–308. Univ. South Carolina Press, Columbia.

Waterman, R. E. (1970). Fine structure and scale development in the teleost, *Brachydanio rerio. Anat. Rec.* **168,** 361–380.

Watson, M. L. (1960). The extracellular nature of enamel in the rat. *J. Biophys. Biochem. Cytol.* **7,** 489–491.

Wattendorff, J., and Meier, H. (1970). Zur Bildung von Kalziumoxalatkristallen in den Zellwänden von Cupressaeeen. *Verhandl. Schweizerischen Naturforschenden Gesellschaft 1970,* 239–242.

Weber, J. N., and Raup, D. M. (1966). Fractionation of the stable isotopes of carbon and oxygen in marine calcareous organisms—the Echinoidea. Part II. Environmental and genetic factors. *Geochim. Cosmochim. Acta* **30,** 705–736.

Weber, J. N., and Woodhead, P.M.J. (1970). Carbon and oxygen isotopic fractionation in the skeletal carbonate of reef-building corals. *Chem. Geol.* **6,** 93–117.

Weber, J. N., and Woodhead, P.M.J. (1972). Stable isotope ratio variations in non-scleractinian Coelenterate carbonates as a function of temperature. *Marine Biol.* **15,** 293–297.

Wefer, G., and Berger, W. H. (1980). Stable isotopes in benthic foraminifera: seasonal variation in large tropical species. *Science* **209,** 803–805.

Wefer, G., and Killingley, J. S. (1980). Growth histories of strombid snails from Bermuda recorded in their O-18 and C-13 profiles. *Marine Biol.* **60,** 129–135.

Weinberg, P., Maroudas, A., Schneiderman, R., Katz, E., and Wachtel, E. (1986). Intrafibrillar water in cartilaginous tissues. *Curr. Adv. Skeletogen.* **3,** 127–131.

Weiner, S. (1979). Aspartic acid-rich proteins: Major components of the soluble organic matrix of mollusk shells. *Calcif. Tissue Int.* **29,** 163–167.

Weiner, S. (1984). Organization of organic matrix components in mineralized tissues. *Amer. Zool.* **24,** 945–951.

Weiner, S. (1985). Organic matrixlike macromolecules associated with the mineral phase of sea urchin skeletal plates and teeth. *J. Exp. Zool.* **234,** 7–15.

Weiner, S. (1986). Organization of extracellularly mineralized tissues: A comparative study of biological crystal growth. *CRC Crit. Rev. Biochem.* **20,** 365–408.

Weiner, S., and Erez, J. (1984). Organic matrix of the shell of the foraminifer, *Heterostegina depressa. J. Foraminf. Res.* **14,** 206–212.

Weiner, S., and Hood, L. (1975). Soluble protein of the organic matrix of mollusk shells: A potential template for shell formation. *Science* **190,** 987–989.

Weiner, S., and Price, P. A. (1986). Disaggregation of bone into crystals. *Calcif. Tissue Res.* **39,** 365–375.

Weiner, S., and Traub, W. (1980). X-Ray diffraction study of the insoluble organic matrix of mollusk shells. *FEBS Lett.* **111,** 311–316.

Weiner, S., and Traub, W. (1981). Organic matrix-mineral relationships in mollusk shell nacreous layers. In *Structural Aspects of Recognition and Assembly in Biological Macromolecules* (eds. M. Balaban, J. L. Sussman, W. Traub and A. Yonath), pp. 467–482. Balaban ISS, Rehovot, Philadelphia.

Weiner, S., and Traub, W. (1984). Macromolecules in mollusc shells and their functions in biomineralization. *Phil. Trans. R. Soc. London Ser. B* **304,** 421–438.

Weiner, S., and Traub, W. (1986). Organization of hydroxyapatite crystals within collagen fibrils. *FEBS Lett.* **206,** 262–266.

Weiner, S., Traub, W., and Lowenstam, H. A. (1983a). Organic matrix in calcified exoskeletons. In *Biomineralization and Biological Metal Accumulation* (eds. P. Westbroek and E. W. de Jong), pp. 205–224. Reidel, Dordrecht.

Weiner, S., Talmon, Y., and Traub, W. (1983b). Electron diffraction of mollusk shell organic matrices and their relationship to the mineral phase. *Int. J. Biol. Macromol.* **5,** 325–328.

Weinstock, M., and Leblond, C. P. (1973). Radioautographic visualization of the deposition of a phosphoprotein at the mineralization front in the dentin of the rat incisor. *J. Cell. Biol.* **56,** 838–845.

Wells, G. P. (1950). Spontaneous activity cycles in polychaeta worms. *Symp. Soc. Exp. Biol.* **4,** 127–142.

Wells, J. W. (1956). Scleractinia. In *Treatise of Invertebrate Paleontology. Part F. Coelenterata* (ed. R. C. Moore), pp. F328–F444. Geol. Soc. America, Lawrence, Kansas.

Wells, J. W. (1963). Coral growth and geochronometry. *Nature (London)* **197,** 948–950.

West, C. D. (1937). Note on the crystallography of the echinoderm skeleton. *J. Paleontol.* **11,** 458–459.

Westbroek, P., de Jong, E. W., Dam, W., and Bosch, L. (1973). Soluble intracrystalline polysaccharides from coccoliths of *Coccolithus huxleyi* (Lohmann) Kamptner (I). *Calcif. Tissue Res.* **12,** 227–238.

Westbroek, P., de Jong, E. W., van der Wal, P., Borman, A. H., de Vrind, J.P.M., Kok, D., de Bruijn, W. C., and Parker, S. B. (1984). Mechanism of calcification in the marine alga *Emiliania huxleyi. Phil. Trans. R. Soc. London Ser. B* **304,** 435–444.

Westbroek, P., van der Wal, P., van Emburg, P. R., de Vrind-de Jong, E. W., and de Bruijn, W. C. (1986). Calcification in the coccolithophorids *Emiliania huxleyi* and *Pleurochrysis carterae.* I. Ultrastructural aspects. In *Biomineralization in*

Lower Plants and Animals (eds. B.S.C. Leadbeater and R. Riding), pp. 189–203. Clarendon Press, Oxford.

Wetz, K., and Crichton, R. R. (1976). Chemical modification as a probe of the topography and reactivity of horse spleen ferritin. *Eur. J. Biochem.* **61,** 545–550.

White, S. W., Hulmes, D.J.S., Miller, A., and Timmins, P.A. (1977). Collagen-mineral axial relationship in calcified turkey leg tendon by X-ray and neutron diffraction. *Nature (London)* **266,** 421–425.

Whitfield, M., and Watson, A. J. (1983). The influence of biomineralisation on the composition of seawater. In *Biomineralization and Biological Metal Accumulation* (eds. P. Westbroek and E. W. de Jong), pp. 57–72. Reidel, Dordrecht.

Wiederhold, M. L., Sheridan, C. E., and Smith, N.K.R. (1986). Statoconia formation in molluscan statocysts. *Scanning Electron Microsc.* 1986 II, 781–792.

Wilbur, K. M. (1964). Shell formation and regeneration. In *Physiology of Mollusca* (eds. K. M. Wilbur and C. M. Yonge), pp. 243–282. Academic Press, New York.

Wilbur, K. M. (1984). Many minerals, several phyla, and a few considerations. *Am. Zool.* **24,** 839–845.

Wilbur, K. M., and Jodrey, L. H. (1952). Studies on shell formation. I. Measurement of the rate of shell formation using Ca^{45}. *Biol. Bull.* **103,** 269–276.

Wilbur, K. M., and Simkiss, K. (1968). Calcified shells. In *Comprehensive Biochemistry* (eds. M. Florkin and E. H. Stotz), pp. 229–295. Elsevier, Amsterdam.

Wilbur, K. M., and Watabe, N. (1963). Experimental studies on calcification in molluscs and the alga *Coccolithus huxleyi. Ann. N.Y. Acad. Sci.* **109,** 82–112.

Wilbur, K. M., Colinvaux, L. H., and Watabe, N. (1969). Electron microscope study of calcification in the alga *Halimeda* (order Siphonales). *Phycologia* **8,** 27–35.

Wilfert, M., and Peters, W. (1969). Vorkommon von Chitin bei Coelenteraten. *Z. Morphol. Tiere* **64,** 77–84.

Williams, R.J.P. (1986). Introduction to silicon chemistry and biochemistry. In *Silicon Biochemistry (Ciba Foundation Symposium 121),* pp. 24–39. Wiley, Chichester.

Wilkinson, B. H. (1979). Biomineralization, paleooceanography and the evolution of calcareous marine organisms. *Geology* **7,** 524–527.

Wilson, M. J., and Jones, D. (1984). The occurrence and significance of manganese oxalate in *Pertusaria corallina* (Lichenes). *Pedobiologia* **26,** 373–379.

Wilson, M. J., Jones, D., and Russel, J. D. (1980). Glushinskite, a natural occurring magnesium oxalate. *Mineralog. Mag.* **43,** 837–840.

Wilt, F. H. (1987). Determination and morphogenesis in the sea urchin embryo. *Development* **100,** 559–575.

Wilt, F. H., Benson, S., and Uzman, J. A. (1985). The origin of the micromeres and formation of the skeletal spicules in developing sea urchin embryos. In *The Cellular and Molecular Biology of Invertebrate Development* (eds. R. H. Sawyer, and R. M. Showman), pp. 297–310. South Carolina Press, Columbia.

Wise, S. W. (1970). Microarchitecture and mode of formation of nacre (mother-of-pearl) in pelecypods, gastropods and cephalopods. *Eclogae Geol.* **63,** 775–797.

Witty, J. E., and Knox, E. G. (1964). Grass opal in some chestnut and forest soils in north central Oregon. *Proc. Soil Sci. Soc. Am.* **28,** 685–688.

Wolff, T. (1960). The hadal community: An introduction. *Deep Sea Res.* **6,** 95–124.

Wolpert, L., and Gustafson, T. (1961). Studies on the cellular basis of morphogenesis of the sea urchin embryo. *Exp. Cell. Res.* **25,** 311–325.

Wood, A. (1949). The structure of the wall of the test in Foraminifera; its value in classification. *Quart. J. Geol. Soc. London* **104,** 229–255.

Wood, N. V. (1947). Specific surfaces of bone, apatite, enamel and dentine. *Science* **105,** 531–532.

Woodland, W. (1907). Studies in spicule formation. VI. *Quart. J. Microsc. Sci.* **51,** 45–53.

Worms, D., and Weiner, S. (1986). Mollusk shell organic matrix: Fourier transform infrared study of the acidic macromolecules. *J. Exp. Zool.* **237,** 11–20.

Wray, J. L. (1977). *Calcareous Algae.* Elsevier, Amsterdam.

Wuthier, R. E. (1984). Calcification of vertebrate hard tissues. In *Calcium and Its Role in Biology* (ed. H. Sigel), pp. 411–472. Marcel Dekker, New York.

Yaari, A. M., Boskey, A. L., and Shapiro, I. M. (1984). Phosphate modulation of calcium transport by a calcium-phospholipid-phosphate complex of calcifying tissues. *Calcif. Tissue Int.* **36,** 317–319.

Yamada, J., and Watabe, N. (1979). Studies on fish scale formation and resorption. 1. *J. Morphol.* **159,** 49–66.

Yamanouchi, T. (1929). Notes on the behavior of the holothurian *Caudina chilensis. Sci. Rpts. Tohoku Univ. ser. 4,* 73–115.

Yamauchi, M., Katz, E. P., and Mechanic, G. L. (1986). Intermolecular cross-linking and sterospecific molecular packing in type I collagen fibrils of the periodontal ligament. *Biochemistry* **25,** 4907–4913.

Yano, I. (1980). Calcification of crab exoskeleton. In *The Mechanisms of Biomineralization in Animals and Plants* (eds. M. Omori and N. Watabe), pp. 187–196. Tokai Univ. Press, Tokyo.

Yatsu, N. (1902). Notes in the histology of *Lingula. J. College Sci., Univ. Tokyo* **17,** 1–112.

Young, J. D. (1950). The structure and some physical properties of the testudinian eggshell. *Proc. Zool. Soc. Lond.* **120,** 455–469.

Young, S. D. (1971). Organic material from scleractinian coral skeletons—I. Variation in composition between several species. *Comp. Biochem. Physiol.* **40B,** 113–120.

Young, S. D., O'Connor, J. D., and Muscatine, L. (1971). Organic material from scleractinian coral skeletons—II. Incorporation of ^{14}C into protein, chitin and lipid. *Comp. Biochem. Physiol.* **40B,** 945–958.

Zheng, S., Tu, A. Y., Renugopalakrishnan, V., Strawich, E., and Glimcher, M. J. (1987). A mixed β-turn and β-sheet structure for bovine tooth enamel amelogenin: Raman spectroscopic evidence. *Biopolymers* **26,** 1809–1813.

Zylberberg, I., Castanet, J., and de Riegles, A. (1980). Structure of the dermal scales in *Gymophiona* (Amphibia). *J. Morphol.* **165,** 41–54.

Zylstra, U., Boer, H. H., and Sminia, T. (1978). Ultrastructure, histology and innervation of the mantle edge of the freshwater pulmonate snails *Lymnae stagnalis* and *Biomphalaria pfeifferi. Calcif. Tissue Res.* **26,** 271–282.

Index

Acantharia, 53
  dissolution in oceans, 20, 31, 207
  skeletal symmetry, 54
Acetabularia, 27, 53, 209
Acicularia, 209
Acidic matrix macromolecules, 21–24
  bone and dentin, 36, 158–62
  cartilage, 170
  Coccolithophoridae, 71
  conformations, 23, 31
  crystal growth, 38–39
  distribution among phyla, 22
  enamel, 36, 184–85
  enameloid, 182
  evolution, 248
  Foraminiferida, 67
  Gorgoniacea, 78
  Mollusca, 107
  nucleation, 38–39
  otoconia, 191
  scleractinian corals, 83, 87
  sea urchins, 129, 132
Acropora palmata, 84
Acroporidae, 239
Actinopoda, 45, 51, 53
Additives, interactions with crystals, 31–32
Adenosine triphosphate (ATP), 31, 33
Agglutinating foraminifera, 63. *See also* Foraminiferidae
Aglantha, 76, 192
Ahermatypic corals, 81. *See also* Scleractinian corals
Algae. *See also* Calcareous algae; Coralline algae; Magnetotactic algae
  environmental influence on
    magnesium, 218
    mineral, 209
  mineralization properties, 208–9
  minerals, 52–53
  oldest mineralized fossils, 234
Alkaline phosphatase, 31, 173
Allogromia laticollaris, 63
Allogromiina, 61
Allopora, 76, 212
Alopias volpinus, 48
Amblema, 102
Ameloblasts, 182, 185
Amelogenins, 183, 184–85, 186, 187
Amoeboid cells, 140
Amorphous calcium carbonate
  Arthropoda, 112
  Ascidiacea, 136, 141
  Crustacea, gastroliths, 121
  cyanobacteria, 44
  Gorgoniacea, 77
  Mollusca, 91
Amorphous calcium phosphate (ACP)
  Arthropoda, 112–14, 117, 122
  Ascidiacea, 136, 141
  bryozoa, 243
  chiton teeth, 97, 99, 164
  Craniata, 137, 139
  fossil record, 228, 241
  granules, 240
  gravity perception, 193–94
  hard parts, composed of, 240
  matrix vesicles, 163
  mitochondria, 171
  Nautilus, 49, 241
  precursor of dahllite, 44–45, 164
  shark otoconia, 48
  stabilization of, 31
  uroliths, 93, 241
Amorphous calcium pyrophosphate, 91, 121
Amorphous ferric oxyhydroxite, 229
Amorphous fluorite
  Ascidiacea, 136, 141
  Mollusca, 91
Amorphous hydrous ferric oxide, 44
Amorphous hydrous ferric phosphate
  chiton teeth, 97–98, 210
  holothurian granules, 125, 46–47, 210
Amorphous Mg-Ca-phosphate, 74, 76, 244
  gravity perception, 192, 231
Amorphous minerals, working definition, 17
Amorphous monohydrocalcite, 91
Amorphous silica. *See* Opal
Amphibia
  fossil record, 243
  magnetite, 197
  otoliths, 196
Amphisorus hemprichii, 64
Angiosperms, 245, 247
Anguilla anguilla, 138
Anguispira, 91
Anisodorus, 91
Anisonema, 53
Anlage, 65

Animalia, distribution of biogenic minerals, 10–11
Annelida
  distribution of minerals, 10–11
  macromolecules, 22
  sternal shields, 240
Anodonta cyanea, 92
Anomalodesmata, 93
Anomiids, 106
Anthozoa, 76, 77. *See also* Cnidaria
Antifreeze glycoproteins, 205–6
Anura, 167
Aplacophora, 89, 90
Aplysia, 92, 191, 193
Apoferritin, 28, 202–3. *See also* Ferritin
Aquaspirillum magnetotacticum, 44, 230
Aragonite
  algae, 208–9
  Annelida, 213
  Arthropoda, 112–13
  Ascidiacea, 136, 140, 143
  Cirripedia, 211
  corals, environmental influences, 208
  Cnidaria, 74, 76, 82, 83, 208, 211
  Craniata, 138
  Ectoprocta, 211
  Foraminiferida, 65
  fossil record, 234, 235
  Mollusca, 90–93
  Nautilus, 49
  nucleation of, 35
  otoconia, shark, 48
  otoliths
    amphibia, reptiles, 195, 196
    fish, 195, 196, 198
  oyster larva, 45, 48
  Phanerozoic, increase during, 235–40
  precursor of, 44
  Protoctista, 52–53
  serpulids, 213–14, 237
  trace elements, 218
Aragonite–calcite proportions in skeletons, 210–17
Aragonite versus calcite deposition, 237–38
Arca zebra, 99–100
Archeocyathids, 233, 234, 238
Archeology, season documentation, 225
Archidoris, 91
Arenicola, 193
Argonauta, 93
Argopecten irradians, 93
Aristotle's lantern, 131
Arthropoda, 111–22
  cuticle, 115–16
  mineral types and locations, 112–14
  molting and temporary storage, 120–22
  oldest fossils, 233, 235
Asaphus raniceps, 114
Ascidiacea, 135, 140–44
  Chordata, association to, 140
  larva, 140, 143–44
  mineral types, 136–37
  spicule formation, 140–43
Astacus leptodactylus, 120
Asteroidea, 123, 126
Atdabianian, 233. *See also* Cambrian
Ateocina culcitella, 92
Atmospheric oxygen, Precambrian, 229, 230, 249
ATPase, calcium, 78
Auralia, 76, 192, 196
Aves, magnetite, 197
Avrainvillea, 209

Bacillariophyta. *See* Diatoms
Bacteria. *See also* Magnetotactic bacteria
  ice crystal nucleation, 204–5
  iron-reducing, 229
  manganese encrusting, 229
  sulfate-reducing, 27, 229
  sulfide minerals, 27
Bacterioferritin, 202. *See also* Ferritin
Bacterionema matruchotii, 33
Balanus balanoides, 113
Banded iron formations, 229
Barite, 51
  gravity perception, 190, 192
  Mollusca, 88, 92
  Protoctista, 52–53
Barium, 19
Barium sulfate, 19, 51, 109. *See also* Barite
Barnacles. *See* Cirripedia
Bathypera, 140
Bathypera ovoida, 136
Bathypera splendens, 136
Bats, magnetite, 197
Bethe, 5
Biogenic minerals, 7–18
  amorphous, 17
  calcium minerals, 17
  chemical formulae, 16
  maturation, 42, 44
  methods of identification, 16–17
  organic crystals, 18
  phosphates, 17–18
  phylum distribution, 97–99, 164–65
Biologically controlled mineralization, 26, 27–41
  evolution of, 229–51
Biologically induced mineralization, 26–27, 42
  evolution of, 229
Biomineralization processes
  controlled and uncontrolled, 26
  definitions, 26
  history of field, 4–5
  impact on biosphere, 18–20
  ontogenetic changes, 45–49
  pioneers in field, 4–5
  real world, 41–49
  scope of field, 3–4
Biomineralization, definition of, 3, 6
Blastocoel, 127
Blatella germanica, 113
Blood. *See* Hemocoel
Bøggild, O. B., 5
Bombyx mori, 113

Bone
collagen-crystal relations, 36–37, 155–58, 165–66
collagen structure, 152–55
crystals
aggregates, 166–67
growth, 165
shapes and sizes, 149–51
density, 147
functions, 144
levels of organization, 144–46
major components, 162–63
mineral, 151–52
age changes, 165
chemical formula of, 152
first precipitates, 164–65
molecular organization of, 149
noncollagenous proteins, 158–62
stages of mineralization, 162–63
surface area, 151
types, 145–46
Bone formation
basic processes, 43, 146
relation to cartilage mineralization, 146, 167
stages, 146–47
Bone-GLA-protein (BGP), 59, 160–61. *See also* GLA-proteins
Bornetella, 53, 209
Boron, relation to salinity, 220
Botryllus schlosseri, 144
Brachiodontes, 212
Brachiodontus variabilis, 40
Brachiopoda, 106
fossil record, 237
macromolecules, shell, 22
mineral types, 10–11
oldest fossils, 233, 235, 242
trace elements, 218–19, 220
Briareum, 77
Bronn, 5
Brushite, 27, 49, 93
Bryozoa. *See* Ectoprocta
Buccinum undatum, 102
Bulla, whale, 150, 151, 166
Buoyancy devices, 88, 93
Burgess Shale, 249
Butschli, 5
Bütterfly, magnetite, 197

Calcareous algae, 21, 27, 209. *See also* Algae
mineral types, 52–53
Calcification, 6, 17. *See also* Biomineralization
Calcisponges, 239
Calcite
Annelida, 213
Arthropoda, 112–14, 117
Ascidiacea, 136, 141, 143
c-axis orientation, 65, 67, 69, 77, 115, 119, 129, 131
Cirripedia, 211
cleavage rhombohedron, 106
Cnidaria, 74, 76, 77, 78, 79, 210
Coccolithophoridae, 68, 69, 239
Craniata, 138–39
Ectoprocta, 211
Foraminiferidae, 65, 239
fossil record, 232–40
gravity perception, 193, 195, 196
Mollusca, 90–93
Nautilus, 49
nucleation of, 35
Ostracoda, 243
Protoctista, 52–53
Pseudokephyrion lorica, 51
scleractinian corals, 82
serpulid worms, 237–38
single biogenic crystals
Echinodermata, 124, 129, 131
foraminifer, 61–62
gorgonian, 77
trace elements, 218, 220, 237–38
twinned, 86
Calcite-aragonite problem, 35
Calcituba polymorpha, 63
Calcium
affinity for proteoglycans, 169
evolution of calcium metabolism, 251
macromolecules, association with, 129
mitochondria, regulation of, 171
privileged status in biomineralization, 17
β-sheet conformation, association with, 23
temporary storage of, 120–22, 164, 189
uptake for mineralization, 34, 130
Calcium carbonate. *See also* Aragonite; Calcite
environmental influence on, 210–21
evolution of, 232–40
marine environment, 19
Calcium citrate, 9, 11, 113
Calcium-phospholipid-phosphorus complexes, 173
Calcium sulfate. *See* Gypsum
Calicoblastic epidermis, 75, 81
Callinectes sapidus, 31, 112, 121
Cambrian
fossil reefs, 238
fossil record, first appearances, 234–35, 242–43
fossil taxa in, 233
opal, 245
phosphate, 241–44
phosphatization, 242
trace fossils, 249
Cancellous bone, 145–46
Cancer paqurus, 117
Caprinid mollusks, 239
Carbon isotopic composition, 222, 249
Carbonate apatite. *See* Dahllite
Carbonate cycle in oceans, 19
Carbonate mineralization. *See* Evolution of carbonate mineralization
Carbonates, biogenic, 8, 10, 18, 233
Carbonic anhydrase, 31, 78, 120
Carboniferous
aragonite fossils, 234
nautiloids, 236

Carboniferous (*continued*)
  pulmonate evolution, 240
  reptiles, 243
γ-Carboxyglutamic acid. *See* GLA-proteins
Carboxylates, 21
  calcium binding to, 23
  coccolith polysaccharide, 71
  ferritin, mineral interface, 204
  nucleation, importance in, 38–39
  β-sheet conformation, orientations of, 23
Carcinus maenas, 112, 118, 119
Cartilage. *See also* Endochondral cartilage
  elasmobranchs, 167
  histological description of, 167
  mineralized cartilage
    distribution of, 167
    extracellular matrix, 171–74
    matrix vesicles, 173–74
    mitochondrial mineralization, 171
  unmineralized cartilage, 168–69
    basic function, 168
    major components, 168
Celestite, 48, 51, 207
  Acantharia, 20, 31
  crystallography of, 54
  distribution, 52–53
Cellulose, 140
Cement
  agglutinating foraminifera, 63
  Textulariina, 61
Cement sheath, 145
Cementum, 149
Cenozoic, 235
  Coccolithophoridae, 236
  Foraminiferidae, 236–37
  Hydrozoa, reef builders, 236
  opal formation, 246
  Pteropoda, 236
  reef formation, 239
Centers of calcification, corals, 82, 84
Central dark line, 165
  bone, dentin and enamel crystals, 184
  synthetic apatite crystals, 164–65
Cephalopods. *See also Nautilus*
  mineral types, 93
  vestibulary apparatus, 190, 193
Cephalaspidian gastropods, 240
Cerastroderma edule, 226
Cetacea, magnetite, 197
Chama, 212, 214, 216
Chama macerophylla, 223
Chama gryphoides, 105
Chama arcana, 214, 217
Chara, 27, 51, 52, 192
Charophyta, 240
Cherts, 245, 246
Chione stutchbury, 225
Chiropsalmus, 76, 192
Chitin
  chemical formula, 115
  chitin–protein complexes, structure of, 108, 115–16
  distribution in mineralized tissues, 22
  fossil, 241
  Nautilus, 49
  polychaete gravity sensors, 190
  scleractinian corals, 83
  α-chitin
    arthropod cuticle, 111, 115–20
    chitons, 96, 98
    periostracum, 101
  β-chitin
    mollusk shells, 36–37, 107–8
Chitin sulfate, 140
Chitons, 94–99
  shell plates, 94
  spicules, 94–95
  teeth, 88
    environmental influence on mineralogy, 210
    functions, 95
    mineralization, 96–99
    mineral transformations, 44–45
    preformed matrix, 30
Choanoflagellates, 51
Chordata
  Ascidiacea, 140–44
  Craniata, 144–88
  fossil record, 243
  mineral types, 136–39
  oldest fossils, 242
  vestibulary apparatus, 190, 194–95
Chlamydomonas, 53
Cholesteric structure, cuticle, 115–16
Chondrocytes, 168, 171, 172
Chondroitin sulfate, 168, 169
Chryosaora, 76, 192
Chrysococcus, 53
Chrysophyta, 51, 53, 244
Chrysotila, 72
Cibicides floridanus, 67
Cichlasoma cyanoguttatum, 178–80
Ciona intestinalis, 137
Cirripedia
  aragonite–calcite proportions, 211
  isotopic composition, 222
  mineral types, 113
  molting, 111
  trace elements, 220
Citrates, calcium, 9, 11, 13
Clarke, F. W., 5
Clavarizona hirtosa, 99
Cleavage rhombohedron, calcite, 68
Clinocardium nutalli, 225
Closterium, 53
Cnidaria, 74–87. *See also* Corals; Scleractinian corals
  aragonite–calcite proportions, 210
  body wall structure, 74
  Ediacaran fauna, 231–32
  fossil record, 236, 239
  Leptogorgia mineralization, 77–78
  massive skeletons, 81–87. *See also* Scleractinian corals

matrix macromolecules, 78
mineral types, 76
polypoid body form, 75
spicule aggregates, 79
spicules, 77–79
symbionts, 81
Coccolithophoridae, 67–73. *See also* Coccoliths
cyst silicification, 73
evolution of, 239
extracellular holococcolith formation, 72
fossil record, 236
intracellular coccolith formation, 69–72
major carbonate producers, 236
noncoccolith mineralization, 72
scale structure, 68
terminology, 69
Coccoliths, 68–72. *See also* Coccolithophoridae
crystal growth cessation, 41
crystallography, 71
morphology of, 68, 71
polysaccharide, 41, 71–72
Coccospheres, 68
Coelenteran, 74
Coelenterates. *See* Cnidaria
Coenenchyme, 79
Collagen
enameloid, 178–80, 187
evolution of multicellularity, 249
gorgonian skeleton, 77
Pennatulacea, 78
sea urchin skeleton, absence of, 132
SLS structure, 155
type I
bone, organic framework of, 144, 153–55, 158
cross-links, 157–58
crystal relations, 36–37, 155–58, 165–66
dynamic structural changes, 166
fibrils, 149, 152
fish scales, 148
noncollagenous proteins, association between, 160–61
structure of, 153–55
tendon, turkey, 148
type II
C-propeptide of, 173, 174
cartilage, major component of, 167, 168
fibril structure, 169
Columba livida, 139
Compact bone, 145–46
Composite material, 134
Controlled mineralization, evolution of, 248. *See also* Biologically controlled mineralization
Conulariida, 242, 243–44
Conus, 110
Cooperativity, crystal nucleation, 38–39, 162
Coralline algae. *See also* Algae
oldest fossils, 235
reef builders, 238, 239
Corals. *See also* Cnidaria; Scleractinian corals
crystal morphologies, 39
rugose, 238
tabulate, 238
Coscinodiscus wailesii, 56–58
Corymorpha, 76, 192
Corymorpha palma, 191
C-propeptide of type II collagen, 173, 174
Craniata, 144–88
bone, 144–67
cartilage, 167–75
enamel, 182–88
enameloid, 180–82
mineral types, 137–39
Craniidae, 218–19
Crassostrea virginica, 45, 48
Cretaceous
carbonates in open ocean, 239
Chrysophyta, 246
crinoids, 239
diatoms, 246
dinoflagellate cysts, 239
reef formation, 239
silicoflagellates, 246
Crinoidea, 123, 126, 239
Critical nucleus, 33
Crossed-lamellar structure, 94, 105, 106, 214
Cross-reactivity, antibodies, 132, 227
Crustacea, 11. *See also* Arthropoda
gastroliths, 121
gravity sensors, 190, 194
hemocoel, 122
macromolecules in cuticle, 119–20
magnetite, 197
midgut gland, 121
mineralized cuticle, 117–20
Cryptochiton, 95–98
Cryptochiton stelleri, 90
Cryptodifflugia, 52
Crystal, single biogenic
Echinodermata, 124, 129, 131, 132
foraminifer, 61–62
gorgonian, 77
Crystal ghosts, 170
Crystal growth, cessation of, 39–41
Crystal growth, control over, 38–39
Crystal-matrix relations, 35–37
Crystal nucleation. *See* Nucleation
Crystallolithus hyalinus, 72
Cubozoa, 74, 76
Culeolus, 136
Culeolus murrayi, 136
Cuticle. *See also* Arthropoda; Crustacea
helicoidal structure, 115–16
mineralization, 117–20
sclerotization, 102
Cyanobacteria, 234
Cylichna cylindracea, 92
Cymopolia, 209
Cystodytes, 136, 143
Cystodytes lobatus, 143
Cysts, 51, 73

Dahllite. *See also* Hydroxyapatite
  Ascidiacea, 136
  bone, 152
  chiton teeth, 97–98, 210
  Craniata, 135, 137–39
  enamel and enameloid, 175, 176, 183
  fish scale, 209–10
  fossil record, 241
  hard parts composed of, 240
  Mollusca, 90, 92
  precursors of, 44–45
  Protoctista, 52–53
Demineralization, laboratory procedures, 21
Density fractionation, bone, 165
Dentin
  biosynthesis of macromolecules, 29–30
  noncollagenous proteins, 158–59
  preformed matrix framework, 148
  similarity to bone, 144
  structure of, 144–45
Dentinoenamel junction, 180, 183, 184, 188
Depth habitat and mineralization, 210–11
Desert varnish, 19
Desulfovibrio, 27, 42, 43
Desulfovibrio desulfuricans, 44
Detoxification
  Crustacea, 121
  Gastropoda, 189
Development. *See* Ontogeny
Devonian, 243
Diagenesis, 234
  carbonates, 234
  phosphates, 240, 242
Diamox (acetazolimide), 78, 120
Diamysis bahirensis, 113
Diatoms, 54–60
  cessation of mineralization, 40–41
  freshwater species, 244
  fossil record, 245–46
  girdle bands, 55–57
  mineral types, 53
  seawater undersaturation induced by, 244
  silicon uptake, 58–59
  valve formation, 56–58
  valve macromolecules, 59–60
Didemnidae, 140
Didymium, 52
Dinoflagellates, 81, 239, 245, 246
Dissolution, mollusk shells, 109–10
Distichopora, 76
Distribution coefficients, Sr and Mg, 218
Dolomite, 125
Donnan effect, 169
DOPA, 101–2

Earth's rotation rate, 224–25
Ebria, 53
Ebridians, 245, 246
Ecdysis. *See* Molt cycle
β-Ecdysone, 122
Echinodermata, 123–34. *See also* Sea urchin
  adult skeleton, 130–32
  larval skeleton, 127–30
  mineral properties, 123–24, 132–34
  oldest fossils, 235
  trace elements, 220
Ectoprocta
  aragonite–calcite proportions, 211
  fossil record, 242–43
  macromolecules, skeletons, 22
  phosphate in skeletons, 243
  skeletal mineralogy, 211
  reef-builders, 238
Ediacaran fauna, 230–32
EDTA, 21, 107
Eggshells
  Craniata, 35, 144
  Mollusca, 88
Elasmobranchs
  enameloid, 176, 180–82
  gravity sensors, 48, 49, 190, 194
  mineralized cartilage, 167
  ontogenetic changes in mineralogy, 48
  vestibulary apparatus, 49
Electron diffraction, mineral identification, 17, 82, 86, 107
Elpidia glacialis, 220–21
Emiliania huxleyi, 33, 41, 68–71
Endochondral growth plate
  developmental stages, 170–72
  ion transport, 171
Endochondral cartilage
  major constituents, 169–70
  molecular organization, 170–71
  selective introduction of ions, 30
Endochondral ossification, 146, 167
Endolymphatic sac, 196
Enamel, 175–76, 182–88
  amelogenins, 185
  comparison with crossed-lamellar structure, 107
  crystals, 183–84
  matrix-crystal relations, 36–37
  maturation, 185–86
  mineralization processes, 43
  nucleation, 36
  organ, 182
  reptile enamel, 176
  tubules, 185
  ultrastructure, 176, 177
Enamelins
  enamel, 36, 183, 184–85, 186, 187
  enameloid, 182
Enameloid, 175–82
  collagen, 180
  crystal morphology, 181
  distribution, 176
  fish teeth, 176–80
Endocrinal control of mineralization, 49
  Crustacea, 122
  mollusks, 101
  sea urchins, 131
Energy dispersive X-ray spectrometry (EDS), mineral identification, 17
Eniwetok atoll, 238

Environmental influences on biomineralization
algal mineralization, 208–9
amount of mineral versus temperature, 208–10
depth, changes in opal content, 210–11
extent of phenomenon, 207–8
mineralogical changes, 210–17
skeletal growth rate, 223–26. *See also* Growth lines
trace element changes, 217–21
oxygen isotopic composition, 221–23
Epiphysis, 146, 147
Epiphyseal growth plate, 146, 147, 172
cartilage compared to bone, 170
premature closure, 160
Epipyrops anomala, 113
Epitaxy
definition, 35
foraminifera, 65
mollusks, 65, 108
hydroxyapatite on octacalcium phosphate, 184
Epithelial cells
Ascidiacea, 140
chiton teeth, 30, 98–99
enamel, 107
enameloid, 178, 180
mollusks, 100
sea urchins, 130
Equisitum, 247
Errinopora, 76
Erwinia herbicola, 204
Esthetes, 95
Eukaryotes, evolution of, 230
Eunicea, 77
Eupomatus gracilis, 212, 213
Evolution of biomineralization, 227–51. *See also* Evolution of carbonate mineralization; Evolution of phosphate mineralization; Evolution of silicification
biologically controlled mineralization, 229–51
biologically induced mineralization, 229
composite skeletons, 232
covergent versus divergent evolution, 25, 227
eukaryote mineralization, 230–51
fossil record of, interpreting, 227–28
Precambrian-Cambrian boundary, 232, 247–51
prokaryote mineralization, 229–30
strontium discrimination, 228
sulfur isotopes, 229
Evolution of carbonate mineralization
amounts formed, 232–34
aragonite versus calcite, 235–40
carbonates of living organisms, 232–34
deposition in open ocean, 239–40
oldest fossils, 234
Precambrian-Cambrian boundary, 233, 234
trends, 235–40
Evolution of phosphate mineralization, 240–44
distribution of phosphatic skeletons, extant and fossil taxa, 241–43
phosphatic skeletons, amorphous and crystalline, 240
Precambrian-Cambrian boundary zone, 233, 241–42
preserved fossil ACP, 241
vertebrate evolution, 243
Evolution of silicification, 244–47
freshwater fossil record, 246
Precambrian-Cambrian boundary zone, 233, 245
preservation in present day environments, 244
seawater saturation, opal, past variations, 246
stratigraphic ranges, 245
terrestrial fossil record, 246–47
Extrapallial fluid, 109
Exuviae, 120
Eyes
Crustacea, 112, 113
trilobites, 114, 115

Ferric oxides, 52–53
Ferrihydrite
chiton teeth, 97–98
Craniata, 137
ferritin, mineral component of, 28, 203
lichens, 27
precursor form, 42, 44
Ferritin
chiton teeth, 98
distribution, 202
mineral, 202–3
protein structure, 202–3
space delineation, 28
Fibrolamellar bone, 145, 146
Fish. *See also* Elasmobranchs
age from otoliths, 196
bone, 166
cartilage, mineralized, 167
fossil record, 243
ice crystal inhibition, polar fish, 205–6
oldest fossils, 242
otocyst, 196
scales, 148, 209–10
teeth, 30, 176–80
Fluorapatite. *See* Francolite
Fluoride, 180
Fluorite
Arthropoda, 20, 113, 194
gravity perception, 194
Mollusca, 20, 88, 92
Fly larva, mineralized puparium, 122
Foliated layers, mollusks, 106, 107
Foraminiferidae, 60–67
agglutinating, 63
chamber formation, 65
evolution of tests, 235
fossil record, 236
magnesium, effect of temperature, 218

Foraminiferidae (*continued*)
  miliolids, 63
  mineralization, 43, 53
  oldest fossils, 235
  opalline test, 61
  planktonic, evolution of, 239
  reef-builders, 238
  rotaline, 64
  taxonomy, 61
  test proteins, 67
Fossilized hard parts
  environment, monitor of, 208
  growth lines in, 224–26
  Precambrian-Cambrian boundary, 233
  strontium discrimination, 228
Fossil record, mineralization, 227–28
  carbonates, 232–34
  extent of, 228
  first mineralized skeletons, 232–33
  magnetotactic bacteria, 230
  noncomposite skeletal materials, 230
  oldest biogenic minerals, 230
  opal, 245–47
  phosphates, 240–44
  skeletal mineralization, 228, 232, 233, 247–51
  sulfide minerals, 229
Framework matrix macromolecules, 21, 23
  amelogenins, 185, 187
  collagen, 149, 158, 187
  crustacean cuticle, 117–19
  distribution among phyla, 22
  evolution of, 250
Francolite
  chiton teeth, 97–98, 210
  Craniata, 137–38
  enameloid, 180–81
  fossil bryozoan, 243
Freshwater mineralization
  barium, 19
  carbonates, 240
  crustaceans, 120, 122
  fossil record, 240, 246
  opal, 19, 244, 246
  impact on water chemistry, 19
Frogs. *See* Anura
Functions of mineralized hard parts
  Arthropoda, 112–14
  Chordata, 136–39
  Cnidaria, 76
  detect magnetic field, 197, 200
  Echinodermata, 126
  gravity perception, 190–96
  iron storage, 202–4
  Mollusca, 90–93
  nonskeletal functions, various, 189
  temporary storage, 120–22
  tooth hardening, 197
Fungae
  common mineralization processes, 26
  distribution, biogenic minerals, 10–11
  lichens, 27
Fusulinina, 61

Galaxaura, 27, 52, 209
Gammarus setosus, 112
Gamophyta, 53, 74
Gastroliths, 112, 121
Genes
  amelogenins, 185
  bacterium, ice crystal nucleation, 162, 204
  sea urchin spicule protein, 129
Geotropism, 190
Gill supports, 92, 240
Girdle bands, diatoms, 55–57
Gizzard plates, 88, 92, 240
GLA-proteins, 83, 158–61, 170
Glauconite, 61
Glycosaminoglycans, 167, 168. *See also* Proteoglycans
Goethite, 88, 91, 139
Golgi, 100, 101
Golgi cisternae, 69, 72
Gorgoniacea, 77. *See also Leptogorgia*
  matrix macromolecules, 78
  spicule aggregates, 79
  spicule formation, 77–78
Granules
  Arthropoda, 112, 113, 121–22
  holothurians, 45–47, 125–27
  mineralization of, 33–34
  Mollusca, 91, 92, 189
  phosphatic, 240
  polychaete worms, 189
Gravity perception
  basic mechanism, 190
  distribution, 192–95
  Ediacaran fauna, 231–32
  locations of receptors, 192–95
  mineralization sites, 192–95
  mineral types, 192–95
  modes of life of organisms with, 190, 192–95
  types of sensors, 191, 196
Greigite, 27, 44
Grobben, 4
Growth hormone, 101
Growth lines
  absence in *Pecten*, 224
  corals, earth's rotation, 224–25
  definition of, 224
  environmental fluctuations recorded, 224–25
  environmental stimuli, 224
  monitoring thermal effluent, 225–26
  seasonal dating, 225
  structural basis of, 224, 226
  theory of formation, 226
Gypsum, 51
  Cnidaria, 74, 76
  gravity perception, 190, 192, 196, 231
  Protoctista, 52–53

Haeckel, 4
Halimeda, 21, 27, 209
Haliotis, 212
Halocynthia roretzi, 140
Haptoneme, 67

Haptophyta. *See* Coccolithophoridae
Hatschek, 4
Haversian system
  manatee bone, 148
  penguin bone, 148
  stages of formation, 145, 147
Helicoidal structure, cuticle, 115–19
Helix, 44, 91
Helix aspersa, 91
Helix pomatia, 91
Hemocoel, 112, 122
Hepatopancreas. *See* Midgut gland
Herdmania momus, 136, 141, 142, 143
Hermatypic corals, 81. *See also Scleractinian corals*
Hermesinum, 53
Heteropoda, 239
Heterostegina depressa, 65–67
Hexactinellid sponges, 245
History of biomineralization field, 4–5
Holococcoliths, 69, 72
Holothuroidea, 123, 126
  mineralization during ontogeny, 45–47
  Molpadiidae, 45–47, 125
  opal content of granules, depth effect, 210–11
  trace elements, hydrostatic pressure effect, 220–21
Holoxonia, 210–12, 214–15
Holthuisana transversa, 112, 122
Homotrema rubrum, 238
Honeybee, magnetite, 197
Hormones, 49, 122
Horse spleen ferritin, 202–3
Horses, teeth abrasion, 247
Humans, magnetite, 197
Huxley, 4
Hyaluronic acid, 168
Hydrocorallina, 77
Hydromedusae, Ediacaran fauna, 231
Hydrostatic pressure
  mineralogy, effect of, 210–11
  Nautilus shell, 24
  trace elements, effect of, 220–21
Hydrotroilite, 44
Hydroxyapatite. *See also* Dahllite
  biogenic, mostly dahllite, 18
  bone, 152, 164
  enamel, 164
  enameloid, 180
  matrix vesicles, 163, 173
Hydroxylysine, 249
Hydroxyproline, 249
Hydrozoa, 74, 76, 190, 191, 192, 231, 244

Ice crystal formation
  inhibition of, 205–6
  nucleation of, 162, 204–5
Ichneumon fly, 115
Infrared spectroscopy, mineral identification, 17
Inhibitors of mineralization
  cessation of mineralization by, 39–41
  glycoproteins, polar fish, 205–6
  mucus, algal, 209
  noncollagenous proteins, 161
  polyelectrolytes, 175
  proteoglycans, 174
  salivary proteins, 129
  sea urchin proteins, 132
  sulfated glycoproteins, 175
  tissue comparison of, 43
Inner dental epithelium, 178–80, 182
Integument. *See* Cuticle
Interactions, crystal-protein
  crystal growth, 38–39, 134
  ice crystal formation, 205
  nucleation, 38, 39
Intracrystalline proteins, 32, 129, 134
Ion transport, 30–31
  cartilage, 171
  ferritin, 203
  matrix vesicles, 173
Ionophore
  calcium, matrix vesicles, 173
  silicate, diatoms, 59
Iron oxides, biogenic, 9, 11, 18
Isotopic composition. *See* Carbon isotopic composition; Oxygen isotopic composition; Sulfur isotopic composition

Janthina, 239
Jurassic
  birds, 243
  carbonate deposition, open oceans, 239
  diatoms, 246
  dinoflagellate cysts, 239
  reef formation, 239

Keratin, 139, 167
Keratin sulfate, 169
Kidneys, *Nautilus,* 49
Kükenthalia borealis, 137

Lagena, 196
Lamarck, 4
Lamellar bone, 145
Lankester, 4
Larval mineralization
  insects, 122
  mollusks, 223
  scleractinian corals, 82
  sea urchins, 127–30
Le Gros Clark, 29, 149
Lepidocrocite
  chiton teeth, 96–98, 210
  mollusks, 88, 90
Leptogorgia virgulata, 77–78, 87
Leucothoe spinicarpa, 113
Liagora, 27, 52, 209
Lichens, 27
Ligumia subrastrata, 92
Lima, 212

Limnaria lignorum, 112
Limpets. *See* Patellacean gastropods
Lingula, 192
Lipids in skeletons, 83
Lirceus brachyurus, 112
Lithophaga antillarum, 92
Lithophaga nigra, 92
Littorina, 212
Lizards, 167, 196
Lorica, 51, 54
Loxodes, 19, 52
Love dart, 88, 91
Lovenella, 76, 192
Lymnae stagnalis, 101
Lytechnus variegatus, 132

Mackinawite, 27, 44
Macromolecules in mineralized tissues, 20–24. *See also* Acidic matrix macromolecules; Framework matrix macromolecules
Magnesium content
  Brachiopoda, 218–19
  calcite versus aragonite, 217
  coralline algae, 218
  Foraminiferidae, 218
  salinity, effect on, 220
  stabilization of ACP, 31
Magnesium calcite. *See also* Calcite
  Cnidaria, 76, 78
  sea urchins
    distribution, 126
    protodolomite, 124–25
    skeleton, spines, 32, 123
    teeth, 132
Magnetite
  algae, magnetotactic, 197–98, 201, 230
  Arthropoda, 114
  bacteria, iron-reducing, 229
  bacteria, magnetotactic, 196, 230. *See also* Magnetotactic bacteria
  chiton teeth, 95–99, 196
  Craniata, 138–39
  discovery of, 196
  distribution of, 18, 197
  fossil record, 248
  Mollusca, 88, 90
  precursor of, 44
  Protoctista, 52–53
Magnetometer, 197
Magnetosome, 197, 200, 248
Magnetotactic algae, 197–98, 201, 230
Magnetotactic bacteria, 197, 200
  discovery of, 196
  fossil record, 230, 248
Malpighian tubules, 113, 114, 122
Manatee bones, 148
Mandibles, *Nautilus,* 49
Manganese encrusting bacteria, 229
Manganese oxalates, 27
Manganese oxides, 9, 11, 52–53
Mantle, mollusks, 45, 99–101
Mantle-shell zone, 109
Marine mammals, bone, 148
Matrix, the term, 149
Matrix degredation in vivo
  bone, 167
  corals, 83
  enamel, 183, 185–86
  enameloid, 180, 182
Matrix-mediated mineralization, 29–30
  bone, 149
  chiton teeth, 99
  enameloid, 178, 180
  gorgonian spicules, 78
  mollusk shell, 103
  terminology, 26
Matrix–mineral relations, 35–37
  bone, 36–37, 155–58
  crustacean cuticle, 117–19
  enamel, 36–37, 184–85
  enameloid, 177, 181
  ferritin, 203
  nacreous layer, mollusks, 36–37, 106, 108
Matrix vesicles, 33, 163–64. *See also* Vesicle mineralization
  bone, 163–64
  cartilage, 173–74
  dentin, 163
  fish scales, 148
  mineral content of, 163
  tendon, 163
Maturation of biogenic minerals, 42–45
Medullary bone, 144
Medusae, 74
Membranous ossification, 146
Melithaea ochracea, 79–81
Menippe mercenaria, 119
Mesozoic, 235
  Coccolithophoridae, 236
  Foraminiferidae, 60, 236
  scleractinian coral reefs, 236
Mercenaria mercenaria, 93, 109, 225–26
Mesenchyme, 127, 128, 130, 140
Mesoderm, 130
Mesoglea, 77, 78
Metaphysis, 146, 147
Metazoans
  composite skeletal formation, fossil record, 232
  Ediacaran fauna, 230–31
  fossil record, first appearance, 230
Micromeres, 127
Microtubules, 128
Midgut gland, 31, 112–14, 121, 189
Miliolids, 63. *See also* Foraminiferidae
Miliolina, 61
Milliporina, 77, 236
Minerals. *See* Biogenic minerals
Mineral, definition of, 6
Mineral identification, 5, 16–17
Mineral–matrix relations. *See* Matrix–mineral relations
Mitochondrial mineralization
  cartilage, 171
  crustaceans, 31, 121
  mollusks, 100

Mola mola, 198
Molgula manhattensis, 137, 143
Mollusca. *See also* Shell, mollusks
 Aplocophora, 89
 chiton teeth, 95–99
 mineral types, 90–93
 Monoplacophora, 89
 oldest fossils, 233, 234–35
 Polyplacophora, chitons, 94–99
 Scaphopoda, 94
Molpadia, 194
Molpadia intermedia, 46–47, 210
Molpadia musculus, 210
Molpadiidae, 45–47, 240
Molt cycle, 120
Monera. *See also* Bacteria
 common mineralization processes, 26
 mineral types, 8–9
Monohydrocalcite, 48, 138, 194
Monolayers, 34
Monoplacophora, 89, 94, 234, 236
Mycobiont, 27
Myostracum, 103, 215
Mysid shrimps, statoliths, 5, 20, 113, 194
Mytilus californianus shell
 calcite–aragonite proportions, 212, 214, 215
 macromolecules of, 107
 matrix–mineral relations, 35, 37, 106
 nacreous layer of, 104
 prismatic layer of, 104
 seasonal deposition, 223
Mytilus edulis, 101–2, 214
Müllers law, 54
Musca autumnalis, 113, 122
Mussa angulosa, 82, 85, 86

Nacre. *See* Nacreous layer
Nacreous layer
 crystal growth cessation, 41
 crystal shape, 104–5
 crystallography of, 106
 forming crystals, morphology of, 39, 40
 location in shell, 103
 matrix–crystal relations, 36–37, 108
 matrix structure, 107–8
 mineralization processes, 43
 Monoplacophora, 94
 nucleation, 108
 variations in *Mytilus*, 214, 215
NADH, 171
Nautilus
 mandible minerals, 27, 93
 mineral types, 88, 93
 mineralization sites, 49, 93
 shell, mechanical properties of, 24
 statoconia, 191, 193
 uroliths, 93, 241
Nautilus pompilius, 38
Nautilus repertus, 35, 106
Nautiloids, 235, 236, 241
Navigation, 200
Neomeris, 209
Nemertean worms, 240
Neomysis intiger, 113
Neomysis rayi, 113
Neopilina, 89
Neotrigonia, 240
Neotrigonia margaritacea, 35, 40, 92, 106
Nerita, 110
Nitzschia alba, 59
Noncollagenous proteins, 152, 158–62
 bone GLA-protein (osteocalcin), 159, 160, 161
 breakdown during maturation, 167
 distribution among tissues, 158–60
 functions, 160–62
 phosphophoryn, 159, 160–61
 sequence conservation, 160
 serum proteins, 158
 types, 159
Nucleation, controlled, 32–38. *See also* Epitaxy
 acidic macromolecules, 34–38
 bone, 36
 calcium-phospholipid-phosphorus complexes, 173
 cooperativity, 38–39
 C-propeptide of type II collagen, 174
 foraminifera, 65
 ice crystals, 204–5
 lipid membranes, 33–34
 sites of, 34, 35, 38, 106
 stereochemical control over, 10
 sulfated glycoproteins, mollusks, 38, 175
 tissue comparisons, 43
 trigger of, 31
Nummulites, 61

Obelia, 76, 192
Octacalcium phosphate, 44, 45, 164–65, 184
Octocorallia, 77, 79, 231
O-diphenyloxidase, 102
Odontella aurita, 55
Odontoblasts
 dentin, 148
 enameloid, 178–80
Oncorhyncus tshanytcha, 200
Oneirophonta matabilis, 221
Ontogeny
 all known examples, 48
 minerals formed during, 45–49
Oocardium, 53
Oolites, 237
Ophiuroidea, 123, 126
Opal, 17, 18. *See also* Evolution of silicification
 Arthropoda, 113
 chemical formula of, 55
 chiton teeth, 97–98
 cysts, 51
 dissolution in laboratory, 21
 dissolution in oceans, 20
 Echinodermata, 125
 Mollusca, 91
 Molpadiidae, 210–11
 precursors of, during ontogeny, 45–48
 Protoctista, speciality, 51

Opal-CT, 244
Opaque boundary, 186
Orchestia cavimana, 112
Orconectes virilis, 112, 121, 122
Ordovician
  first appearances in, 235
  nautiloids, 236
  oldest preserved aragonite, 234
  siliceous skeletal deposits, 245
  vertebrates, 243
Organic crystals, 9, 11, 18
Organic matrix, 29, 30. *See also* Acidic matrix macromolecules; Framework matrix macromolecules; Matrix
Ossicles, 123
Ossification. *See* Bone formation
Osteoblasts, 147
Osteocalcin, 59, 160. *See also* GLA-proteins
Osteoclasts, 145, 147
Osteocytes, 147
Osteonectin, 159, 161
Ostracods, 119, 242, 243
Ostwald-Lussac law, 42, 45
Otic chambers, 3 types, 196
Otoconia, 191, 194–96. *See also* Gravity perception
Otoliths, 14, 143–44. *See also* Gravity perception
Ototyphlonemertes, 192
Oxalates, 9, 11, 27. *See also* Weddelite; Whewellite
Oxygen isotopic composition
  carbonates
    disequilibrium skeletal deposits, 27, 222
    equilibrium skeletal deposits, 222
    paleotemperatures, 208, 221, 222
    seasons of shell deposition, 223
    water isotopic composition, 222
    whale migration, 221, 222
  phosphates, 221, 222
Oxygen, past atmospheric levels, 51, 230

Padina, 27, 53, 209
Paleotemperature analysis, 208, 223
  basis for, 221
  historical anecdote, 221
  nonequilibrium precipitates, 222
Paleozoic, 235, 236
Pallial line, 100
Paracentrotus lividus, 38, 124, 125, 132, 133
Parafins, 111, 113
Parallel-fibered bone, 145
Paraquadrula, 52
Parathyroid hormone, 49, 122
Patella vulgata, 91
Patellacean gastropods, 95, 96, 244
Patellina corrugata, 62
Pearls, 93, 224
Pecten diegensis, 224
Pecten inflexus, 93
Pedalion alatum, 212
Penecillus, 27, 52, 209
Penguin bones, 148
Penium, 53
Pennatulacea, 27, 78, 231. *See also* Cnidaria
Periodontal ligament, 157
Periosteal surface, 146
Periostracin, 28, 101
Periostracum, 101–3
  function in space delineation, 28
  Monoplacophora, 94
Periplaneta americana, 113
Periplasts, 68
Peritubular dentin, 148
Pertusaria corallina, 27
Phagocytes, 131
Phanerozoic, 228, 232
Philline, 92
Phosphate isotopic composition, 222
Phosphate mineralization. *See* Evolution of phosphate mineralization
Phosphates, biogenic, 8, 10, 17–18
Phosphatic hard parts. *See also* Dahllite
  distribution of, 240–41
  fossil record of, 240–44
  phosphate minerals, 17–18
Phosphatidylserine, 173
Phosphatization, 242
Phospholipids
  cartilage, matrix vesicles, 173
  multimolecular complexes with calcium, 34
  scleractinian corals, 83, 87
Phosphophoryn, 159, 160–61
Phosphoproteins, matrix
  bone and dentin, 159–61
  fish scales, 148
  mollusks, 109
  sea urchins, 132
Phyalidium, 76, 192
Phytoliths, 247
Pigeons, magnetite, 197
Pinctada margaratifera, 35
Pinctada radiata, 106
Planktotrophic mollusk larvae, 223
Plantae
  distribution of minerals, 10–11
  fossil record, 245, 246–47
  gravity perception, 190, 191
  opal, 244, 245
  silicification processes, 60
Plasmalemma, 59, 68
Pleiotropy, 237
Pleistocene, 239
Pleurobrachia, 192
Pleurochrysis carterae, 69, 71
Pleurotaenium, 53
Plexaura flexuasa, 212
Plusiotis, 113
Pocillopora damicornis, 82, 83
Podarcis sicula, 196
Polycarpa, 136
Polychaete worms, 189, 190, 193, 240
Polycitoridae, 140
Polyethylene glycol, 155
Polyplacophora. *See* Chitons
Pomacea paludosa, 91

Porcelanite, 244
Porcellaneous shells, foraminifera, 61, 63
Porifera
  fossil record, 233, 234–35. *See also* Archeocyathids
  macromolecules, skeletons, 22
  oldest fossil opal, 245
  spicules, 246
  tetractinellids, 246
Praunus flexuosus, 113
Precambrian, mineralization during
  biologically induced mineralization, 229
  controlled mineralization, 230, 248
  magnetite crystals, 230–31, 248
  noncomposite skeletons, 230
Precambrian–Cambrian boundary zone, 232, 233, 247–51
Precursor minerals. *See* Maturation of biogenic minerals
Predators, protection against, 250
Predentin, 160, 178, 180
Preformed organic matrix framework, 29–30
  bone, 162
  chiton teeth, 98
  crustacean gastroliths, 121
  enamel, 182
  enameloid, 180, 182
  mollusk shell, 103, 107
  predentin, 148, 160, 178, 180
  tissues, comparison of, 43
Primary organic membrane, 65, 67
Primates, magnetite, 197
Prionace glauca, 48, 182
Prismatic layer, mollusks
  Monoplacophora, 94
  Mytilus californianus, 104
  varying proportions in *Mytilus,* 215
Problematica, 232, 234, 235, 241–42, 250
Processes. *See* Biomineralization processes
Prociphilus tesselatus, 113
Proteoglycans
  bone and dentin, 158–59, 162
  cation affinities, 169
  distribution in skeletal hard parts, 22
  structure of, 168–69
Proteolipid, multimolecular complex with calcium, 34
Protoctista, 50–73. *See also* Algae
  distribution of biogenic minerals, 8–9
  evolution of siliceous protoctists, 244–46
  impact of mineralization on seawater, 19
  invasion of ocean surface waters, 239
Protodolomite, 124, 126, 132
Prymnesiophyceae. *See* Coccolithophoridae
Prymnesium, 52
Prymnesium parvum, 73
Prymnesium patellifera, 73
Psammechinus miliaris, 131
Pseudokephyrion, 48, 51, 52
Pseudomonas fluorescens, 204
Pseudomonas syringae, 204
Pseudoplexaura flagellosa, 78
Pseudopodia, 63, 65, 78, 127
Pseudosquilla bigelowi, 112
Pteropoda, 236, 239
Pulmonate gastropods, 240
Pyrite, 44
Pyura australis, 136, 141
Pyura bradleyi, 136, 140
Pyura cataphracta, 136
Pyura littoralis, 136
Pyura sacciformis, 136
Pyura stolonifera, 136
Pyuridae, 135
  mineral types, 136, 140–41
  spicules, 140–41
  vertebrate ancestor, 141

Quartz, 244
Quinone-tanning, 107

Radial tests, foraminifera, 65
Radiolaria
  fossil record, 245–46
  isospore, celestite, 51
  macromolecules, skeleton, 21, 22
  minerals formed, 53
  ontogenetic changes in mineralogy, 48
  seawater undersaturation, 244
  skeletal symmetry, 50
Radula, 95–96. *See also* Chitons
Raia clavata, 137
Raja erinacae, 181
Rangia cuneata, 92, 109
Reefs
  algae, red, 209
  carbonate sink, 19
  formation processes, 238
  fossil record, 238–39
  oldest reefs, fossil record, 235
  organisms, reef-building, 238
Regeneration, sea urchins, 127, 131
Remodeling
  bone, 147
  mollusks, 109–10
Renal concretions
  Ascidiacea, 143
  fossil *Nautilus,* 241
  Nautilus, 93
Reptilia
  fossil record, 243
  magnetite, 197
Resorption. *See* Remodeling
Reticular body, 69
Rhipocephalus, 27, 52, 209
Rhizomes, 190
Rhizopoda, 50, 52, 54, 244
Rhythmic deposition of mineral. *See* Growth lines
Rickets, 167
Rivularia, 44
Robertinacea, 236
Robertinina, 65. *See also* Foraminiferidae
Rotaline foraminifera, 65–67. *See also* Foraminiferidae
Rotaliina, 61

Rudistid mollusks, 239
Rutile, 61

Sacculus, 196
Salinity, influence on mineralized hard parts, 214, 220
Salivary proteins, 129
Salmon, magnetite, 200
Saturated solution for mineralization, 30–32
Scales, fish, 148. *See also* Fish
Scaphander, 92
Scaphopoda, 94
Schizoporella unicornis, 211, 212
Schmidt, W. J., 5
Scleractinian corals, 81–87. *See also* Cnidaria; Corals
  adult skeleton, 82–83
  fossil reefs of, 238–39
  growth lines, 224–25, 226
  increase in aragonite in tropics, 208
  larval skeleton, 82
  mineralization processes, 83–87
  mode of life, 81
  reef-builders, 238, 239
Scleroblasts, 78
Sclerocytes
  Ascidiacea, 143
  sea urchins, 130
Sclerosponges, 238, 244
Sclerotization
  Arthropoda, 111
  Mollusca, 102
Scotoplanus globosa, 221
Scrippsiella, 52
Scyphomedusae, 231
Scyphozoa, 74, 76, 196, 231, 242, 244
Sea urchin mineralization, 127–34. *See also* Echinodermata
  adult test formation, 38, 41, 43, 130–32
  evolution of test, 235
  growth of skeletal elements, 131
  intracrystalline proteins, 32, 134
  larval spicule formation, 127–30
  mineral phase, 123–24, 132–34
  spine structure, 123, 125, 131
  teeth, 124, 127, 131–32
  test ultrastructure, 124
Seawater
  impact of biomineralization on, 18–20
  isotopic composition of, 222
  Mg/Ca ratio of, 237
  opal saturation, fossil record, 244–46
  phosphate concretions, fossil record, 251
Secondary matrix protein conformation, 23. *See also* β-Sheet conformation
Secondary osteon. *See* Haversian system
Sepia, 93
Serpulid polychaete worms, 213, 237, 238
Sharks. *See* Elasmobranchs
β-Sheet conformation, matrix proteins, 23
  acidic macromolecules, 175
  amelogenins, 185
  collagen telopeptide, 154
  enamelins, 184
  mollusk shell acidic proteins, 38, 39, 107
  polyaspartate, 38
Shell apex theory, 223
Shell gland, 45
Shell, mollusks, 99–110. *See also* Mollusca
  calcite–aragonite proportions, 214–17
  dissolution and remodeling, 109–10
  extrapallial fluid, 109
  growth lines, 225–26
  isotopic composition of, 222, 223
  mantle, 99–101
  matrix–mineral relations, 106
  matrix structure, 108
  ontogenetic changes, 45
  organic components of, 107–9
  periostracum, 101–3
  trace element contents of, 220
  ultrastructural types, 103–5
Silica, 9, 10, 56. *See also* Opal
Silica cycle, oceans, 20
Silica-depositing vesicle, 39–40
Silica mineralization. *See* Diatoms; Silicification
Silicification. *See also* Evolution of silicification
  cysts, Haptophyta, 73
  diatoms, 58–60
  plants, 60, 244
  terminology, 55
Silicified fossil deposits, taxonomic distribution, 244
Silicoflagellates
  fossil record, 245, 246
  minerals formed, 53
Silicalemma, 56, 58, 59–60
Silicoloculina, 52
Silicon
  bone, 251
  definition of, 55
  diatoms, essential nutrient, 59, 251
  Molpadiidae granules, 210
Silurian
  conularids, 242
  first massive reefs, 238
  monoplacophorans, 236
  nautiloids, 236
Sodium chloride crystals, 34
Sodium hypochlorite, 118, 124, 134, 150–51, 166
Somalinautilus libanoticus, 241
Sound reception, fish, 196
Space delineation, mineralization, 28–29, 43
Sparry calcite, 234
Sphenophyta, 244, 247
Spherulites
  algae, red, 209
  chiton tegmentum, 95
  Crustacea, 118–19
  crystal growth cessation, 41
  distribution among mineralized tissues, 35

Haptophyta, 72
Pennatulacea, 79
scleractinian corals, 82, 85
Spicules
Aplacophora, 89
Ascidiacea, 140–43
chitons, 94
Cnidaria, 77–79
Ediacaran Cnidaria, 231
functions, not understood, 189
holothurians, 45–46, 220–21
sclerosponges, 244
sea urchins, 127–30
spicular skeletons, Cambrian, 234
sponge, 244
Spirillinacea, 61, 62
Spirobranchus gigantheus, 237
Spirogyra, 53, 192
Spiroloculina hyalina, 63
Spirorbis, 212, 214
Spirostomum, 51, 52
Spirula, 93
Sponges. *See* Porifera
Starch granules, 190, 191, 192
Statoconia
gravity receptors, distribution, 191–96
Hydrozoa, 74
Statocyst
Cnidaria medusae, 74
gravity receptors, distribution, 191–94
Statoliths
gravity receptors, distribution, 191–96
mysids, 5
Stephanoeca diplocostata, 54
Stereochemical interactions, additives and crystals, 32
Stereochemical rules governing protein–crystal interactions, 38
Stereom, 123, 124, 130, 235
Sternaspis, 240
Strabo, 61
Stromatolites, 228
Stromatoporoid sponges, 238
Strombus, 24
Strongylocentrotus purpuratus, 132
Strontianite, 220
Strontium
Brachiopoda, 218–20
evolution of Sr discrimination, 228
salinity, effect on, 220
seawater, effect of biomineralization on, 19, 20, 31
temperature, effect on, 220
Strontium sulfate. *See* Celestite
Struvite, 139
Styella partita, 143
Stylaster, 76
Stylasterina, 236
Stylopoma, 212
Sulfated glycoproteins. *See also* Proteoglycans
Ascidiacea, 140
bone and dentin, 159
cartilage, 168–69
EDTA soluble fraction, 21
mollusk shells, 38, 39, 107, 175
nucleation site, 38
Sulfated polysaccharides, 38, 71. *See also* Chitin sulfate; Proteoglycans
Sulfate-reducing bacteria, 27, 229
Sulfates, biogenic, 8, 10
Sulfides, biogenic, 9, 11
Sulfur isotopic composition, Precambrian, 229
Surface area, bone, 166
Symbiodinium, 81
Symbionts
Ascidiacea, fungal, 143
foraminifera, 61–63
scleractinian corals, 81
Symbiodinium, 81
Syncytium, 127, 128, 130, 132
Synura, 53
Synura petersenii, 41

Tarletonbeania crenularis, 198
Tegmentum, chiton, 95
Temporary storage of mineral
Arthropoda, 112–14, 120–22
Chordata
bones, 144
matrix vesicles, 164
medullary bones, 144
milk, 139
mitochondria, 139
ferritin, 98, 202–4
Mollusca
intracellular granules, 91, 92
uroliths, 93
Telememorus, 53
Teleost fish. *See* Fish
Telopeptides of collagen, 153
Temperature, effect on biogenic minerals
basis for effect, 207–8
calcite–aragonite proportions, 210–17
isotopic fractionation, 221–23
trace elements, 217–20
Terrestrial mineralization processes
crustaceans, 120, 122
fossil record, 240, 246–47
lichens, 19, 27
plants, 19. *See also* Plantae
weathering, biologic, 19
Tetraclita, 211, 212
Tetractinellid sponges, 246
Tetracycline, 131
Textulariina, 61
Thermal shock breaks, mollusk shells, 226
Thiopneutes, 27
Thompson, D'Arcy, 50
Thyroglutus malayus, 114
Tomes processes, 182
Tommotian. *See* Cambrian
Tomocerus minor, 114

Tooth formation
  Chordata
    dentin, 29–30, 147–48. *See also* Bone
      enamel and enameloid, 175–88
  continuously growing, 29, 96, 127
  Mollusca
    chitons, 30, 94–99
    patellid gastropods, 96, 244
  sea urchins, 127, 131–33
Trace elements, effects of
  hydrostatic pressure, 220–21
  redistribution in oceans by biomineralization, 20
  salinity, 220
  temperature, 218–20
  water chemistry, 220
Trace fossils, 230, 249
Transformation of minerals. *See* Maturation of biogenic minerals
Trap doors, 88, 91
Trematomus borchgrevinki, 205
Triassic
  crinoids, planktonic, 239
  mammals, 243
  reef formation, 238
Tridacna, 24
Trididemnum cereum, 141
Trimerellacea, 237
Trilobites, 115, 233, 235, 239, 242
Triradiate spicule, sea urchin, 128–29
Tubipora, 76
Tubipora musica, 79
Tunic, 140, 141, 143
Tunicate, 140. *See also* Ascidiacea
Tunicin, 140
Turbellaria, 192
Turbo, 110
Turkey tendon, mineralized, 148–49
  collagen fibrils, 156–57
  crystal dimensions, 151
  crystal growth, 165
  disaggregation of, 166
  matrix-mineral relations, 36–37
Turtle, magnetite, 197, 200
Twinning, calcite, 82, 86
Tympanic bulla. *See* Bulla

Udotea, 52, 209
Urate, 137, 141
Urey, H. C., 221
Uric acid, 111–14
Urochordata. *See* Ascidiacea
Uroliths, 93, 241
Utriculus, 196

Vaterite
  Arthropoda, 112–13
  Ascidiacea, 136, 141–43
  Craniata, 138
  gravity perception, 194–95
  otoliths, fish, 195, 196, 199
  precursor of aragonite, 42, 45
Vendian
  adaptive radiation, 249
  algae, mineralized, 234
  carbon-13, isotopic shift, 249
  cyanobacteria, calcified, 234
  mineralized skeletons, fossil taxa, 233
  phosphatic hard parts, 241
Vertebrates, 135. *See also* Chordata
Vesicle mineralization, 33–34. *See also* Matrix vesicles
  Ascidiacea, 141
  chitons, 94
  choanoflagellates, 54
  Coccolithophoridae, 69
  Gorgoniacea, 78
  magnetotactic bacteria, 197
  miliolids, 63
  sea urchins, 127–28, 130
Vestibulary apparatus, 49, 190, 191. *See also* Gravity perception
Vital effect, 222
Vitamin D, 49
  Echinodermata, 131
  Mollusca, 101, 122
Vitamin K, 160
Veretillum cyanomorium, 78
Viviparus, 91
Viviparus viviparus, 44

Warfarin, 160
Water chemistry, trace elements, 220
Wax, insects, 111, 113
Weddelite
  Arthropoda, 113
  Ascidiacea, 137, 141, 143
  Craniata, 139
  lichens, 27
  Mollusca, 88, 92, 93
  Nautilus mandible, 27, 49
Whales
  migration, 223
  tympanic bulla, 139, 151
Wheeler, W. C., 5
Whewellite
  Arthropoda, 113
  Craniata, 139
  Protoctista, 52–53
Woven bone, 145

Xenophoridae, 63
X-ray diffraction, mineral identification, 5, 16–17

Zinc phosphate, 113
Zoomastigina, 45, 51, 52
Zooxanthellae. *See also* Symbionts
  Foraminiferidae, 61–63
  scleractinian corals, 81